THE WORLD'S CLASSICS

THE ORIGIN OF SPECIES

CHARLES DARWIN was born on 12 February 1809 to Robert Darwin, a highly successful doctor, and his wife Susannah Wedgwood, a member of the famous pottery family. His paternal grandfather was the free-thinking scientist and poet Erasmus Darwin, who was formulating evolutionary ideas at the end of the eighteenth century.

Initially training as a medical student, then intending to become a clergyman, Darwin found his life and thinking profoundly altered by a five-year round-the-world trip on HMS *Beagle*. Twenty-odd years lay between his initial formulation of the idea of natural selection and the publication of *The Origin of Species* in 1859, in the course of which he painstakingly drew together materials from an extraordinary range of sources and intellectual domains to confirm his observations and intuitions, which he understood from the start would challenge fundamentally both society's and science's assumptions. However, by 1871, when he published *The Descent of Man*, much of the first rancour of the debate concerning 'man's place in nature' had passed.

Throughout his life, Darwin was beset by intermittent debilitating illness, which nevertheless had the advantage of exempting him from public life and allowing him to concentrate on writing, reading, and observation. He was also profoundly affected by the death of several of his children in infancy. The *Autobiography* that he wrote late in life for his family makes it clear that he ceased relatively early in his career to have any religious faith and that he resented the punitive element in orthodox Christianity. He died on 19 April 1882 and, after controversy, was buried in Westminster Abbey.

GILLIAN BEER has written extensively about Darwin and about scientific writing in its cultural context. Her *Darwin's Plots* (1983) has been followed by a number of essays on related topics, collected in the (forthcoming) volume *Open Fields*, and she is at present completing a book on the English fascination with islands in the later nineteenth and earlier twentieth century. She is King Edward VII Professor of English at the University of Cambridge and President of Clare Hall.

THE WORLD'S CLASSICS

THE ORIGIN OF SPECIES

CHARLES DARWIN was born on 12 February 1809 to Robert Darwin, a highly successful doctor, and his wife Susannah Wedgwood, a member of the famous pottery family. His paternal grandfather was the free-thinking scientist and poet Erasmus Darwin, who was formulating evolutionary ideas at the end of the eighteenth century.

Initially training as a medical student, then intending to become a clergyman, Darwin found his life and thinking profoundly altered by a five-year round-the-world trip on HMS *Beagle*. Twenty-odd years lay between his initial formulation of the idea of natural selection and the publication of *The Origin of Species* in 1859, in the course of which he painstakingly drew together materials from an extraordinary range of sources and intellectual domains to confirm his observations and intuitions, which he understood from the start would challenge fundamentally both society's and science's assumptions. However, by 1871, when he published *The Descent of Man*, much of the first rancour of the debate concerning 'man's place in nature' had passed.

Throughout his life, Darwin was beset by intermittent debilitating illness, which nevertheless had the advantage of exempting him from public life and allowing him to concentrate on writing, reading, and observation. He was also profoundly affected by the death of several of his children in infancy. The *Autobiography* that he wrote late in life for his family makes it clear that he ceased relatively early in his career to have any religious faith and that he regretted the punitive element in orthodox Christianity. He died on 19 April 1882 and, after controversy, was buried in Westminster Abbey.

GILLIAN BEER has written extensively about Darwin and about scientific writing in its cultural context. Her *Darwin's Plots* (1983) has been followed by a number of essays on related topics, collected in the (forthcoming) volume *Open Fields*, and she is at present completing a book on the English fascination with islands in the later nineteenth and earlier twentieth century. She is King Edward VII Professor of English at the University of Cambridge and President of Clare Hall.

THE WORLD'S CLASSICS

CHARLES DARWIN

The Origin of Species

Edited with an Introduction by
GILLIAN BEER

Oxford New York
OXFORD UNIVERSITY PRESS
1996

Oxford University Press, Walton Street, Oxford OX2 6DP

Oxford New York
Athens Auckland Bangkok Bombay
Calcutta Cape Town Dar es Salaam Delhi
Florence Hong Kong Istanbul Karachi
Kuala Lumpur Madras Madrid Melbourne
Mexico City Nairobi Paris Singapore
Taipei Tokyo Toronto

and associated companies in
Berlin Ibadan

Oxford is a trade mark of Oxford University Press

First published as a World's Classics paperback 1996

British Library Cataloguing in Publication Data
Data available

Library of Congress Cataloging in Publication Data
Darwin, Charles, 1809–1882.
[On the origin of species]
The origin of species / Charles Darwin ; edited with an introduction by Gillian Beer.
Includes bibliographical references (p.) and index
1. Evolution (Biology) 2. Natural selection. I. Beer, Gillian.
II. Title. III. Series.
QH365.02 1996 575.01'62—dc20 95–8377
ISBN 0–19–281783–3

1 3 5 7 9 10 8 6 4 2

Typeset by Pure Tech India Ltd, Pondicherry
Printed in Great Britain by
BPC Paperbacks Ltd.
Aylesbury, Bucks.

CONTENTS

INTRODUCTION

WHEN Charles Darwin published *The Origin of Species* in November 1859 he presented it as a hasty introduction to his ideas, for which he would present further evidence in the future. So it may have seemed to him: the text was written in thirteen months after a gestation of more than twenty years. It was written in the anxiety of knowing that Alfred Russel Wallace, like Darwin himself earlier, had recently conceived of a process that Darwin would name 'natural selection'. Instead of all species having been created together at the beginning of time, or even at punctuated intervals through time, the present array of kinds throughout the world had come into being by a gradual process of genetic differentiation and selection under environmental pressures. Slight mutations could advantage individual organisms, and such mutations might then be enhanced over generations. This insight involved extinction as well as proliferation; it was disquieting in a great number of ways, however much each man later sought to palliate the disturbance.

The idea grew in both minds through extensive travel as natural historians, through detailed observation of natural phenomena around the world (not always the same parts), and through dream and reflection. In each case it seemed to the thinker that the full force of the theory seized him after reading Thomas Malthus's *An Essay on the Principle Of Population* (first edition 1798), probably in the edition of 1826. Malthus argued that human population and population growth will always outrun resources of nutrition and space; therefore competition between those occupying common environments will control population.[1] Scholars have since discriminated the differences between the theories of Wallace and of Darwin and have demurred at, or emphasized, Malthus's role. In 1858, though, the insight the two men shared seemed close enough to drive Darwin at last into a steady frenzy of composition.

[1] See Appendix I.

Darwin wished to marshal sufficient evidence to convince scientists adept in a variety of fields, from geology to botany to taxonomy and morphology. At the same time he wanted to address a very broad spectrum of readers, thus acting as fundamental initiator and popularizer at once. That is, the *Origin* was to speed up the process of reception, so that the ideas it contained could become available simultaneously to Darwin's fellow-workers in science and to any educated person. In this double task the book was quite remarkably successful, perhaps the more so as Darwin himself made no stable discrimination between the diverse audiences he was addressing. The language of the text is accessible and non-mathematical. Yet evidence is piled on evidence for colleagues to pursue, as the 'Register of writers referred to in the text of the *Origin*' at the end of this edition makes abundantly clear. The continuing fertility of Darwin's work for scientists is evident in the degree to which it is still a fundamental prerequisite for work in genetics and still a source of controversy in taxonomy.[2]

The extraordinary creativity of the writing, and its capacity to lend itself to contradictory social programmes, comes out of this liberal amalgamation of audiences. Indeed, this openness plays into—is part of—the vigour of Darwin's argument. It declares itself in the novelty of his associative power, and in his ability to pursue small discrepancies to large effect. Different readers can find their hopes and fears confirmed by extending the implications of Darwin's thought in one direction or another; and, it would later prove, those readers might be individualists, Fascists, Marxists, imperialists, or anarchists—or indeed, quietists. There is something fascinating and perturbing in a text that, while pursuing, in Darwin's words, 'one long argument', ballasted by multiple evidences, can generate such a variety of ideological potentialities.

To understand its impact in Western cultures over the past one hundred and fifty years, it is necessary to track the history of the work and its context. This will help the reader to analyse

[2] For a useful non-technical account by a variety of authors see Laurie Rohde Godfrey (ed.), *What Darwin Began: Modern Darwinian and Non-Darwinian Perspectives on Evolution* (Boston, 1985).

the process by which Darwin's ideas (or those associated with his name, not always to be found within the covers of this book) have come to have so dominant a role in the construction of social domains apparently remote from the biological. Darwin himself insisted always on constraining the extra-scientific implications of his work and resisted any overt politicization (itself, of course, a political position). Wallace, on the contrary, became an active socialist who saw evolutionism as caught into that enterprise. But it was Wallace who uncoupled the human from all other species development so as to preserve a place for the soul, and Darwin who, more radically, faced the complete integration of the human into the natural order. In this story no simple contraries survive.

Above all, the *Origin* made its impact because it raised questions fundamental to the life of humankind without making humankind the centre of its enquiry. That shift away from the centre was a silent and intense challenge to the reader's assumptions. Survival and descent, extinction and forgetfulness, being briefly alive and struggling to stay so, living in an environment composed of multiple other needs, coupling and continuing, ceasing to be: all these pressures, desires, and fears are alerted in this work without any particular attention being granted to the human person. This is a history of a world in which the human has a place but has not always been present, and where other kinds have each their own lost and fitfully recorded histories: in the strata of rocks, in reproduction, in the silt of the deep ocean, on remote islands where conditions have not changed, in parthenogenesis or hermaphroditism (modes of production more stable than the two-sexed system humans share).

How did the work come to be? It began, as most projects do, before the author recognized its existence; perhaps before the author existed. To take that second prehistory first: there was a long hinterland of attempts to understand how the diversity of species had been established and whether changes had occurred through time.[3] Some held to the view that all

[3] From the 3rd edn. on (1861) Darwin prefaced the *Origin* with 'An Historical Sketch of the Recent Progress of Opinion on the Origin of Species'; in the 6th edn. (1872), in response to criticisms that he had been ungenerous to his

kinds had been stable from the start of the world; others that different species had been introduced at intervals in a 'progressive' sequence. The Mosaic version of creation presented Adam with a complete roster of living kinds come into being at one time, to name at his own will. That strong myth certainly continued to command belief, or institutional acceptance, in England up to and beyond the generation in the late eighteenth century of Darwin's innovative grandfather, Erasmus Darwin, who himself produced a proto-evolutionary treatise, *Zoonomia*:

> all warm-blooded animals have arisen from one living filament, which THE GREAT FIRST CAUSE endued with animality, with the power of acquiring new parts, attended with new propensities . . . and thus possessing the faculty of continuing to improve by its own inherent activity, and of delivering down those improvements by generation to its posterity, world without end![4]

Darwin later distanced himself from his grandfather's theory by making it clear that there is no steady movement across all species towards complexity; it all depends on the relations between individual organism and medium:

> As the variability of each species is an independent property, and will be taken advantage of by natural selection, only so far as it profits the individual in its complex struggle for life, so the degree of modification in different species will be no uniform quantity. (p. 284)

The biblical account had also been under persistent pressure, particularly since Linnaeus set out a new scientific taxonomy of the relations between extant plant and animal kinds. Various versions of natural change had woven in and out for centuries. 'Transmutation' was not a new idea; it had been explored in the early nineteenth century by French scientists such as Jean-Baptiste Lamarck and Étienne Geoffroy Saint-Hilaire and, long before them, in literary texts such as Edmund Spenser's *Faerie Queene* in the late sixteenth century. And

predecessors, he added the phrase 'Previously to the Publication of the First Edition of this Work'. See Morse Peckham (ed.), *The Origin of Species By Charles Darwin: A Variorum Text* (Philadelphia, 1959).

[4] Erasmus Darwin, *Zoonomia; or, the Laws of Organic Life*, 1 (2 vols.; London, 1794–6), 1, 505.

metamorphosis, with its sideways motion, had been a familiar imaginative resource since Ovid and Apuleius. More directly, transformation within the individual life cycle was a familiar phenomenon among insects (butterflies were the most delightful and frequently cited example). This seemed to raise the possibility of natural change at a species level. Nor was the close relation of the human to other primates altogether ignored: Lord Monboddo insisted (to much mockery) that the orang-utan, though mute, was a brother to the human at an earlier point of development.[5] And in discussions across a wide range of fields in the years before Darwin, the idea of 'development' was often figured as 'progress', or even—at the end of the eighteenth century in France—as revolution.

So these diverse elements, fruitful for evolutionary thinking, were already available and under discussion in intellectual and activist circles, some from long before Darwin was born and all before he was a grown man. Indeed, a number of recent historians have argued that, so widespread were these discussions at both a popular and a more technical level, that it is possible to construct a history of nineteenth-century evolutionary thinking almost without Darwin.[6] And it is certainly the case, as James A. Secord has convincingly demonstrated, that the anonymous publication of Robert Chambers's *Vestiges of the Natural History of Creation* had a galvanizing effect on public discussion of evolutionist claims, from its first appearance in 1844 through its fourteen editions.[7] Even those who despised, or affected to despise, the work could not ignore its popular success, nor the degree to which it had stirred up and brought into question a settled cosmos. Moreover, the Darwin Correspondence[8] and, modestly, the glossary of information I have

[5] James Burnett Monboddo, *Of the Origin and Progress of Language* (6 vols.; Edinburgh, 1773–92).

[6] For example, Peter Bowler, *The Non-Darwinian Revolution: Reinterpreting a Historical Myth* (London and Baltimore, 1988); Adrian Desmond, *The Politics of Evolution: Morphology, Medicine, and Reform in Radical London* (Chicago, 1989).

[7] Robert Chambers, *Vestiges of the Natural History of Creation and Other Evolutionary Writings*, pub. anonymously (Edinburgh, 1845); ed. James A. Secord (Chicago, 1994).

[8] *The Correspondence of Charles Darwin*, ed. Frederick Burkhardt *et al.* (Cambridge, 1985–).

put together in this edition about the men (and they all are men) whose work Darwin alludes to and absorbs bring out how densely he was involved in the scientific community of the time, both through reading and through letters and conversation. They show, too, how much research was under way that could be turned to Darwin's theoretical purposes.

Yet the reception and continuing influence of the *Origin* indicate how much remained to be accomplished. Major originality may be a matter, first of all, of condensing materials that would otherwise remain discrete observations. Darwin drew his evidences from an extraordinary range of domains, many of which he had studied at first hand. He brought to his task a career manifestly devoted to empirical research and observation. His notebooks also show how daring, quirky, radical, his questions to himself were: 'have plants any notion of cause & effect'; 'Has the oyster necessary notion of space'; '?Is the shortness of life of *species* in certain orders connected with gaps in the *series of connections*?'[9]

That curiosity and assurance allowed him to pursue hypotheses with immense energy and to take them on beyond the bounds of current data. His project moved back and forth along a trajectory from the present into a newly construed and imagined past. His invention of the term 'natural selection', with (among other things) its inhering time-span, gave him an analytical tool that was to prove an extremely powerful discriminator, though it brought with it also a set of philosophical and ideological questions that Darwin himself was puzzled to resolve, as I shall later discuss. Retrospect had all the narrative excitement of discovery. Darwin allows the reader to participate in a copious world, a voyage into the past as exuberant as his voyage round the world in the *Beagle*.[10] This double motion had much to do with the process of composition

[9] Howard Gruber, *Darwin on Man: A Psychological Study of Scientific Creativity*, with Paul Barrett (ed.), *Darwin's Early and Unpublished Notebooks* (London, 1974) 332, 333, 443.

[10] The work we now know as *The Voyage of the Beagle* was originally called *Journal of Researches into the Geology and Natural History of the Various Countries Visited by HMS Beagle* (London, 1839). It first appeared as the 3rd vol. of a *Narrative* controlled by Robert Fitzroy.

and its deep involvement over many years with Darwin's entire life. Despite his intellectual caution, Darwin retained and communicated in his writing the open vulnerability of the enthusiast.

Darwin saw himself during his youth as intellectually unremarkable. He began to study medicine at Edinburgh, left there, and went to Cambridge where he took the then-common pass BA. Even at that time he probably underestimated his proficiency, since he ranked tenth out of 178 candidates in the final examinations. Moreover, he had quite remarkably impressed the Professor of Botany, John Stevens Henslow, who spent much time with him and pursued questions he raised. Indeed, shortly after, Henslow transformed Darwin's life and future by recommending him (when the initial candidate dropped out) to Fitzroy, Captain of His Majesty's Ship, the *Beagle*, as captain's companion and natural historian. When Darwin set out on the *Beagle* he was not quite 23, about the same age as a present-day Ph.D. student. He was active, enthusiastic, inquisitive. The five years of the *Beagle* voyage were the transformative period of Darwin's life. During that time he undertook extensive journeys by land, often through little-known territory; he saw rapine and slaughter and the decimation of the Indians by the Spaniards, and of the aboriginal inhabitants of Van Diemen's Land by the British; he collected, preserved, and catalogued specimens of many kinds; he kept field-notes and a diary; he wrote frequent long letters home and to friends and fellow-scientists such as Henslow and Charles Lyell, the geologist.

Still considering himself to be a geologist despite the wide experience he was gaining in botany and zoology, Darwin made his first important theoretical intervention, establishing the difference between oceanic and continental islands, and, as he said, 'deductively', working out a new view on the formation of coral islands before he ever saw a reef. He learnt to look, to interpret, and, above all, to write and to use writing as part of the interpretive process. Through the interplay of sensory experience, record, collection, revision, reassessment, he gained the confidence to think theoretically, to risk large hypotheses, to subject them to close-grained instances without yielding them to a narrow empiricism.

In recent years the tendency among historians of science has been to emphasize the 'belatedness' of Darwin's recognition of many of the materials he gathered, including the implications of his visit to the Galapagos. But this emphasis, though in its way sensible, can be needlessly exaggerated. It needs to be understood how the many phases of writing and revision, even while on board ship, played in with his physical exploration to generate thought. This interplay allowed him to reserve as well as to articulate, and to specify as well as to suspend ideas. He did not wait until he got home among his scientific peers to brood and, as he put it, 'build castles in the air'—a process which he declared to be quite 'as hard work as the closest train of reasoning in geology'.[11]

The public outcome of this voyage has an important bearing on the later reception of the *Origin*. It helps to account for a phenomenon that has sometimes puzzled commentators: the very rapid recognition by his contemporaries of how Darwin's ideas bore on the descent of man, despite what he called his 'diplomatic' attempt to exclude it from the argument. When Darwin published the *Origin* he was not only a distinguished scientist but a well-known popular author. The work we now know as *The Voyage of the Beagle* had twenty years earlier had a remarkable success, rapidly moving from being the third volume of a narrative of voyages dominated by the authorship of Fitzroy to a free-standing publication which went into a second edition on its own. And the *Voyage* was full of accounts of the variety of indigenous peoples, their struggle for life, their imminent extinction, the disturbance in their social patterns as the result of colonizing invasion. Far from being purely a 'natural historical' account, if such an account excludes the human, Darwin offered a contribution to the debate then current concerning the place of ethnography in natural history. The debate was active at the British Association for the Advancement of Science in 1839, and Darwin was there made a member of a committee to draw up questionnaires for

[11] Gruber, *Darwin on Man*, 272.

travellers who might contact remote peoples. In several instances in the *Voyage*, for example in his account of the Fuegians, Darwin emphasized the very close approximation between the human and the animal if physical conditions prevent anything but the round of minute-by-minute survival in adverse circumstances. Darwin, that is, was already well known as a vivid recorder of human as well as other kinds.

Why did more than twenty years elapse between Darwin's initial insights recorded in his notebooks in the late 1830s and the publication of the *Origin*? The question can be answered at very different levels, not necessarily mutually exclusive of each other. Howard Gruber, and in his wake many others, has compellingly argued that Darwin resisted his own insight because of the social and familial upheaval it would bring upon him. Whether or not theology could, strictly, accommodate the new version of how things came to be, it would undoubtedly cause disturbance and grief to many believers, including his own wife. Moreover, Darwin's passionate attachment to his widowed father included also an Oedipal struggle which intermittently froze his activities. That is one strain of explanation proferred with twentieth-century hindsight.

Equally, more empirically minded commentators have emphasized the necessity for him to test his ideas against as many different kinds of evidence as he could gather. Hence, the argument goes, he needed the delay to augment the first-hand materials of the *Beagle* journey with others' reports and comments. And certainly the work as finally composed is studded with frequent and often generous reference to the works of others, especially Alphonse de Candolle, Edward Forbes, Asa Gray, Linnaeus, Charles Lyell, Geoffroy Saint-Hilaire, and now lesser-known contemporaries such as Kirby, Milne Edwards, Wollaston, and H. C. Watson. The section on hybridism draws directly and extensively on Köllreuter and Gärtner. Intriguingly, Malthus is mentioned twice only, and Wallace (though acknowledged in the introduction as having 'arrived at almost exactly the same general conclusions that I have on the origin of species') is also referred to twice only in the body of the work.

The work Darwin did during the 1840s on cirripedes was certainly not just a delaying tactic but a highly relevant investigation concerned with barnacles which exist on the borders of hermaphroditism, parasitism, and a two-sex system. They therefore raise fascinating questions of taxonomy and of reproductive systems, adaptation, and descent. (One can see to what good effect Darwin used such material in his argument in the *Origin* by turning to page 121 of this edition.) Darwin realized with mounting enthusiasm that cirripedes could be accurately studied *only* in the light of his new arguments. This implied empirical support for his theory: he wrote to Joseph Hooker in 1848:

> I never shd. have made this out, had not my species theory convinced me, that an hermaphrodite species must pass into a bisexual species by insensible small stages, & here we have it, for the male organs in the hermaphrodite are beginning to fail, & independent males ready formed.[12]

So the years passed on, well spent in intellectual enquiry and experiment. Further, Darwin was not pressed for funds and was not in good health; the effort to bring his work to completion seemed less attractive than thorough, ongoing enquiry. The retired life he led at Down House did not bring to crisis any demand that he make his work public. Thus runs a third explanation.

Each of these interpretations is coherent, none is sufficient on its own, and—in their simplified form—none can quite include the extent to which Darwin *was* engaged in advancing and writing his own specific evolutionary programme during the intervening years. Darwin wrote two essays, one in 1842 and one in 1844, in which he first set out his 'species work'. Though he judged both as insufficient sketches, he yet considered the 1844 essay sufficiently important to be published in case of his early death. In a formal letter to his wife he wrote:

> . . . I have just finished my sketch of the species theory. If, as I believe, my theory in time be accepted even by one competent judge, it will be a considerable step in science.
>
> I therefore write this in case of my sudden death, as my most solemn and last request, which I am sure you will consider the same

[12] *Correspondence*, 4, p. 159.

as legally entered in my will, that you will devote £400 to its publication.[13]

Darwin knew that, for his theory to be fulfilled, it could not remain without an audience for ever. Despite his reluctance to publish (as he saw it, prematurely) he yet could not bear to think of his idea drowned in his own death. By 1844 he was confident that the theory was sufficiently watertight to be floated in his absence.

In the event, though, it was not until 1856 that he undertook the massive labour of sorting and collating his portfolios of notes and, under the urging of Charles Lyell, began the work now known as the 'Big Book'—an immense study, *Natural Selection*, never to be finished. Instead, alarm over Wallace's insights drove him to compose what he initially held to be another 'sketch', *The Origin of Species*. What started out as an 'abstract' (the 'Big Book' severely pruned and re-ordered) then opened out into new work written between August 1858 and September 1859 and published that November. The account by Adrian Desmond and James Moore in their *Darwin* (London, 1991) conveys a vivid impression of the clutter, the groaning of the spirit, the sickness, the impatience and anxiety, that went into this urgent composition. Without Wallace's intervention it is possible that Darwin could have remained a Mr Casaubon, weltering in research grown old. Instead, the *Origin*, as *Middlemarch*'s author remarked shortly after its appearance, 'makes an epoch'.

It is important to remember that, at the time that Darwin was writing, the laws of genetics had not been established. The descent of characteristics and their distribution within a family, tribe, or species could be observed, but not technically accounted for. Gregor Mendel's 1866 paper on plant hybridism did not become generally known until 1900, well after Darwin's death. In the 1920s, population geneticists such as Ronald Fisher, J. B. S. Haldane, and Sewall Wright, put together

[13] *The Foundation of the Origin of Species: Two Essays written in 1842 and 1844 (The Works of Charles Darwin*, ed. P. H. Barrett and R. B. Freeman, 10; London, 1986), p. xxi.

Darwin's work on natural selection and Mendel's genetics to produce the 'modern' or 'neo-Darwinian' synthesis. That movement took Darwin's work forward into the present day, particularly in molecular genetics and the work that has flowed from the elucidation of DNA. With our hindsight the extraordinary importance of Darwin's thinking is manifest.

But what is it that accounts for the strength of the impact that Darwin's work had for his contemporaries, few of whom were during his lifetime convinced of the absolute centrality of natural selection as the central mechanism of evolutionary change? And how did he react to the criticisms mounted? Clearly his contemporaries were most disconcerted by the implications of his theory for religion—and this goes for many of the scientists among them, such as Lyell and Asa Gray, as well as professional clerics like the Bishop of Oxford, who neatly invoked traditional scientific methodology to bolster his theological criticisms: 'when subjected to the stern Baconian law of observation of facts, the theory breaks down utterly.'[14]

Despite Darwin's own later claims in his *Autobiography* that he had followed Baconian principles, such principles would have been quite inadequate to his purposes. Those purposes were to describe what could no longer be seen in the present world and to produce a history of what had vanished that would sufficiently explain what is now present. This could be accomplished at best by by a process of backward induction (what Huxley called 'the method of Zadig') using what Darwin himself described as the slight fragments surviving:

> For my part, following out Lyell's metaphor, I look at the natural geological record, as a history of the world imperfectly kept, and written in a changing dialect; of this history we possess the last volume alone, relating only to two or three countries. Of this volume, only here and there a short chapter has been preserved; and of each page, only here and there a few lines. Each word of the slowly-changing language, in which the history is supposed to be written, being more or less different in the interrupted succession of chapters, may represent the apparently abruptly changed forms of

[14] Quoted in Alvar Ellegård, *Darwin and the General Reader: The Reception of Darwin's Theory of Evolution in the British Periodical Press, 1859–1872* (Chicago, 1990) 186. From *Quarterly Review*, 108 (1860), 239.

life, entombed in our consecutive, but widely separated, formations. (p. 251)

Darwin here and elsewhere used metaphors from language and from history. Indeed, his reading in language theory and in the new ideas of comparative grammarians and etymologists gave him a thought-model at a crucial time in the precipitation of his theories, as I have demonstrated elsewhere. In this passage he emphasizes the elements of time and of decay, and suggests how impossible it is to retrieve a full roster of evidence. Principles of change must serve. The living forms of the past studied in geology have been crumbled, striated, homogenized, ignited, compacted: rendered all but uninterpretable. Darwin implies an imagined synchronicity between writing the record and being part of the experience, as any historian must. But the passage also admits how fitful and slight is any access to the forms of the past. And now, in his own new and peculiar mind-world, far more was this so, since those forms have changed and changed again over aeons, and whole species have vanished and are vanishing. Darwin's theories demand recognition of loss, irrecuperable.

Darwin always denied that he was studying how life began. In his title he emphasized that he was concerned with process rather than stable origins, although the shortened version we usually now adopt reverses that emphasis: *The Origin of Species* was originally entitled *On the Origin of Species by means of Natural Selection, or the Preservation of Favoured Races in the Struggle for Life.* 'By Means of': a sequence of change is suggested in the main title, though stabilization is recognized in the subtitle with its keyword 'Preservation'. His position was very like that described by Foucault in 'Nietzsche, Genealogy, History':

> Genealogy . . . depends on a vast accumulation of source material . . . and does not oppose itself to history as the lofty and profound gaze of the philosopher might compare to the molelike perspective of the scholar; on the contrary, it rejects the metahistorical deployment of ideal significations and indefinite teleologies. It opposes itself to the search for 'origins'.[15]

[15] Michel Foucault (1971) in P. Rabinow (ed.), *The Foucault Reader* (London, 1984), 76–7.

Darwin studied the originating sequences, not the stable origin (if such existed). That paradox released him from the final task of challenging first causes, and yet usefully destabilized any insistence on design or plan.

Darwin placed three ideas in interaction to form his novel insight: those ideas were hyperproductivity, variability, and selection. The first two necessary ideas tend towards largesse: (1) superabundance of individuals and kinds; (2) variety potentiating a wide diversity of ecological niches. The third idea—selection—is more frugal. It drives ruthlessly across the populous ideal landscape of the first two, culling, controlling, weeding out. The tensions among these ideas are both indicative of Darwin's originality and damagingly difficult to keep in equal play. Perhaps in an attempt to do so, Darwin discriminated three forms of selection set oddly askance from each other: artificial selection, natural selection, and sexual selection. I shall return to these distinctions below.

Darwin's thinking drew fruitfully on nineteenth-century individualism in its insistence that the potentiality for change is borne by the individual, and that it is by means of the individual organism that evolutionary change comes about. Moreover, he wittily observed that 'it should be remembered that systematists are far from pleased at finding variability in important characters' and that variation was declared not to occur in 'important parts of the structure'. This circular argument had closed discussion because where variation did occur it was deemed that this was mere 'individual difference' in 'unimportant parts'. In contrast, Darwin asserted: 'These individual differences are highly important for us, as they afford materials for natural selection to accumulate' (p. 39); 'Hence I look at individual differences, though of small interest to the systematist, as of high importance for us, as being the first step towards such slight varieties as are barely thought worth recording in works on natural history' (p. 44). Such variability is the ground condition on which natural selection can fruitfully work; without such slight variations natural selection could make no headway. This emphasis on slight differences, deviations, rather than approximation to the parent-type, is among the most profound of Darwin's challenges to conventional thinking. Not the normative but difference proves to be the generative principle. Even now, in the light of the human genome project with its emphasis on normalization, that insight needs emphasis.

Before Darwin, J. B. Lamarck's 1815 theory of the descent of acquired characteristics, in *Histoire naturelle des animaux sans vertèbres*, seemed to promise some efficacy to self-help and self-betterment. It had its appeal for a society that emphasized individual effort and the learning process. According to the Lamarckian view, descendants could draw on the learning process of the parent type. The popular example was giraffes reaching up for the topmost leaves that other animals could not get to and 'therefore' gradually acquiring longer and longer necks.

Darwin's view was bleaker: from among a population of giraffes those individuals who happened to have longer necks would be advantaged, would survive, breed, and leave descendants, some of whom would have necks as long or even longer than the parent. No virtue or effort is involved, simply a sorting process that favours those individuals most apt to a particular contemporary environment. But sheltered within this amoral and non-teleological world are assumptions about 'improvement' that draw on the practices of human breeders. Gardeners, pigeon-fanciers, stock-breeders can all refine chosen characteristics within a few short generations from among the animals or plants under their control. The advantage there is not to the animals or plants, but to those who own them. That procedure Darwin called 'artificial selection' and he saw it as morally ambiguous. Natural selection worked to the betterment of the individual and its species, never to the advantage of another kind. Thus he avoided instating the master–servant relation implicit in artificial selection as natural.

But the hope that 'improvement' will be the outcome of the process he names 'natural selection' and sets in opposition to artificial selection still haunts his vocabulary and argument. 'Natural Selection, it should never be forgotten, can act on each part of each being, solely through and for its advantage' (p. 122). In the last paragraph of the whole work he enunciates the laws he has laid down, the last of them being: 'a Ratio of Increase so high as to lead to a Struggle for Life, and as a consequence to Natural Selection, entailing Divergence of Character and the Extinction of less improved forms'. This will to emphasize *improvement* eventually resulted, in the fifth, 1869, edition, in his retitling the crucial fourth chapter, 'Natural Selection', as 'Natural Selection; or the Survival of the Fittest'. Thus he acquiesced in an idea not of his initial making, an idea derived from the writing of the philosopher Herbert Spencer; it proved to be an unfortunate move. 'The survival of the fittest' implies, tautologically, that those who survive are the 'fit', the 'superior' examples. Having survived proves them fit to survive; that other signification of 'fitness' as 'aptness', also crucial to Darwin's argument, is lost.

So the way was cleared for some of the more venomous appropriations of his theories to crude 'Social Darwinism', with its emphasis on competition, profit, and exclusion of the weaker members of society from its good. 'Social Darwinism' tends to identify current society with natural law and ignores the degree to which any environment is composed of many, conflicting, needs and capacities. An environment is not a static entity surrounding an organism. Nor can responsiveness to environmental demands guarantee moral worth: at any time, long arms or short, dun colouring, bright colouring, the ability to co-operate, or the capacity for uninhibited violence, may happen to favour survival.

In the 1859 and 1860s editions, natural selection is represented as a benign, slow-acting, unconscious process that relies on the relations between organisms in terms of every kind of interaction ('a web of complex relations' (p. 61)) rather than competition alone. Indeed, in the light of the present ecological movement, the opening pages of Chapter IV of the *Origin* read in a style very different from that seized on by his contemporaries. Darwin here emphasizes 'the intimate and complex manner in which the inhabitants of each country are bound together', and he urges the reader to remember 'how powerful the influence of a single introduced tree or mammal has been shown to be' (p. 68). He then offers the thought-experiment of an island 'into which new and better adapted forms could not freely enter'. In that closed system, he suggests, natural selection would work at its most pacific, without niches being 'seized on by intruders':

In such case, every slight modification, which in the course of ages chanced to arise, and which in any way favoured the individuals of any of the species, by better adapting them to their altered conditions, would tend to be preserved; and natural selection would thus have free scope for the work of improvement. (p. 68)

This passage does not indicate, necessarily, that a current reading is correct and the Victorian emphasis on competition incorrect. Both are feasible interpretations; each is put under stress in different passages in the book, because Darwin himself moves between what we may call different moods of interpretation. What is fascinating is to see the extent to which diverse societies can read in these pages those elements most justifying their own ideals and practices.

Darwin himself could not rest content with the *Origin* as first published. He used later editions to respond to criticisms, sometimes, by the end of the 1860s, moving towards a more Lamarckian view under the pressure of objections that, lacking genetics, he could not satisfactorily answer. The second edition, which forms the copy-text here, follows the first edition in most particulars, but even here some intriguing shifts take place, perhaps the most striking being the re-admission of the Creator and his works, as when we are told that 'natural selection [acting on some originally created form] will' (p. 352).[16] The phrase in square brackets is added in this edition. The second edition was issued only a very few weeks after the first, so here we can observe Darwin under the stress of the very first responses to his work. The opening section of the book's final sentence includes a reparation or climb-down—or perhaps a sense at this time that, since he was not concerned to describe the beginning of life, family peace could be bought without loss of honesty. In any event, agency is here assigned to a Creator rather than left vague as in the first edition. Again, the phrase here placed in square brackets is the addition:

[16] Peckham, *Variorum Text*, 678.

There is grandeur in this view of life, with its several powers, having been originally breathed [by the Creator] into a few forms or into one. (p. 396)[17]

Similarly, Darwin may be making either a biological or a theological exception, or both, when he adds a passage in parenthesis:

In the case of animals and plants with separated sexes, it is of course obvious that two individuals must always [(with the exception of the curious and not well-understood cases of parthenogenesis)] unite for each birth. (p. 79)[18]

Virgin birth (parthenogenesis) returns. His changes tend to avoid confrontation, or evade uncertain claims. Thus, he changes the description 'most of our domestic animals have descended from two or more *aboriginal* species' (my italics) into 'two or more wild species'. The contraries he needs for his argument are domesticated and wild; how long they have existed is uncertain. He does not wish to suggest that these two or more species were there from the start: ab-original.

Above all, Darwin had trouble with questions of *agency*. 'Natural selection' might still imply an active selector, albeit a selector called Nature, not God. How can one get away from the idea of creation to that of production, in which diversity of individual forms produces diversity of futures without initiation or conscious intervention? Darwin, after all, in contrast to some modern biologists, is not describing a process of genetic drift. Some shaping and patterning is performed by the play of forces he summarizes as 'natural selection'. In the later editions, long paragraphs agglomerate around the terms 'natural selection' and 'nature', paradoxically in an attempt to cut back their superplus of meaning. In the third edition, for instance, he complains that

Several writers have misapprehended or objected to the term Natural Selection. Some have even imagined that natural selection induces variability, whereas it implies only the preservation of such variations as occur and are beneficial to the being under its conditions of life.[19]

[17] Peckham, *Variorum Text*, 759 [18] Ibid. 185 [19] Ibid. 164

This new formulation denies any active role to natural selection, but problems remain; Darwin tries to solve them by appealing to the dead metaphors of other scientific fields and invoking the authority of Newton: 'who objects to an author speaking of the attraction of gravity as ruling the movements of the planets?' A note of irritation creeps in: 'Every one knows what is meant and is implied by such metaphorical expressions; and they are almost necessary for brevity.'[20] But Darwin had invented a new concept, and 'every one' did *not* know how to read it, as metaphor or as force.

Nor did he himself quite know clearly how far a transcendental level surrounded his invocation of nature. Certainly there is something extra, above and beyond the technical, in his insistence on the *naturalness* of natural selection. Nature authorizes and is internal at once. Whereas artificial selection implies an outside agency, natural selection brings agency inside production. In this second edition Darwin added to the passages from Whewell and Bacon a third epigraph, from Bishop Butler's *Analogy of Revealed Religion*, a text of constant resource for theologians at the time, and of particular value to Darwin, whose method of argument often has recourse to analogy. Nature here is figured as at once settled and continuously active, but also as the agent of an agent:

> The only distinct meaning of the word 'natural' is *stated, fixed*, or *settled*; since what is natural as much requires and pre-supposes an intelligent agent to render it so, *i.e.*, to effect it continually or at stated times, as what is supernatural or miraculous does to effect it for once.

As so often with epigraphs, the statement does not solve a problem; rather it suggests further ones. Who or what, in Darwin's argument, can this 'intelligent agent' be? The addition of Butler's epigraph seems to be part of the move in the second edition to mollify religious readers.

Darwin had a hard task, internally as well as externally, in giving up the idea of design with which the language and values of his culture were so deeply imbued. The problem of how to move from a theory of creation to one of production continued to beset him. He wavered. There is in the language

[20] Ibid. 165

of the *Origin* an intermittent suggestion that Nature's processes are, if not always manifestly benign, at least directed towards some ultimate shared good. He welcomed Asa Gray's pamphlet 'Natural Selection Not Inconsistent with Natural Theology' (1861). It is, then, not quite easy to believe him when, in the third edition, he insists parsimoniously:

> So again it is difficult to avoid personifying the word Nature; but I mean by Nature, only the aggregate action and product of many natural laws, and by laws the sequence of events as ascertained by us. With a little familiarity such superficial objections will be forgotten.[21]

The sentence reels back recursively into an empirical abysm: 'by laws the sequence of events as ascertained by us'. Reflexivity rules. Darwin favours but does not perform frugal one-to-one relations between signifier and signified. His dense, enthusiastic writing gave scope to others to press out counter-meanings. So from time to time he was, as here, obliged to curtail signification. But it is a process he finds difficult to sustain.

Equally, natural selection might be read as a satire on natural theology, with its comfortable assumption that all manifestations of nature are aspects of a relationship between God and Man. That view Darwin refuted, demonstrating that natural selection produced an organization in which the human had no separate or central place. Moreover, this organization produced a world in which its denizens were 'modified but not perfected' and where (writing of bees) 'maternal love or maternal hatred . . . is all the same to the inexorable principle of natural selection' (p. 165).

Natural selection is the central category among his three selective forms: artificial, natural, and sexual selection. He intriguingly partitions off 'sexual' from 'natural' selection, thereby seeming to recognize the degree to which sexual selection is *social* selection. Sexual selection works at its fullest in two-sex systems, and such systems produce the greatest degree of variation in offspring (as opposed, for example, to splitting or 'scission', parthenogenesis, or hermaphroditism). Sexual selection also has the effect of opening up questions of aesthetics:

[21] Peckham, *Variorum Text*, 165

I can see no good reason to doubt that female birds, by selecting, during thousands of generations, the most melodious or beautiful males, according to their standard of beauty, might produce a marked effect. (p. 74)

Darwin came to feel that he had underestimated, or under-explained, the power of sexual selection, and in *The Descent of Man, and Selection in Relation to Sex* (1872) he devoted an entire volume to its action. There he emphasized that, though in almost all species the female was the choice-maker, and the males competed among themselves, in human societies the privilege of choice-making had passed to the male, with deleterious effects. Darwin's representation of Nature as female in the *Origin* (though in a common enough rhetorical tradition) seems to lead his thought forward into this later phase of his argument. Curiously, although his *theory* emphasized male–female distinctions, his *description* in the *Origin* always placed the primordial ancestor, or 'the ancient progenitor', outside, or previous to, gender or sexed systems: an 'it', never a 'he' or 'she'. In the *Descent*, on the other hand, he misconstrued the differences between the capacities of men and women as an outcome of sexual selection, accepting that 'separate spheres' had advanced and would irrevocably advance men intellectually beyond women.

Darwin evokes and needs a more personified Nature than his disclaimer in the third edition allows. More typical of nature's maternal presence in the *Origin*'s argument are passages such as this: 'as man can certainly produce great results by adding up in any given direction mere individual differences, so could Nature, but far more easily, from having incomparably longer time at her disposal' (p. 69). But that evocation of an ordering and active Nature allows others to naturalize the current organization of society, a point Marx was quick to see:

It is remarkable how Darwin recognizes among beasts and plants his English society with its division of labour, competition, opening-up of

[22] Marx to Engels, 18 June 1862, *Selected Correspondence*, ed. S. W. Ryazanskaya (Moscow, 1965), 128.

new markets, 'inventions', and the Malthusian 'struggle for existence'.[22]

Marx here pinpoints a sanguine tendency in Darwin's summaries which might justify the operations of a society based on raw competition:

> When we reflect on this struggle, we may console ourselves with the full belief, that the war of nature is not incessant, that no fear is felt, that death is generally prompt, and that the vigorous, the healthy, and the happy survive and multiply. (p. 66)

Even here, the clause 'we may console ourselves with the full belief' undermines its own apparent certainty.

What Marx did *not* see was that Darwin's theory also potentiated an alternative to current society, one in which all organisms are bound together in 'an intimate and complex manner' and 'the relation of organism to organism the most important of all relations' (p. 51). 'Let it be borne in mind how infinitely complex and close-fitting are the mutual relations of all organic beings to each other and to their physical conditions of life' (p. 67). It would be false and sentimental to suggest that Darwin denied the power of competition for resources in evolutionary change. He viewed extinctions as inevitable in evolutionary process. But he emphasized also that survival depends upon variability, and that diversity is essential to the island of the world. Those emphases speak more strongly than ever now.

NOTE ON THE TEXT

THIS edition follows Darwin's second edition of *On The Origin of Species By Means of Natural Selection, or the Preservation of Favoured Races in the Struggle for Life* (London, 1859) first published on 24 November 1859. The second edition has type, paper, and binding identical to the first edition but corrects misprints. It might therefore be considered to be a reprint. However, despite Darwin's intention of making only a few important corrections, Morse Peckham's *Variorum* edition notes that 9 sentences are dropped, 30 added, and 483 are rewritten or repunctuated. The edition was published by Murray on 7 January 1860, a matter of only about six weeks from the first edition. It therefore represents Darwin's immediate urgent response to the private and public reception of the *Origin*. Six editions of the *Origin* were issued between 1859 and 1872 and these editions introduced considerable revisions, amplifications, corrections, and new material. The third (1861), fourth (1866), fifth (1869), and sixth (1872) editions move the text increasingly far away from many of Darwin's initial arguments in an attempt to assimilate the responses and critiques of his contemporaries, particularly palaeontologists and physicists.

SELECT BIBLIOGRAPHY

Beer, Gillian, *Darwin's Plots: Darwin, George Eliot and Nineteenth Century Fiction* (London, 1983).

—— 'Darwin and the Growth of Language Theory', in John Christie and Sally Shuttleworth (eds.), *Nature Transfigured: Science and Literature* (Manchester, 1989).

Burrow, J. W., *Evolution and Society: A Study in Victorian Social Theory* (Cambridge, 1966).

Bowler, Peter J., *The Non-Darwinian Revolution: Reinterpreting a Historical Myth* (London and Baltimore, 1988).

Chambers, Robert, *Vestiges of the Natural History of Creation and Other Evolutionary Writings*, ed. James A. Secord (Chicago, 1994) [*Vestiges* was originally published in 1844].

Darwin, Charles, *The Correspondence of Charles Darwin*, ed., Burkhardt, Frederick, *et al.*, 9 vols., further vols. forthcoming (Cambridge, 1985–).

—— *The Works of Charles Darwin*, eds. P. H. Barratt and R. B. Freeman, 29 vols. (London, 1986).

Desmond, Adrian, *The Politics of Evolution: Morphology, Medicine, and Reform in Radical London* (Chicago, 1989).

Desmond, Adrian and Moore, James, *Darwin* (London, 1991).

Ellegård, Alvar, *Darwin and the General Reader: The Reception of Darwin's Theory of Evolution in the British Periodical Press, 1859–1872* (Chicago, 1990).

Godfrey, Laurie Rohde (ed.), *What Darwin Began: Modern Darwinian and Non-Darwinian Perspectives on Evolution* (Boston, 1985).

Gould, Stephen Jay, *The Panda's Thumb: More Reflections in Natural History* (New York, 1980).

Gruber, Howard, *Darwin on Man: A Psychological Study of Scientific Creativity; together with Darwin's Early and Unpublished Notebooks*, ed. P. H. Barrett (London, 1974).

Leakey, Richard E. (ed.), *The Illustrated Origin of Species* [abridged, but with supplementary material illustrating twentieth-century evolutionary and genetic research] (London, 1979, 1986).

Peckham, Morse (ed.), *The Origin of Species By Charles Darwin: A Variorum Text* (Philadelphia, 1959).

Young, Robert, *Darwin's Metaphor: Nature's Place in Victorian Culture* (Cambridge, 1985).

A CHRONOLOGY OF CHARLES DARWIN

1809 Born 12 February in Shrewsbury, UK.
1813 Dated 'some of my earliest recollections' from a summer sea-bathing holiday at Gros near Abergele.
1817 Starts at day school, run by Unitarian minister Mr Case.
1817 Mother Susannah dies on 15 July.
1818 In September becomes a boarder at Shrewsbury School.
1825 In September begins medical studies at Edinburgh.
1827 Leaves Edinburgh in April without completing degree.
1828 In January goes up to Christ's College, Cambridge.
1830 Comes to know and work with the Professor of Botany, James Henslow.
1831 Gains the BA in January, passing tenth in a list of 178 candidates. On 27 December, after delays, the *Beagle* sets sail under its Captain, Robert Fitzroy, with Darwin as gentleman companion and naturalist.
1831–2 Parliamentary Reform Bills finally enacted in England.
1835 In September visits the Galapagos Islands.
1836 On 2 October the *Beagle* docks at Falmouth after a tour of the globe that included visits to the Tropics, South America, the Pacific Islands, Australia and New Zealand, South Africa, and innumerable islands, including the Falklands and St Helena.
1838 Manifestations of the illness which is to dog him at intervals for the rest of his life (possibly the result of a bite by the Chagas Beetle, or M.E., compounded with psychosomatic effects).
1838 Reads Thomas Malthus, *On Population.*
1839 Elected a Fellow of the Royal Society on 24 January.
1839 Marries Emma Wedgwood on 29 January.
1839 *Journal of Researches into the Geology and Natural History of the Various Countries visited by H.M.S. Beagle* first published by Henry Colburn as volume iii of Fitzroy's *Narrative*, reissued as a separate volume without permission by Colburn. The first version of what is now called *The Voyage of the Beagle.*
1842 Writes his first 'Sketch' of his theory.

1842 *The Structure and Distribution of Coral Reefs* published by Smith, Elder.

1844 Writes a more developed 'Essay' which, during the following years, he shows to Hooker, Lyell, and Asa Gray. Asks his wife to prepare the 'Essay' for publication in the event of his death.

1844 *Geological Observations on the Volcanic Islands visited during the Voyage of HMS Beagle* published by Smith, Elder.

1846 *Geological Observations on South America* published by Smith, Elder.

1851–4 *A Monograph on the Sub-Class Cirripedia*, 2 vols., published by Ray Society.

1850s Work on his big book *Natural Selection* progressing.

1858 On 18 June receives Alfred Russel Wallace's letter and paper enunciating an evolutionary mechanism parallel to Darwin's natural selection.

1858 On 30 June a joint Darwin–Wallace paper presented to the Linnean Society.

1859 Publication of *On the Origin of Species* on 24 November; Samuel Smiles, *Self-Help* published.

1860 At meeting of British Association for the Advancement of Science in Oxford, Thomas Henry Huxley clashes with Bishop Samuel Wilberforce over Darwin's theory and gets the better of him with a joke.

1863 Thomas Henry Huxley, *Man's Place in Nature and Other Anthropological Lectures* published.

1871 *The Descent of Man, and Selection in Relation to Sex* (2 vols.) published by John Murray.

1872 *The Expression of the Emotions in Man and Animals* published by Murray.

1875 *The Movements and Habits of Climbing Plants* published by Murray.

1875 *The Variation of Animals and Plants Under Domestication* revised, published by Murray.

1876 *The Effects of Cross and Self Fertilization in the Vegetable Kingdom* published by Murray.

1877 *The Various Contrivances by which Orchids are Fertilized by Insects* published by Murray.

1877 *The Different Forms of Flowers on Plants of the Same Species* published by Murray.

1881 *The Formation of Vegetable Mould, through the Action of Worms, with Observations on their Habits* published by Murray.

1882 Dies on 19 April.

THE ORIGIN OF SPECIES

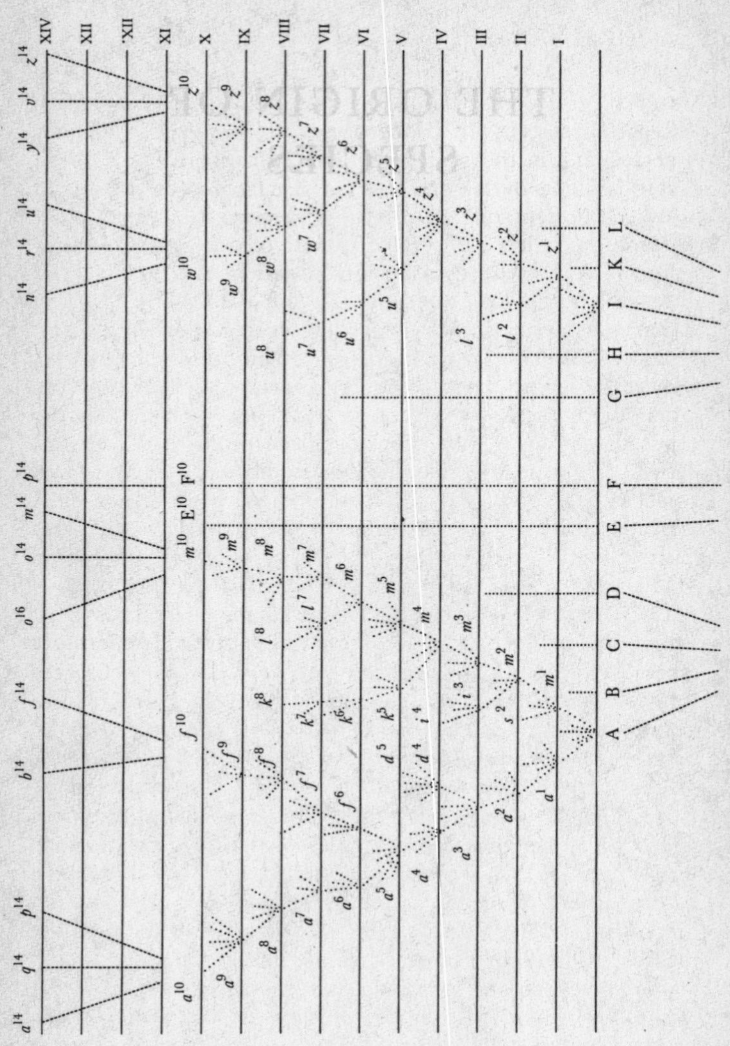

For a discussion of this diagram, see pp. 96 ff.

INTRODUCTION

When on board HMS *Beagle*, as naturalist, I was much struck with certain facts in the distribution of the inhabitants of South America, and in the geological relations of the present to the past inhabitants of that continent. These facts seemed to me to throw some light on the origin of species—that mystery of mysteries, as it has been called by one of our greatest philosophers. On my return home, it occurred to me, in 1837, that something might perhaps be made out on this question by patiently accumulating and reflecting on all sorts of facts which could possibly have any bearing on it. After five years' work I allowed myself to speculate on the subject, and drew up some short notes; these I enlarged in 1844 into a sketch of the conclusions, which then seemed to me probable: from that period to the present day I have steadily pursued the same object. I hope that I may be excused for entering on these personal details, as I give them to show that I have not been hasty in coming to a decision.

My work is now nearly finished; but as it will take me two or three more years to complete it, and as my health is far from strong, I have been urged to publish this Abstract. I have more especially been induced to do this, as Mr Wallace, who is now studying the natural history of the Malay archipelago, has arrived at almost exactly the same general conclusions that I have on the origin of species. Last year he sent me a memoir on this subject, with a request that I would forward it to Sir Charles Lyell, who sent it to the Linnean Society, and it is published in the third volume of the Journal of that Society. Sir C. Lyell and Dr Hooker, who both knew of my work—the latter having read my sketch of 1844—honoured me by thinking it advisable to publish, with Mr Wallace's excellent memoir, some brief extracts from my manuscripts.

This Abstract, which I now publish, must necessarily be imperfect. I cannot here give references and authorities for my several statements; and I must trust to the reader reposing some confidence in my accuracy. No doubt errors will have crept in, though I hope I have always been cautious in trusting to good authorities alone. I can here give only the general conclusions at which I have arrived, with a few facts in illustration, but which, I hope, in most cases will suffice. No one can feel more sensible than I do of the necessity of hereafter publishing in detail all the facts, with references, on which my conclusions have been grounded; and I hope in a future work to do this. For I am well aware that scarcely a single point is discussed in this volume on which facts cannot be adduced, often apparently leading to conclusions directly opposite to those at which I have arrived. A fair result can be obtained only by fully stating and balancing the facts and arguments on both sides of each question; and this cannot possibly be here done.

I much regret that want of space prevents my having the satisfaction of acknowledging the generous assistance which I have received from very many naturalists, some of them personally unknown to me. I cannot, however, let this opportunity pass without expressing my deep obligations to Dr Hooker, who for the last fifteen years has aided me in every possible way by his large stores of knowledge and his excellent judgment.

In considering the Origin of Species, it is quite conceivable that a naturalist, reflecting on the mutual affinities of organic beings, on their embryological relations, their geographical distribution, geological succession, and other such facts, might come to the conclusion that each species had not been independently created, but had descended, like varieties, from other species. Nevertheless, such a conclusion, even if well founded, would be unsatisfactory, until it could be shown how the innumerable species inhabiting this world have been modified, so as to acquire that perfection of structure and coadaptation which most justly excites our admiration. Naturalists continually refer to external conditions, such as climate, food, etc., as the only possible cause of variation. In one very limited sense, as we shall hereafter see, this may be true; but it is preposterous to attribute to mere external conditions the structure, for instance, of the woodpecker, with its feet, tail, beak, and tongue, so admirably adapted to catch insects under the bark of trees. In the case of the mistletoe, which draws its nourishment from certain trees, which has seeds that must be transported by certain birds, and which has flowers with separate sexes absolutely requiring the agency of certain insects to bring pollen from one flower to the other, it is equally preposterous to account for the structure of this parasite, with its relations to several distinct organic beings, by the effects of external conditions, or of habit, or of the volition of the plant itself.

The author of the *Vestiges of Creation* would, I presume, say that, after a certain unknown number of generations, some bird had given birth to a woodpecker, and some plant to the mistletoe, and that these had been produced perfect as we now see them; but this assumption seems to me to be no explanation, for it leaves the case of the coadaptations of organic beings to each other and to their physical conditions of life, untouched and unexplained.

It is, therefore, of the highest importance to gain a clear insight into the means of modification and coadaptation. At the commencement of my observations it seemed to me probable that a careful study of domesticated animals and of cultivated plants would offer the best chance of making out this obscure problem. Nor have I been disappointed; in this and in all other perplexing cases I have invariably found that our knowledge, imperfect though it be, of variation under domestication, afforded the best and safest clue. I may venture to express my conviction of the high value of such studies, although they have been very commonly neglected by naturalists.

From these considerations, I shall devote the first chapter of this Abstract to Variation under Domestication. We shall thus see that a large amount of hereditary modification is at least possible; and, what is equally or more important, we shall see how great is the power of man in accumulating by his Selection successive slight variations. I will then pass on to the variability of species in a state of nature; but I shall, unfortunately, be compelled to treat this subject far too briefly, as it can be treated properly only by giving long catalogues of facts. We shall, however, be enabled to discuss what circumstances are most favourable to variation. In the next chapter the Struggle for Existence amongst all organic beings throughout the world, which inevitably follows from the high geometrical ratio of their increase, will be treated of. This is the doctrine of Malthus, applied to the whole animal and vegetable kingdoms. As many more individuals of each species are born than can possibly survive; and as, consequently, there is a frequently recurring struggle for existence, it follows that any being, if it vary however slightly in any manner profitable to itself, under the complex and sometimes varying conditions of life, will have a better chance of surviving, and thus be *naturally selected*. From the strong principle of inheritance, any selected variety will tend to propagate its new and modified form.

This fundamental subject of Natural Selection will be treated at some length in the fourth chapter; and we shall then see how Natural Selection almost inevitably causes much Extinction of the less improved forms of life, and leads to what I have called Divergence of Character. In the next chapter I shall discuss the complex and little known laws of variation and of correlation of growth. In the four succeeding chapters, the most apparent and gravest difficulties on the theory will be given: namely, first, the difficulties of transitions, or in understanding how a simple being or a simple organ can be changed and perfected into a highly developed being or elaborately constructed organ; secondly, the subject of Instinct, or the mental powers of animals; thirdly, Hybridism, or the infertility of species and the fertility of varieties when intercrossed; and fourthly, the imperfection of the Geological Record. In the next chapter I shall consider the geological succession of organic beings throughout time; in the eleventh and twelfth, their geographical distribution throughout space; in the thirteenth, their classification or mutual affinities, both when mature and in an embryonic condition. In the last chapter I shall give a brief recapitulation of the whole work, and a few concluding remarks.

No one ought to feel surprise at much remaining as yet unexplained in regard to the origin of species and varieties, if he makes due allowance for our profound ignorance in regard to the mutual relations of all the beings which live around us. Who can explain why one species ranges widely and is very numerous, and why another allied species has a narrow range and is rare? Yet these relations are of the highest importance, for they determine the present welfare, and, as I believe, the future success and modification of every inhabitant of this world. Still less do we know of the mutual relations of the innumerable inhabitants of the world during the many past geological epochs in its history. Although much remains obscure, and will long remain obscure, I can entertain no doubt, after the most deliberate study and dispassionate judgment of which I am capable, that the view which most naturalists entertain, and which I formerly entertained—namely, that each species has been independently created—is erroneous. I am fully convinced that species are not immutable; but that those belonging to what are called the same genera are lineal descendants of some other and generally extinct species, in the same manner as the acknowledged varieties of any one species are the descendants of that species. Furthermore, I am convinced that Natural Selection has been the main but not exclusive means of modification.

CHAPTER I

VARIATION UNDER DOMESTICATION

Causes of Variability—Effects of Habit—Correlation of Growth—Inheritance—Character of Domestic Varieties—Difficulty of distinguishing between Varieties and Species—Origin of Domestic Varieties from one or more Species—Domestic Pigeons, their Differences and Origin—Principle of Selection anciently followed, its Effects—Methodical and Unconscious Selection—Unknown Origin of our Domestic Productions—Circumstances favourable to Man's power of Selection.

When we look to the individuals of the same variety or sub-variety of our older cultivated plants and animals, one of the first points which strikes us is, that they generally differ more from each other than do the individuals of any one species or variety in a state of nature. When we reflect on the vast diversity of the plants and animals which have been cultivated, and which have varied during all ages under the most different climates and treatment, I think we are driven to conclude that this great variability is simply due to our domestic productions having been raised under conditions of life not so uniform as, and somewhat different from, those to which the parent-species have been exposed under nature. There is also, I think, some probability in the view propounded by Andrew Knight, that this variability may be partly connected with excess of food. It seems pretty clear that organic beings must be exposed during several generations to the new conditions of life to cause any appreciable amount of variation; and that when the organisation has once begun to vary, it generally continues to vary for many generations. No case is on record of a variable being ceasing to be variable under cultivation. Our oldest cultivated plants, such as wheat, still often yield new varieties: our oldest domesticated animals are still capable of rapid improvement or modification.

It has been disputed at what period of life the causes of variability, whatever they may be, generally act; whether during the early or late period of development of the embryo, or at the instant of conception. Geoffroy St Hilaire's experiments show that unnatural treatment of the embryo causes monstrosities; and monstrosities cannot be separated by any clear line of distinction from mere variations. But I am strongly inclined to suspect that the most frequent cause of variability may be attributed to the male and female reproductive elements having been affected prior to the act of conception. Several reasons make me believe in this; but the chief one is the remarkable effect which confinement or cultivation has on the function of the reproductive system; this system appearing to be far more susceptible than any other part of the organisation, to the action of any change in the conditions of life. Nothing is more easy than to tame an animal, and few things more difficult than to get it to breed freely under confinement, even in the many cases when the male and female unite. How many animals there are which will not breed, though living long under not very close confinement in their native country! This is generally attributed to vitiated instincts; but how many cultivated plants display the utmost vigour, and yet rarely or never seed! In some few such cases it has been discovered that very trifling changes, such as a little more or less water at some particular period of growth, will determine whether or not the plant sets a seed. I cannot here enter on the copious details which I have collected on this curious subject; but to show how singular the laws are which determine the reproduction of animals under confinement, I may just mention that carnivorous animals, even from the tropics, breed in this country pretty freely under confinement, with the exception of the plantigrades or bear family; whereas carnivorous birds, with the rarest exceptions, hardly ever lay fertile eggs. Many exotic plants have pollen utterly worthless, in the same exact condition as in the most sterile hybrids. When, on the one hand, we see domesticated animals and plants, though often weak and sickly, yet breeding quite freely under confinement; and when, on the other hand, we see individuals, though taken young from a state of nature, perfectly tamed, long-lived, and healthy (of which I could give numerous instances), yet having their reproductive system so seriously affected by unperceived causes as to fail in acting, we need not be surprised at this system, when it does act under confinement, acting not quite regularly, and producing offspring not perfectly like their parents.

Sterility has been said to be the bane of horticulture; but on this view we owe variability to the same cause which produces sterility; and variability is the source of all the choicest productions of the garden. I may add, that as some organisms will breed freely under the most unnatural conditions (for instance, the rabbit and ferret kept in hutches), showing that their reproductive system has not been thus affected; so will some animals and plants withstand domestication or cultivation, and vary very slightly—perhaps hardly more than in a state of nature.

A long list could easily be given of 'sporting plants'; by this term gardeners mean a single bud or offset, which suddenly assumes a new and sometimes very different character from that of the rest of the plant. Such buds can be propagated by grafting, etc., and sometimes by seed. These 'sports' are extremely rare under nature, but far from rare under cultivation; and in this case we see that the treatment of the parent has affected a bud or offset, and not the ovules or pollen. But it is the opinion of most physiologists that there is no essential difference between a bud and an ovule in their earliest stages of formation; so that, in fact, 'sports' support my view, that variability may be largely attributed to the ovules or pollen, or to both, having been affected by the treatment of the parent prior to the act of conception. These cases anyhow show that variation is not necessarily connected, as some authors have supposed, with the act of generation.

Seedlings from the same fruit, and the young of the same litter, sometimes differ considerably from each other, though both the young and the parents, as Müller has remarked, have apparently been exposed to exactly the same conditions of life; and this shows how unimportant the direct effects of the conditions of life are in comparison with the laws of reproduction, of growth, and of inheritance; for had the action of the conditions been direct, if any of the young had varied, all would probably have varied in the same manner. To judge how much, in the case of any variation, we should attribute to the direct action of heat, moisture, light, food, etc., is most difficult: my impression is, that with animals such agencies have produced very little direct effect, though apparently more in the case of plants. Under this point of view, Mr Buckman's recent experiments on plants are extremely valuable. When all or nearly all the individuals exposed to certain conditions are affected in the same way, the change at first appears to be directly due to such conditions; but in some cases it can be shown that quite opposite conditions produce similar changes of structure. Nevertheless some slight amount of change may, I think, be attributed to the direct action of the conditions of life—as, in some cases, increased size from amount of food, colour from particular kinds of food or from light, and perhaps the thickness of fur from climate.

Habit also has a decided influence, as in the period of flowering with plants when transported from one climate to another. In animals it has a more marked effect; for instance, I find in the domestic duck that the bones of the wing weigh less and the bones of the leg more, in proportion to the whole skeleton, than do the same bones in the wild-duck; and I presume that this change may be safely attributed to the domestic duck flying much less, and walking more, than its wild parent. The great and inherited development of the udders in cows and goats in countries where they are habitually milked, in comparison with the state of these organs in other countries, is another instance of the effect of use. Not a single domestic animal can be named which has not in some country drooping ears; and the view suggested by some authors, that the drooping is due to the disuse of the muscles of the ear, from the animals not being much alarmed by danger, seems probable.

There are many laws regulating variation, some few of which can be dimly seen, and will be hereafter briefly mentioned. I will here only allude to what may be called correlation of growth. Any change in the embryo or larva will almost certainly entail changes in the mature animal. In monstrosities, the correlations between quite distinct parts are very curious; and many instances are given in Isidore Geoffroy St Hilaire's great work on this subject. Breeders believe that long limbs are almost always accompanied by an elongated head. Some instances of correlation are quite whimsical: thus cats with blue eyes are invariably deaf; colour and constitutional peculiarities go together, of which many remarkable cases could be given amongst animals and plants. From the facts collected by Heusinger, it appears that white sheep and pigs are differently affected from coloured individuals by certain vegetable poisons. Hairless dogs have imperfect teeth; long-haired and coarse-haired animals are apt to have, as is asserted, long or many horns; pigeons with feathered feet have skin between their outer toes; pigeons with short beaks have small feet, and those with long beaks large feet. Hence, if man goes on selecting, and thus augmenting, any peculiarity, he will almost certainly unconsciously modify other parts of the structure, owing to the mysterious laws of the correlation of growth.

The result of the various, quite unknown, or dimly seen laws of variation is infinitely complex and diversified. It is well worth while carefully to study the several treatises published on some of our old cultivated plants, as on the hyacinth, potato, even the dahlia, etc.; and it is really surprising to note the endless points in structure and constitution in which the varieties and sub-varieties differ slightly from each other. The whole organisation seems to have become plastic, and tends to depart in some small degree from that of the parental type.

Any variation which is not inherited is unimportant for us. But the number and diversity of inheritable deviations of structure, both those of slight and those of considerable physiological importance, is endless. Dr Prosper Lucas's treatise, in two large volumes, is the fullest and the best on this subject. No breeder doubts how strong is the tendency to inheritance: like produces like is his fundamental belief: doubts have been thrown on this principle by theoretical writers alone. When any deviation of structure often appears, and we see it in the father and child, we cannot tell whether it may not be due to the same cause having acted on both; but when amongst individuals, apparently exposed to the same conditions, any very rare deviation, due to some extraordinary combination of circumstances, appears in the parent—say, once amongst several million individuals—and it reappears in the child, the mere doctrine of chances almost compels us to attribute its reappearance to inheritance. Every one must have heard of cases of albinism, prickly skin, hairy bodies, etc., appearing in several members of the same family. If strange and rare deviations of structure are truly inherited, less strange and commoner deviations may be freely admitted to be inheritable. Perhaps the correct way of viewing the whole subject, would be, to look at the inheritance of every character whatever as the rule, and non-inheritance as the anomaly.

The laws governing inheritance are quite unknown; no one can say why a peculiarity in different individuals of the same species, or in individuals of different species, is sometimes inherited and sometimes not so; why the child often reverts in certain characters to its grandfather or grandmother or other more remote ancestor; why a peculiarity is often transmitted from one sex to both sexes, or to one sex alone, more commonly but not exclusively to the like sex. It is a fact of some little importance to us, that peculiarities appearing in the males of our domestic breeds are often transmitted either exclusively, or in a much greater degree, to males alone. A much more important rule, which I think may be trusted, is that, at whatever period of life a peculiarity appears, it tends to appear in the offspring at a corresponding age, though sometimes earlier. In many cases this could not be otherwise: thus the inherited peculiarities in the horns of cattle could appear only in the offspring when nearly mature; peculiarities in the silkworm are known to appear at the corresponding caterpillar or cocoon stage. But hereditary diseases and some other facts make me believe that the rule has a wider extension, and that when there is no apparent reason why a peculiarity should appear at any particular age, yet that it does tend to appear in the offspring at the same period at which it first appeared in the parent. I believe this rule to be of the highest importance in explaining the laws of embryology. These remarks are of course confined to the first *appearance* of the peculiarity, and not to its primary cause, which may have acted on the ovules or male element; in nearly the same manner as in the crossed offspring from a short-horned cow by a long-horned bull, the greater length of horn, though appearing late in life, is clearly due to the male element.

Having alluded to the subject of reversion, I may here refer to a statement often made by naturalists—namely, that our domestic varieties, when run wild, gradually but certainly revert in character to their aboriginal stocks. Hence it has been argued that no deductions can be drawn from domestic races to species in a state of nature. I have in vain endeavoured to discover on what decisive facts the above statement has so often and so boldly been made. There would be great difficulty in proving its truth: we may safely conclude that very many of the most strongly-marked domestic varieties could not possibly live in a wild state. In many cases we do not know what the aboriginal stock was, and so could not tell whether or not nearly perfect reversion had ensued. It would be quite necessary, in order to prevent the effects of inter-crossing, that only a single variety should be turned loose in its new home. Nevertheless, as our varieties certainly do occasionally revert in some of their characters to ancestral forms, it seems to me not improbable, that if we could succeed in naturalising, or were to cultivate, during many generations, the several races, for instance, of the cabbage, in very poor soil (in which case, however, some effect would have to be attributed to the direct action of the poor soil), that they would to a large extent, or even wholly, revert to the wild aboriginal stock. Whether or not the experiment would succeed, is not of great importance for our line of argument; for by the experiment itself the conditions of life are changed. If it could be shown that our domestic varieties manifested a strong tendency to reversion,—that is, to lose their acquired characters, whilst kept under the same conditions, and whilst kept in a considerable body, so that free intercrossing might check, by blending together, any slight deviations in their structure, in such case, I grant that we could deduce nothing from domestic varieties in regard to species. But there is not a shadow of evidence in favour of this view: to assert that we could not breed our cart and race horses, long and short horned cattle, and poultry of various breeds, and esculent vegetables, for an almost infinite number of generations, would be opposed to all experience. I may add, that when under nature the conditions of life do change, variations and reversions of character probably do occur; but natural selection, as will hereafter be explained, will determine how far the new characters thus arising shall be preserved.

When we look to the hereditary varieties or races of our domestic animals and plants, and compare them with closely-allied species, we generally perceive in each domestic race, as already remarked, less uniformity of character than in true species. Domestic races of the same species, also, often have a somewhat monstrous character; by which I mean, that, although differing from each other, and from other species of the same genus, in several trifling respects, they often differ in an extreme degree in some one part, both when compared one with another, and more especially when compared with all the species in nature to which they are nearest allied. With these exceptions (and with that of the perfect fertility of varieties when crossed,—a subject hereafter to be discussed), domestic races of the same species differ from each other in the same manner as, only in most cases in a lesser degree than, do closely-allied species of the same genus in a state of nature. I think this must be admitted, when we find that there are hardly any domestic races, either amongst animals or plants, which have not been ranked by competent judges as mere varieties, and by other competent judges as the descendants of aboriginally distinct species. If any marked distinction existed between domestic races and species, this source of doubt could not so perpetually recur. It has often been stated that domestic races do not differ from each other in characters of generic value. I think it could be shown that this statement is hardly correct; but naturalists differ widely in determining what characters are of generic value; all such valuations being at present empirical. Moreover, on the view of the origin of genera which I shall presently give, we have no right to expect often to meet with generic differences in our domesticated productions.

When we attempt to estimate the amount of structural difference between the domestic races of the same species, we are soon involved in doubt, from not knowing whether they have descended from one or several parent species. This point, if it could be cleared up, would be interesting; if, for instance, it could be shown that the greyhound, bloodhound, terrier, spaniel, and bull-dog, which we all know propagate their kind so truly, were the offspring of any single species, then such facts would have great weight in making us doubt about the immutability of the many very closely-allied natural species—for instance, of the many foxes—inhabiting different quarters of the world. I do not believe, as we shall presently see, that the whole amount of difference between the several breeds of the dog has been produced under domestication; I believe that some small part of the difference is due to their being descended from distinct species. In the case of some other domesticated species, there is presumptive, or even strong evidence, that all the breeds have descended from a single wild stock.

It has often been assumed that man has chosen for domestication animals and plants having an extraordinary inherent tendency to vary, and likewise to withstand diverse climates. I do not dispute that these capacities have added largely to the value of most of our domesticated productions; but how could a savage possibly know, when he first tamed an animal, whether it would vary in succeeding generations, and whether it would endure other climates? Has the little variability of the ass or guinea-fowl, or the small power of endurance of warmth by the reindeer, or of cold by the common camel, prevented their domestication? I cannot doubt that if other animals and plants, equal in number to our domesticated productions, and belonging to equally diverse classes and countries, were taken from a state of nature, and could be made to breed for an equal number of generations under domestication, they would vary on an average as largely as the parent species of our existing domesticated productions have varied.

In the case of most of our anciently domesticated animals and plants, I do not think it is possible to come to any definite conclusion, whether they have descended from one or several wild species. The argument mainly relied on by those who believe in the multiple origin of our domestic animals is, that we find in the most ancient records, more especially on the monuments of Egypt, much diversity in the breeds; and that some of the breeds closely resemble, perhaps are identical with, those still existing. Even if this latter fact were found more strictly and generally true than seems to me to be the case, what does it show, but that some of our breeds originated there, four or five thousand years ago? But Mr Horner's researches have rendered it in some degree probable that man sufficiently civilised to have manufactured pottery existed in the valley of the Nile thirteen or fourteen thousand years ago; and who will pretend to say how long before these ancient periods, savages, like those of Tierra del Fuego or Australia, who possess a semi-domestic dog, may not have existed in Egypt?

The whole subject must, I think, remain vague; nevertheless, I may, without here entering on any details, state that, from geographical and other considerations, I think it highly probable that our domestic dogs have descended from several wild species. Knowing, as we do, that savages are very fond of taming animals, it seems to me unlikely, in the case of the dog-genus, which is distributed in a wild state throughout the world, that since man first appeared one single species alone should have been domesticated. In regard to sheep and goats I can form no opinion. I should think, from facts communicated to me by Mr Blyth, on the habits, voice, and constitution, etc., of the humped Indian cattle, that these had descended from a different aboriginal stock from our European cattle; and several competent judges believe that these latter have had more than one wild parent. With respect to horses, from reasons which I cannot give here, I am doubtfully inclined to believe, in opposition to several authors, that all the races have descended from one wild stock. Mr Blyth, whose opinion, from his large and varied stores of knowledge, I should value more than that of almost any one, thinks that all the breeds of poultry have proceeded from the common wild Indian fowl (Gallus bankiva). In regard to ducks and rabbits, the breeds of which differ considerably from each other in structure, I do not doubt that they have all descended from the common wild duck and rabbit.

The doctrine of the origin of our several domestic races from several aboriginal stocks, has been carried to an absurd extreme by some authors. They believe that every race which breeds true, let the distinctive characters be ever so slight, has had its wild prototype. At this rate there must have existed at least a score of species of wild cattle, as many sheep, and several goats in Europe alone, and several even within Great Britain. One author believes that there formerly existed in Great Britain eleven wild species of sheep peculiar to it. When we bear in mind that Britain has now hardly one peculiar mammal, and France but few distinct from those of Germany and conversely, and so with Hungary, Spain, etc., but that each of these kingdoms possesses several peculiar breeds of cattle, sheep, etc., we must admit that many domestic breeds have originated in Europe; for whence could they have been derived, as these several countries do not possess a number of peculiar species as distinct parent-stocks? So it is in India. Even in the case of the domestic dogs of the whole world, which I fully admit have probably descended from several wild species, I cannot doubt that there has been an immense amount of inherited variation. Who can believe that animals closely resembling the Italian greyhound, the bloodhound, the bull-dog, or Blenheim spaniel, etc.—so unlike all wild Canidæ—ever existed freely in a state of nature? It has often been loosely said that all our races of dogs have been produced by the crossing of a few aboriginal species; but by crossing we can only get forms in some degree intermediate between their parents; and if we account for our several domestic races by this process, we must admit the former existence of the most extreme forms, as the Italian greyhound, bloodhound, bull-dog, etc., in the wild state. Moreover, the possibility of making distinct races by crossing has been greatly exaggerated. There can be no doubt that a race may be modified by occasional crosses, if aided by the careful selection of those individual mongrels, which present any desired character; but that a race could be obtained nearly intermediate between two extremely different races or species, I can hardly believe. Sir J. Sebright expressly experimentised for this object, and failed. The off-spring from the first cross between two pure breeds is tolerably and sometimes (as I have found with pigeons) extremely uniform, and everything seems simple enough; but when these mongrels are crossed one with another for several generations, hardly two of them will be alike, and then the extreme difficulty, or rather utter hopelessness, of the task becomes apparent. Certainly, a breed intermediate between *two very distinct* breeds could not be got without extreme care and long-continued selection; nor can I find a single case on record of a permanent race having been thus formed.

On the Breeds of the Domestic Pigeon.—Believing that it is always best to study some special group, I have, after deliberation, taken up domestic pigeons. I have kept every breed which I could purchase or obtain, and have been most kindly favoured with skins from several quarters of the world, more especially by the Hon. W. Elliot from India, and by the Hon. C. Murray from Persia. Many treatises in different languages have been published on pigeons, and some of them are very important, as being of considerable antiquity. I have associated with several eminent fanciers, and have been permitted to join two of the London Pigeon Clubs. The diversity of the breeds is something astonishing. Compare the English carrier and the short-faced tumbler, and see the wonderful difference in their beaks, entailing corresponding differences in their skulls. The carrier, more especially the male bird, is also remarkable from the wonderful development of the carunculated skin about the head, and this is accompanied by greatly elongated eyelids, very large external orifices to the nostrils, and a wide gape of mouth. The short-faced tumbler has a beak in outline almost like that of a finch; and the common tumbler has the singular inherited habit of flying at a great height in a compact flock, and tumbling in the air head over heels. The runt is a bird of great size, with long, massive beak and large feet; some of the sub-breeds of runts have very long necks, others very long wings and tails, others singularly short tails. The barb is allied to the carrier, but, instead of a very long beak, has a very short and very broad one. The pouter has a much elongated body, wings, and legs; and its enormously developed crop, which it glories in inflating, may well excite astonishment and even laughter. The turbit has a very short and conical beak, with a line of reversed feathers down the breast; and it has the habit of continually expanding slightly the upper part of the œsophagus. The Jacobin has the feathers so much reversed along the back of the neck that they form a hood, and it has, proportionally to its size, much elongated wing and tail feathers. The trumpeter and laughter, as their names express, utter a very different coo from the other breeds. The fantail has thirty or even forty tail feathers, instead of twelve or fourteen, the normal number in all members of the great pigeon family; and these feathers are kept expanded, and are carried so erect that in good birds the head and tail touch; the oil-gland is quite aborted. Several other less distinct breeds might be specified.

In the skeletons of the several breeds, the development of the bones of the face in length and breadth and curvature differs enormously. The shape, as well as the breadth and length of the ramus of the lower jaw, varies in a highly remarkable manner. The number of the caudal and sacral vertebræ vary; as does the number of the ribs, together with their relative breadth and the presence of processes. The size and shape of the apertures in the sternum are highly variable; so is the degree of divergence and relative size of the two arms of the furcula. The proportional width of the gape of mouth, the proportional length of the eyelids, of the orifice of the nostrils, of the tongue (not always in strict correlation with the length of beak), the size of the crop and of the upper part of the œsophagus; the development and abortion of the oil-gland; the number of the primary wing and caudal feathers; the relative length of wing and tail to each other and to the body; the relative length of leg and of the feet; the number of scutellæ on the toes, the development of skin between the toes, are all points of structure which are variable. The period at which the perfect plumage is acquired varies, as does the state of the down with which the nestling birds are clothed when hatched. The shape and size of the eggs vary. The manner of flight differs remarkably; as does in some breeds the voice and disposition. Lastly, in certain breeds, the males and females have come to differ to a slight degree from each other.

Altogether at least a score of pigeons might be chosen, which if shown to an ornithologist, and he were told that they were wild birds, would certainly, I think, be ranked by him as well-defined species. Moreover, I do not believe that any ornithologist would place the English carrier, the short-faced tumbler, the runt, the barb, pouter, and fantail in the same genus; more especially as in each of these breeds several truly-inherited sub-breeds, or species as he might have called them, could be shown him.

Great as the differences are between the breeds of pigeons, I am fully convinced that the common opinion of naturalists is correct, namely, that all have descended from the rock-pigeon (Columba livia), including under this term several geographical races or sub-species, which differ from each other in the most trifling respects. As several of the reasons which have led me to this belief are in some degree applicable in other cases, I will here briefly give them. If the several breeds are not varieties, and have not proceeded from the rock-pigeon, they must have descended from at least seven or eight aboriginal stocks; for it is impossible to make the present domestic breeds by the crossing of any lesser number: how, for instance, could a pouter be produced by crossing two breeds unless one of the parent-stocks possessed the characteristic enormous crop? The supposed aboriginal stocks must all have been rock-pigeons, that is, not breeding or willingly perching on trees. But besides C. livia, with its geographical sub-species, only two or three other species of rock-pigeons are known; and these have not any of the characters of the domestic breeds. Hence the supposed aboriginal stocks must either still exist in the countries where they were originally domesticated, and yet be unknown to ornithologists; and this, considering their size, habits, and remarkable characters, seems very improbable; or they must have become extinct in the wild state. But birds breeding on precipices, and good fliers, are unlikely to be exterminated; and the common rock-pigeon, which has the same habits with the domestic breeds, has not been exterminated even on several of the smaller British islets, or on the shores of the Mediterranean. Hence the supposed extermination of so many species having similar habits with the rock-pigeon seems to me a very rash assumption. Moreover, the several above-named domesticated breeds have been transported to all parts of the world, and, therefore, some of them must have been carried back again into their native country; but not one has ever become wild or feral, though the dovecot-pigeon, which is the rock-pigeon in a very slightly altered state, has become feral in several places. Again, all recent experience shows that it is most difficult to get any wild animal to breed freely under domestication; yet on the hypothesis of the multiple origin of our pigeons, it must be assumed that at least seven or eight species were so thoroughly domesticated in ancient times by half-civilised man, as to be quite prolific under confinement.

An argument, as it seems to me, of great weight, and applicable in several other cases, is, that the above-specified breeds, though agreeing generally in constitution, habits, voice, colouring, and in most parts of their structure, with the wild rock-pigeon, yet are certainly highly abnormal in other parts of their structure; we may look in vain throughout the whole great family of Columbidæ for a beak like that of the English carrier, or that of the short-faced tumbler, or barb; for reversed feathers like those of the Jacobin; for a crop like that of the pouter; for tail-feathers like those of the fantail. Hence it must be assumed not only that half-civilised man succeeded in thoroughly domesticating several species, but that he intentionally or by chance picked out extraordinarily abnormal species; and further, that these very species have since all become extinct or unknown. So many strange contingencies seem to me improbable in the highest degree.

Some facts in regard to the colouring of pigeons well deserve consideration. The rock-pigeon is of a slaty-blue, and has a white rump (the Indian sub-species, C. intermedia of Strickland, having it bluish); the tail has a terminal dark bar, with the bases of the outer feathers externally edged with white; the wings have two black bars; some semi-domestic breeds and some apparently truly wild breeds have, beside the two black bars, the wings chequered with black. These several marks do not occur together in any other species of the whole family. Now, in every one of the domestic breeds, taking thoroughly well-bred birds, all the above marks, even to the white edging of the outer tail-feathers, sometimes concur perfectly developed. Moreover, when two birds belonging to two distinct breeds are crossed, neither of which is blue or has any of the above-specified marks, the mongrel offspring are very apt suddenly to acquire these characters; for instance, I crossed some uniformly white fantails with some uniformly black barbs, and they produced mottled brown and black birds; these I again crossed together, and one grandchild of the pure white fantail and pure black barb was of as beautiful a blue colour, with the white rump, double black wing-bar, and barred and white-edged tail-feathers, as any wild rock-pigeon! We can understand these facts, on the well-known principle of reversion to ancestral characters, if all the domestic breeds have descended from the rock-pigeon. But if we deny this, we must make one of the two following highly improbable suppositions. Either, firstly, that all the several imagined aboriginal stocks were coloured and marked like the rock-pigeon, although no other existing species is thus coloured and marked, so that in each separate breed there might be a tendency to revert to the very same colours and markings. Or, secondly, that each breed, even the purest, has within a dozen or, at most, within a score of generations, been crossed by the rock-pigeon: I say within a dozen or twenty generations, for we know of no fact countenancing the belief that the child ever reverts to some one ancestor, removed by a greater number of generations. In a breed which has been crossed only once with some distinct breed, the tendency to reversion to any character derived from such cross will naturally become less and less, as in each succeeding generation there will be less of the foreign blood; but when there has been no cross with a distinct breed, and there is a tendency in both parents to revert to a character, which has been lost during some former generation, this tendency, for all that we can see to the contrary, may be transmitted undiminished for an indefinite number of generations. These two distinct cases are often confounded in treatises on inheritance.

Lastly, the hybrids or mongrels from between all the domestic breeds of pigeons are perfectly fertile. I can state this from my own observations, purposely made, on the most distinct breeds. Now, it is difficult, perhaps impossible, to bring forward one case of the hybrid offspring of two animals *clearly distinct* being themselves perfectly fertile. Some authors believe that long-continued domestication eliminates this strong tendency to sterility: from the history of the dog I think there is some probability in this hypothesis, if applied to species closely related together, though it is unsupported by a single experiment. But to extend the hypothesis so far as to suppose that species, aboriginally as distinct as carriers, tumblers, pouters, and fantails now are, should yield offspring perfectly fertile, *inter se*, seems to me rash in the extreme.

From these several reasons, namely, the improbability of man having formerly got seven or eight supposed species of pigeons to breed freely under domestication; these supposed species being quite unknown in a wild state, and their becoming nowhere feral; these species having very abnormal characters in certain respects, as compared with all other Columbidæ, though so like in most other respects to the rock-pigeon; the blue colour and various marks occasionally appearing in all the breeds, both when kept pure and when crossed; the mongrel offspring being perfectly fertile;—from these several reasons, taken together, I can feel no doubt that all our domestic breeds have descended from the Columbia livia with its geographical sub-species.

In favour of this view, I may add, firstly, that C. livia, or the rock-pigeon, has been found capable of domestication in Europe and in India; and that it agrees in habits and in a great number of points of structure with all the domestic breeds. Secondly, although an English carrier or short-faced tumbler differs immensely in certain characters from the rock-pigeon, yet by comparing the several sub-breeds of these varieties, more especially those brought from distant countries, we can make an almost perfect series between the extremes of structure. Thirdly, those characters which are mainly distinctive of each breed, for instance the wattle and length of beak of the carrier, the shortness of that of the tumbler, and the number of tail-feathers in the fantail, are in each breed eminently variable; and the explanation of this fact will be obvious when we come to treat of selection. Fourthly, pigeons have been watched, and tended with the utmost care, and loved by many people. They have been domesticated for thousands of years in several quarters of the world; the earliest known record of pigeons is in the fifth Egyptian dynasty, about 3000 BC, as was pointed out to me by Professor Lepsius; but Mr Birch informs me that pigeons are given in a bill of fare in the previous dynasty. In the time of the Romans, as we hear from Pliny, immense prices were given for pigeons; 'nay, they are come to this pass, that they can reckon up their pedigree and race.' Pigeons were much valued by Akber Khan in India, about the year 1600; never less than 20,000 pigeons were taken with the court. 'The monarchs of Iran and Turan sent him some very rare birds'; and, continues the courtly historian, 'His Majesty by crossing the breeds, which method was never practised before, has improved them astonishingly.' About this same period the Dutch were as eager about pigeons as were the old Romans. The paramount importance of these considerations in explaining the immense amount of variation which pigeons have undergone, will be obvious when we treat of Selection. We shall then, also, see how it is that the breeds so often have a somewhat monstrous character. It is also a most favourable circumstance for the production of distinct breeds, that male and female pigeons can be easily mated for life; and thus different breeds can be kept together in the same aviary.

I have discussed the probable origin of domestic pigeons at some, yet quite insufficient, length; because when I first kept pigeons and watched the several kinds, knowing well how true they bred, I felt fully as much difficulty in believing that they could have descended from a common parent, as any naturalist could in coming to a similar conclusion in regard to the many species of finches, or other large groups of birds, in nature. One circumstance has struck me much; namely, that all the breeders of the various domestic animals and the cultivators of plants, with whom I have ever conversed, or whose treatises I have read, are firmly convinced that the several breeds to which each has attended, are descended from so many aboriginally distinct species. Ask, as I have asked, a celebrated raiser of Hereford cattle, whether his cattle might not have descended from long-horns, and he will laugh you to scorn. I have never met a pigeon, or poultry, or duck, or rabbit fancier, who was not fully convinced that each main breed was descended from a distinct species. Van Mons, in his treatise on pears and apples, shows how utterly he disbelieves that the several sorts, for instance a Ribston-pippin or Codlin-apple, could ever have proceeded from the seeds of the same tree. Innumerable other examples could be given. The explanation, I think, is simple: from long-continued study they are strongly impressed with the differences between the several races; and though they well know that each race varies slightly, for they win their prizes by selecting such slight differences, yet they ignore all general arguments, and refuse to sum up in their minds slight differences accumulated during many successive generations. May not those naturalists who, knowing far less of the laws of inheritance than does the breeder, and knowing no more than he does of the intermediate links in the long lines of descent, yet admit that many of our domestic races have descended from the same parents—may they not learn a lesson of caution, when they deride the idea of species in a state of nature being lineal descendants of other species?

Selection.—Let us now briefly consider the steps by which domestic races have been produced, either from one or from several allied species. Some little effect may, perhaps, be attributed to the direct action of the external conditions of life, and some little to habit; but he would be a bold man who would account by such agencies for the differences of a dray and a race horse, a greyhound and bloodhound, a carrier and tumbler pigeon. One of the most remarkable features in our domesticated races is that we see in them adaptation, not indeed to the animal's or plant's own good, but to man's use or fancy. Some variations useful to him have probably arisen suddenly, or by one step; many botanists, for instance, believe that the fuller's teasel, with its hooks, which cannot be rivalled by any mechanical contrivance, is only a variety of the wild Dipsacus; and this amount of change may have suddenly arisen in a seedling. So it has probably been with the turnspit dog; and this is known to have been the case with the ancon sheep. But when we compare the dray-horse and race-horse, the dromedary and camel, the various breeds of sheep fitted either for cultivated land or mountain pasture, with the wool of one breed good for one purpose, and that of another breed for another purpose; when we compare the many breeds of dogs, each good for man in very different ways; when we compare the game-cock, so pertinacious in battle, with other breeds so little quarrelsome, with 'everlasting layers' which never desire to sit, and with the bantam so small and elegant; when we compare the host of agricultural, culinary, orchard, and flower-garden races of plants, most useful to man at different seasons and for different purposes, or so beautiful in his eyes, we must, I think, look further than to mere variability. We cannot suppose that all the breeds were suddenly produced as perfect and as useful as we now see them; indeed, in several cases, we know that this has not been their history. The key is man's power of accumulative selection: nature gives successive variations; man adds them up in certain directions useful to him. In this sense he may be said to make for himself useful breeds.

The great power of this principle of selection is not hypothetical. It is certain that several of our eminent breeders have, even within a single lifetime, modified to a large extent some breeds of cattle and sheep. In order fully to realise what they have done, it is almost necessary to read several of the many treatises devoted to this subject, and to inspect the animals. Breeders habitually speak of an animal's organisation as something quite plastic, which they can model almost as they please. If I had space I could quote numerous passages to this effect from highly competent authorities. Youatt, who was probably better acquainted with the works of agriculturists than almost any other individual, and who was himself a very good judge of an animal, speaks of the principle of selection as 'that which enables the agriculturist, not only to modify the character of his flock, but to change it altogether. It is the magician's wand, by means of which he may summon into life whatever form and mould he pleases.' Lord Somerville, speaking of what breeders have done for sheep, says:—'It would seem as if they had chalked out upon a wall a form perfect in itself, and then had given it existence.' That most skilful breeder, Sir John Sebright, used to say, with respect to pigeons, that 'he would produce any given feather in three years, but it would take him six years to obtain head and beak.' In Saxony the importance of the principle of selection in regard to merino sheep is so fully recognised, that men follow it as a trade: the sheep are placed on a table and are studied, like a picture by a connoisseur; this is done three times at intervals of months, and the sheep are each time marked and classed, so that the very best may ultimately be selected for breeding.

What English breeders have actually effected is proved by the enormous prices given for animals with a good pedigree; and these have now been exported to almost every quarter of the world. The improvement is by no means generally due to crossing different breeds; all the best breeders are strongly opposed to this practice, except sometimes amongst closely allied sub-breeds. And when a cross has been made, the closest selection is far more indispensable even than in ordinary cases. If selection consisted merely in separating some very distinct variety, and breeding from it, the principle would be so obvious as hardly to be worth notice; but its importance consists in the great effect produced by the accumulation in one direction, during successive generations, of differences absolutely inappreciable by an uneducated eye—differences which I for one have vainly attempted to appreciate. Not one man in a thousand has accuracy of eye and judgment sufficient to become an eminent breeder. If gifted with these qualities, and he studies his subject for years, and devotes his lifetime to it with indomitable perseverance, he will succeed, and may make great improvements; if he wants any of these qualities, he will assuredly fail. Few would readily believe in the natural capacity and years of practice requisite to become even a skilful pigeon-fancier.

The same principles are followed by horticulturists; but the variations are here often more abrupt. No one supposes that our choicest productions have been produced by a single variation from the aboriginal stock. We have proofs that this is not so in some cases, in which exact records have been kept; thus, to give a very trifling instance, the steadily-increasing size of the common gooseberry may be quoted. We see an astonishing improvement in many florists' flowers, when the flowers of the present day are compared with drawings made only twenty or thirty years ago. When a race of plants is once pretty well established, the seed-raisers do not pick out the best plants, but merely go over their seed-beds, and pull up the 'rogues,' as they call the plants that deviate from the proper standard. With animals this kind of selection is, in fact, also followed; for hardly any one is so careless as to allow his worst animals to breed.

In regard to plants, there is another means of observing the accumulated effects of selection—namely, by comparing the diversity of flowers in the different varieties of the same species in the flower-garden; the diversity of leaves, pods, or tubers, or whatever part is valued, in the kitchen-garden, in comparison with the flowers of the same varieties; and the diversity of fruit of the same species in the orchard, in comparison with the leaves and flowers of the same set of varieties. See how different the leaves of the cabbage are, and how extremely alike the flowers; how unlike the flowers of the heartsease are, and how alike the leaves; how much the fruit of the different kinds of gooseberries differ in size, colour, shape, and hairiness, and yet the flowers present very slight differences. It is not that the varieties which differ largely in some one point do not differ at all in other points; this is hardly ever, perhaps never, the case. The laws of correlation of growth, the importance of which should never be overlooked, will ensure some differences; but, as a general rule, I cannot doubt that the continued selection of slight variations, either in the leaves, the flowers, or the fruit, will produce races differing from each other chiefly in these characters.

It may be objected that the principle of selection has been reduced to methodical practice for scarcely more than three-quarters of a century; it has certainly been more attended to of late years, and many treatises have been published on the subject; and the result has been, in a corresponding degree, rapid and important. But it is very far from true that the principle is a modern discovery. I could give several references to the full acknowledgment of the importance of the principle in works of high antiquity. In rude and barbarous periods of English history choice animals were often imported, and laws were passed to prevent their exportation: the destruction of horses under a certain size was ordered, and this may be compared to the 'roguing' of plants by nurserymen. The principle of selection I find distinctly given in an ancient Chinese encyclopædia. Explicit rules are laid down by some of the Roman classical writers. From passages in Genesis, it is clear that the colour of domestic animals was at that early period attended to. Savages now sometimes cross their dogs with wild canine animals, to improve the breed, and they formerly did so, as is attested by passages in Pliny. The savages in South Africa match their draught cattle by colour, as do some of the Esquimaux their teams of dogs. Livingstone shows how much good domestic breeds are valued by the negroes of the interior of Africa who have not associated with Europeans. Some of these facts do not show actual selection, but they show that the breeding of domestic animals was carefully attended to in ancient times, and is now attended to by the lowest savages. It would, indeed, have been a strange fact, had attention not been paid to breeding, for the inheritance of good and bad qualities is so obvious.

At the present time, eminent breeders try by methodical selection, with a distinct object in view, to make a new strain or sub-breed, superior to anything existing in the country. But, for our purpose, a kind of Selection, which may be called Unconscious, and which results from every one trying to possess and breed from the best individual animals, is more important. Thus, a man who intends keeping pointers naturally tries to get as good dogs as he can, and afterwards breeds from his own best dogs, but he has no wish or expectation of permanently altering the breed. Nevertheless I cannot doubt that this process, continued during centuries, would improve and modify any breed, in the same way as Bakewell, Collins, etc., by this very same process, only carried on more methodically, did greatly modify, even during their own lifetimes, the forms and qualities of their cattle. Slow and insensible changes of this kind could never be recognised unless actual measurements or careful drawings of the breeds in question had been made long ago, which might serve for comparison. In some cases, however, unchanged, or but little changed individuals of the same breed may be found in less civilised districts, where the breed has been less improved. There is reason to believe that King Charles's spaniel has been unconsciously modified to a large extent since the time of that monarch. Some highly competent authorities are convinced that the setter is directly derived from the spaniel, and has probably been slowly altered from it. It is known that the English pointer has been greatly changed within the last century, and in this case the change has, it is believed, been chiefly effected by crosses with the fox-hound; but what concerns us is, that the change has been effected unconsciously and gradually, and yet so effectually, that, though the old Spanish pointer certainly came from Spain, Mr Borrow has not seen, as I am informed by him, any native dog in Spain like our pointer.

By a similar process of selection, and by careful training, the whole body of English racehorses have come to surpass in fleetness and size the parent Arab stock, so that the latter, by the regulations for the Goodwood Races, are favoured in the weights they carry. Lord Spencer and others have shown how the cattle of England have increased in weight and in early maturity, compared with the stock formerly kept in this country. By comparing the accounts given in old pigeon treatises of carriers and tumblers with these breeds as now existing in Britain, India, and Persia, we can, I think, clearly trace the stages through which they have insensibly passed, and come to differ so greatly from the rock-pigeon.

Youatt gives an excellent illustration of the effects of a course of selection, which may be considered as unconsciously followed, in so far that the breeders could never have expected or even have wished to have produced the result which ensued—namely, the production of two distinct strains. The two flocks of Leicester sheep kept by Mr Buckley and Mr Burgess, as Mr Youatt remarks, 'have been purely bred from the original stock of Mr Bakewell for upwards of fifty years. There is not a suspicion existing in the mind of any one at all acquainted with the subject that the owner of either of them has deviated in any one instance from the pure blood of Mr Bakewell's flock, and yet the difference between the sheep possessed by these two gentlemen is so great that they have the appearance of being quite different varieties.'

If there exist savages so barbarous as never to think of the inherited character of the offspring of their domestic animals, yet any one animal particularly useful to them, for any special purpose, would be carefully preserved during famines and other accidents, to which savages are so liable, and such choice animals would thus generally leave more offspring than the inferior ones; so that in this case there would be a kind of unconscious selection going on. We see the value set on animals even by the barbarians of Tierra del Fuego, by their killing and devouring their old women, in times of dearth, as of less value than their dogs.

In plants the same gradual process of improvement, through the occasional preservation of the best individuals, whether or not sufficiently distinct to be ranked at their first appearance as distinct varieties, and whether or not two or more species or races have become blended together by crossing, may plainly be recognised in the increased size and beauty which we now see in the varieties of the heartsease, rose, pelargonium, dahlia, and other plants, when compared with the older varieties or with their parent-stocks. No one would ever expect to get a first-rate heartsease or dahlia from the seed of a wild plant. No one would expect to raise a first-rate melting pear from the seed of the wild pear, though he might succeed from a poor seedling growing wild, if it had come from a garden-stock. The pear, though cultivated in classical times, appears, from Pliny's description, to have been a fruit of very inferior quality. I have seen great surprise expressed in horticultural works at the wonderful skill of gardeners, in having produced such splendid results from such poor materials; but the art, I cannot doubt, has been simple, and, as far as the final result is concerned, has been followed almost unconsciously. It has consisted in always cultivating the best known variety, sowing its seeds, and, when a slightly better variety has chanced to appear, selecting it, and so onwards. But the gardeners of the classical period, who cultivated the best pear they could procure, never thought what splendid fruit we should eat; though we owe our excellent fruit, in some small degree, to their having naturally chosen and preserved the best varieties they could anywhere find.

A large amount of change in our cultivated plants, thus slowly and unconsciously accumulated, explains, as I believe, the well-known fact, that in a vast number of cases we cannot recognise, and therefore do not know, the wild parent-stocks of the plants which have been longest cultivated in our flower and kitchen gardens. If it has taken centuries or thousands of years to improve or modify most of our plants up to their present standard of usefulness to man, we can understand how it is that neither Australia, the Cape of Good Hope, nor any other region inhabited by quite uncivilised man, has afforded us a single plant worth culture. It is not that these countries, so rich in species, do not by a strange chance possess the aboriginal stocks of any useful plants, but that the native plants have not been improved by continued selection up to a standard of perfection comparable with that given to the plants in countries anciently civilised.

In regard to the domestic animals kept by uncivilised man, it should not be overlooked that they almost always have to struggle for their own food, at least during certain seasons. And in two countries very differently circumstanced, individuals of the same species, having slightly different constitutions or structure, would often succeed better in the one country than in the other; and thus by a process of 'natural selection,' as will hereafter be more fully explained, two sub-breeds might be formed. This, perhaps, partly explains what has been remarked by some authors, namely, that the varieties kept by savages have more of the character of species than the varieties kept in civilised countries.

On the view here given of the all-important part which selection by man has played, it becomes at once obvious how it is that our domestic races show adaptation in their structure or in their habits to man's wants or fancies. We can, I think, further understand the frequently abnormal character of our domestic races, and likewise their differences being so great in external characters and relatively so slight in internal parts or organs. Man can hardly select, or only with much difficulty, any deviation of structure excepting such as is externally visible; and indeed he rarely cares for what is internal. He can never act by selection, excepting on variations which are first given to him in some slight degree by nature. No man would ever try to make a fantail, till he saw a pigeon with a tail developed in some slight degree in an unusual manner, or a pouter till he saw a pigeon with a crop of somewhat unusual size; and the more abnormal or unusual any character was when it first appeared, the more likely it would be to catch his attention. But to use such an expression as trying to make a fantail, is, I have no doubt, in most cases, utterly incorrect. The man who first selected a pigeon with a slightly larger tail, never dreamed what the descendants of that pigeon would become through long-continued, partly unconscious and partly methodical selection. Perhaps the parent bird of all fantails had only fourteen tail-feathers somewhat expanded, like the present Java fantail, or like individuals of other and distinct breeds, in which as many as seventeen tail-feathers have been counted. Perhaps the first pouter-pigeon did not inflate its crop much more than the turbit now does the upper part of its œsophagus,—a habit which is disregarded by all fanciers, as it is not one of the points of the breed.

Nor let it be thought that some great deviation of structure would be necessary to catch the fancier's eye: he perceives extremely small differences, and it is in human nature to value any novelty, however slight, in one's own possession. Nor must the value which would formerly be set on any slight differences in the individuals of the same species, be judged of by the value which would now be set on them, after several breeds have once fairly been established. Many slight differences might, and indeed do now, arise amongst pigeons, which are rejected as faults or deviations from the standard of perfection of each breed. The common goose has not given rise to any marked varieties; hence the Thoulouse and the common breed, which differ only in colour, that most fleeting of characters, have lately been exhibited as distinct at our poultry-shows.

I think these views further explain what has sometimes been noticed—namely, that we know nothing about the origin or history of any of our domestic breeds. But, in fact, a breed, like a dialect of a language, can hardly be said to have had a definite origin. A man preserves and breeds from an individual with some slight deviation of structure, or takes more care than usual in matching his best animals and thus improves them, and the improved individuals slowly spread in the immediate neighbourhood. But as yet they will hardly have a distinct name, and from being only slightly valued, their history will be disregarded. When further improved by the same slow and gradual process, they will spread more widely, and will get recognised as something distinct and valuable, and will then probably first receive a provincial name. In semi-civilised countries, with little free communication, the spreading and knowledge of any new sub-breed will be a slow process. As soon as the points of value of the new sub-breed are once fully acknowledged, the principle, as I have called it, of unconscious selection will always tend,—perhaps more at one period than at another, as the breed rises or falls in fashion,—perhaps more in one district than in another, according to the state of civilisation of the inhabitants—slowly to add to the characteristic features of the breed, whatever they may be. But the chance will be infinitely small of any record having been preserved of such slow, varying, and insensible changes.

I must now say a few words on the circumstances, favourable or the reverse, to man's power of selection. A high degree of variability is obviously favourable, as freely giving the materials for selection to work on; not that mere individual differences are not amply sufficient, with extreme care, to allow of the accumulation of a large amount of modification in almost any desired direction. But as variations manifestly useful or pleasing to man appear only occasionally, the chance of their appearance will be much increased by a large number of individuals being kept; and hence this comes to be of the highest importance to success. On this principle Marshall has remarked, with respect to the sheep of parts of Yorkshire, that 'as they generally belong to poor people, and are mostly *in small lots*, they never can be improved.' On the other hand, nurserymen, from raising large stocks of the same plants, are generally far more successful than amateurs in getting new and valuable varieties. The keeping of a large number of individuals of a species in any country requires that the species should be placed under favourable conditions of life, so as to breed freely in that country. When the individuals of any species are scanty, all the individuals, whatever their quality may be, will generally be allowed to breed, and this will effectually prevent selection. But probably the most important point of all is, that the animal or plant should be so highly useful to man, or so much valued by him, that the closest attention should be paid to even the slightest deviation in the qualities or structure of each individual. Unless such attention be paid nothing can be effected. I have seen it gravely remarked, that it was most fortunate that the strawberry began to vary just when gardeners began to attend closely to this plant. No doubt the strawberry had always varied since it was cultivated, but the slight varieties had been neglected. As soon, however, as gardeners picked out individual plants with slightly larger, earlier, or better fruit, and raised seedlings from them, and again picked out the best seedlings and bred from them, then, there appeared (aided by some crossing with distinct species) those many admirable varieties of the strawberry which have been raised during the last thirty or forty years.

In the case of animals with separate sexes, facility in preventing crosses is an important element of success in the formation of new races,—at least, in a country which is already stocked with other races. In this respect enclosure of the land plays a part. Wandering savages or the inhabitants of open plains rarely possess more than one breed of the same species. Pigeons can be mated for life, and this is a great convenience to the fancier, for thus many races may be kept true, though mingled in the same aviary; and this circumstance must have largely favoured the improvement and formation of new breeds. Pigeons, I may add, can be propagated in great numbers and at a very quick rate, and inferior birds may be freely rejected, as when killed they serve for food. On the other hand, cats, from their nocturnal rambling habits, cannot be matched, and, although so much valued by women and children, we hardly ever see a distinct breed kept up; such breeds as we do sometimes see are almost always imported from some other country, often from islands. Although I do not doubt that some domestic animals vary less than others, yet the rarity or absence of distinct breeds of the cat, the donkey, peacock, goose, etc., may be attributed in main part to selection not having been brought into play: in cats, from the difficulty in pairing them; in donkeys, from only a few being kept by poor people, and little attention paid to their breeding; in peacocks, from not being very easily reared and a large stock not kept; in geese, from being valuable only for two purposes, food and feathers, and more especially from no pleasure having been felt in the display of distinct breeds.

To sum up on the origin of our Domestic Races of animals and plants. I believe that the conditions of life, from their action on the reproductive system, are so far of the highest importance as causing variability. I do not believe that variability is an inherent and necessary contingency, under all circumstances, with all organic beings, as some authors have thought. The effects of variability are modified by various degrees of inheritance and of reversion. Variability is governed by many unknown laws, more especially by that of correlation of growth. Something may be attributed to the direct action of the conditions of life. Something must be attributed to use and disuse. The final result is thus rendered infinitely complex. In some cases, I do not doubt that the intercrossing of species, aboriginally distinct, has played an important part in the origin of our domestic productions. When in any country several domestic breeds have once been established, their occasional intercrossing, with the aid of selection, has, no doubt, largely aided in the formation of new sub-breeds; but the importance of the crossing of varieties has, I believe, been greatly exaggerated, both in regard to animals and to those plants which are propagated by seed. In plants which are temporarily propagated by cuttings, buds, etc., the importance of the crossing both of distinct species and of varieties is immense; for the cultivator here quite disregards the extreme variability both of hybrids and mongrels, and the frequent sterility of hybrids; but the cases of plants not propagated by seed are of little importance to us, for their endurance is only temporary. Over all these causes of Change I am convinced that the accumulative action of Selection, whether applied methodically and more quickly, or unconsciously and more slowly, but more efficiently, is by far the predominant Power.

CHAPTER II

VARIATION UNDER NATURE

Variability—Individual differences—Doubtful species—Wide ranging, much diffused, and common species vary most—Species of the larger genera in any country vary more than the species of the smaller genera—Many of the species of the larger genera resemble varieties in being very closely, but unequally, related to each other, and in having restricted ranges.

BEFORE applying the principles arrived at in the last chapter to organic beings in a state of nature, we must briefly discuss whether these latter are subject to any variation. To treat this subject at all properly, a long catalogue of dry facts should be given; but these I shall reserve for my future work. Nor shall I here discuss the various definitions which have been given of the term 'species.' No one definition has as yet satisfied all naturalists; yet every naturalist knows vaguely what he means when he speaks of a species. Generally the term includes the unknown element of a distinct act of creation. The term 'variety' is almost equally difficult to define; but here community of descent is almost universally implied, though it can rarely be proved. We have also what are called monstrosities; but they graduate into varieties. By a monstrosity I presume is meant some considerable deviation of structure in one part, either injurious to or not useful to the species, and not generally propagated. Some authors use the term 'variation' in a technical sense, as implying a modification directly due to the physical conditions of life; and 'variations' in this sense are supposed not to be inherited: but who can say that the dwarfed condition of shells in the brackish waters of the Baltic, or dwarfed plants on Alpine summits, or the thicker fur of an animal from far northwards, would not in some cases be inherited for at least some few generations? and in this case I presume that the form would be called a variety.

Again, we have many slight differences which may be called individual differences, such as are known frequently to appear in the offspring from the same parents, or which may be presumed to have thus arisen, from being frequently observed in the individuals of the same species inhabiting the same confined locality. No one supposes that all the individuals of the same species are cast in the very same mould. These individual differences are highly important for us, as they afford materials for natural selection to accumulate, in the same manner as man can accumulate in any given direction individual differences in his domesticated productions. These individual differences generally affect what naturalists consider unimportant parts; but I could show by a long catalogue of facts, that parts which must be called important, whether viewed under a physiological or classificatory point of view, sometimes vary in the individuals of the same species. I am convinced that the most experienced naturalist would be surprised at the number of the cases of variability, even in important parts of structure, which he could collect on good authority, as I have collected, during a course of years. It should be remembered that systematists are far from pleased at finding variability in important characters, and that there are not many men who will laboriously examine internal and important organs, and compare them in many specimens of the same species. I should never have expected that the branching of the main nerves close to the great central ganglion of an insect would have been variable in the same species; I should have expected that changes of this nature could have been effected only by slow degrees: yet quite recently Mr Lubbock has shown a degree of variability in these main nerves in Coccus, which may almost be compared to the irregular branching of the stem of a tree. This philosophical naturalist, I may add, has also quite recently shown that the muscles in the larvæ of certain insects are very far from uniform. Authors sometimes argue in a circle when they state that important organs never vary; for these same authors practically rank that character as important (as some few naturalists have honestly confessed) which does not vary; and, under this point of view, no instance of an important part varying will ever be found: but under any other point of view many instances assuredly can be given.

There is one point connected with individual differences which seems to me extremely perplexing: I refer to those genera which have sometimes been called 'protean' or 'polymorphic,' in which the species present an inordinate amount of variation; and hardly two naturalists can agree which forms to rank as species and which as varieties. We may instance Rubus, Rosa, and Hieracium amongst plants, several genera of insects, and several genera of Brachiopod shells. In most polymorphic genera some of the species have fixed and definite characters. Genera which are polymorphic in one country seem to be, with some few exceptions, polymorphic in other countries, and likewise, judging from Brachiopod shells, at former periods of time. These facts seem to be very perplexing, for they seem to show that this kind of variability is independent of the conditions of life. I am inclined to suspect that we see in these polymorphic genera variations in points of structure which are of no service or disservice to the species, and which consequently have not been seized on and rendered definite by natural selection, as hereafter will be explained.

Those forms which possess in some considerable degree the character of species, but which are so closely similar to some other forms, or are so closely linked to them by intermediate gradations, that naturalists do not like to rank them as distinct species, are in several respects the most important for us. We have every reason to believe that many of these doubtful and closely-allied forms have permanently retained their characters in their own country for a long time; for as long, as far as we know, as have good and true species. Practically, when a naturalist can unite two forms together by others having intermediate characters, he treats the one as a variety of the other, ranking the most common, but sometimes the one first described, as the species, and the other as the variety. But cases of great difficulty, which I will not here enumerate, sometimes occur in deciding whether or not to rank one form as a variety of another, even when they are closely connected by intermediate links; nor will the commonly-assumed hybrid nature of the intermediate links always remove the difficulty. In very many cases, however, one form is ranked as a variety of another, not because the intermediate links have actually been found, but because analogy leads the observer to suppose either that they do now somewhere exist, or may formerly have existed; and here a wide door for the entry of doubt and conjecture is opened.

Hence, in determining whether a form should be ranked as a species or a variety, the opinion of naturalists having sound judgment and wide experience seems the only guide to follow. We must, however, in many cases, decide by a majority of naturalists, for few well-marked and well-known varieties can be named which have not been ranked as species by at least some competent judges.

That varieties of this doubtful nature are far from uncommon cannot be disputed. Compare the several floras of Great Britain, of France, or of the United States, drawn up by different botanists, and see what a surprising number of forms have been ranked by one botanist as good species, and by another as mere varieties. Mr H. C. Watson, to whom I lie under deep obligation for assistance of all kinds, has marked for me 182 British plants, which are generally considered as varieties, but which have all been ranked by botanists as species; and in making this list he has omitted many trifling varieties, but which nevertheless have been ranked by some botanists as species, and he has entirely omitted several highly polymorphic genera. Under genera, including the most polymorphic forms, Mr Babington gives 251 species, whereas Mr Bentham gives only 112,—a difference of 139 doubtful forms! Amongst animals which unite for each birth, and which are highly locomotive, doubtful forms, ranked by one zoologist as a species and by another as a variety, can rarely be found within the same country, but are common in separated areas. How many of those birds and insects in North America and Europe, which differ very slightly from each other, have been ranked by one eminent naturalist as undoubted species, and by another as varieties, or, as they are often called, as geographical races! Many years ago, when comparing, and seeing others compare, the birds from the separate islands of the Galapagos Archipelago, both one with another, and with those from the American mainland, I was much struck how entirely vague and arbitrary is the distinction between species and varieties.

On the islets of the little Madeira group there are many insects which are characterised as varieties in Mr Wollaston's admirable work, but which it cannot be doubted would be ranked as distinct species by many entomologists. Even Ireland has a few animals, now generally regarded as varieties, but which have been ranked as species by some zoologists. Several most experienced ornithologists consider our British red grouse as only a strongly-marked race of a Norwegian species, whereas the greater number rank it as an undoubted species peculiar to Great Britain. A wide distance between the homes of two doubtful forms leads many naturalists to rank both as distinct species; but what distance, it has been well asked, will suffice? if that between America and Europe is ample, will that between the Continent and the Azores, or Madeira, or the Canaries, or Ireland, be sufficient? It must be admitted that many forms, considered by highly-competent judges as varieties, have so perfectly the character of species that they are ranked by other highly-competent judges as good and true species. But to discuss whether they are rightly called species or varieties, before any definition of these terms has been generally accepted, is vainly to beat the air.

Many of the cases of strongly-marked varieties or doubtful species well deserve consideration; for several interesting lines of argument, from geographical distribution, analogical variation, hybridism, etc., have been brought to bear on the attempt to determine their rank. I will here give only a single instance,—the well-known one of the primrose and cowslip, or Primula vulgaris and veris. These plants differ considerably in appearance; they have a different flavour, and emit a different odour; they flower at slightly different periods; they grow in somewhat different stations; they ascend mountains to different heights; they have different geographical ranges; and lastly, according to very numerous experiments made during several years by that most careful observer Gärtner, they can be crossed only with much difficulty. We could hardly wish for better evidence of the two forms being specifically distinct. On the other hand, they are united by many intermediate links, and it is very doubtful whether these links are hybrids; and there is, as it seems to me, an overwhelming amount of experimental evidence, showing that they descend from common parents, and consequently must be ranked as varieties.

Close investigation, in most cases, will bring naturalists to an agreement how to rank doubtful forms. Yet it must be confessed that it is in the best-known countries that we find the greatest number of forms of doubtful value. I have been struck with the fact, that if any animal or plant in a state of nature be highly useful to man, or from any cause closely attract his attention, varieties of it will almost universally be found recorded. These varieties, moreover, will be often ranked by some authors as species. Look at the common oak, how closely it has been studied; yet a German author makes more than a dozen species out of forms, which are very generally considered as varieties; and in this country the highest botanical authorities and practical men can be quoted to show that the sessile and pedunculated oaks are either good and distinct species or mere varieties.

When a young naturalist commences the study of a group of organisms quite unknown to him, he is at first much perplexed to determine what differences to consider as specific, and what as varieties; for he knows nothing of the amount and kind of variation to which the group is subject; and this shows, at least, how very generally there is some variation. But if he confine his attention to one class within one country, he will soon make up his mind how to rank most of the doubtful forms. His general tendency will be to make many species, for he will become impressed, just like the pigeon or poultry fancier before alluded to, with the amount of difference in the forms which he is continually studying; and he has little general knowledge of analogical variation in other groups and in other countries, by which to correct his first impressions. As he extends the range of his observations, he will meet with more cases of difficulty; for he will encounter a greater number of closely-allied forms. But if his observations be widely extended, he will in the end generally be enabled to make up his own mind which to call varieties and which species; but he will succeed in this at the expense of admitting much variation,— and the truth of this admission will often be disputed by other naturalists. When, moreover, he comes to study allied forms brought from countries not now continuous, in which case he can hardly hope to find the intermediate links between his doubtful forms, he will have to trust almost entirely to analogy, and his difficulties rise to a climax.

Certainly no clear line of demarcation has as yet been drawn between species and sub-species—that is, the forms which in the opinion of some naturalists come very near to, but do not quite arrive at the rank of species; or, again, between sub-species and well-marked varieties, or between lesser varieties and individual differences. These differences blend into each other in an insensible series; and a series impresses the mind with the idea of an actual passage.

Hence I look at individual differences, though of small interest to the systematist, as of high importance for us, as being the first step towards such slight varieties as are barely thought worth recording in works on natural history. And I look at varieties which are in any degree more distinct and permanent, as steps leading to more strongly marked and more permanent varieties; and at these latter, as leading to sub-species, and to species. The passage from one stage of difference to another and higher stage may be, in some cases, due merely to the long-continued action of different physical conditions in two different regions; but I have not much faith in this view; and I attribute the passage of a variety, from a state in which it differs very slightly from its parent to one in which it differs more, to the action of natural selection in accumulating (as will hereafter be more fully explained) differences of structure in certain definite directions. Hence I believe a well-marked variety may be called an incipient species; but whether this belief be justifiable must be judged of by the general weight of the several facts and views given throughout this work.

It need not be supposed that all varieties or incipient species necessarily attain the rank of species. They may whilst in this incipient state become extinct, or they may endure as varieties for very long periods, as has been shown to be the case by Mr Wollaston with the varieties of certain fossil land-shells in Madeira. If a variety were to flourish so as to exceed in numbers the parent species, it would then rank as the species, and the species as the variety; or it might come to supplant and exterminate the parent species; or both might co-exist, and both rank as independent species. But we shall hereafter have to return to this subject.

From these remarks it will be seen that I look at the term species, as one arbitrarily given for the sake of convenience to a set of individuals closely resembling each other, and that it does not essentially differ from the term variety, which is given to less distinct and more fluctuating forms. The term variety, again, in comparison with mere individual differences, is also applied arbitrarily, and for mere convenience' sake.

Guided by theoretical considerations, I thought that some interesting results might be obtained in regard to the nature and relations of the species which vary most, by tabulating all the varieties in several well-worked floras. At first this seemed a simple task; but Mr H. C. Watson, to whom I am much indebted for valuable advice and assistance on this subject, soon convinced me that there were many difficulties, as did subsequently Dr Hooker, even in stronger terms. I shall reserve for my future work the discussion of these difficulties, and the tables themselves of the proportional numbers of the varying species. Dr Hooker permits me to add, that after having carefully read my manuscript, and examined the tables, he thinks that the following statements are fairly well established. The whole subject, however, treated as it necessarily here is with much brevity, is rather perplexing, and allusions cannot be avoided to the 'struggle for existence,' 'divergence of character,' and other questions, hereafter to be discussed.

Alph. de Candolle and others have shown that plants which have very wide ranges generally present varieties; and this might have been expected, as they become exposed to diverse physical conditions, and as they come into competition (which, as we shall hereafter see, is a far more important circumstance) with different sets of organic beings. But my tables further show that, in any limited country, the species which are most common, that is abound most in individuals, and the species which are most widely diffused within their own country (and this is a different consideration from wide range, and to a certain extent from commonness), often give rise to varieties sufficiently well marked to have been recorded in botanical works. Hence it is the most flourishing, or, as they may be called, the dominant species,—those which range widely over the world, are the most diffused in their own country, and are the most numerous in individuals,—which oftenest produce well-marked varieties, or, as I consider them, incipient species. And this, perhaps, might have been anticipated; for, as varieties, in order to become in any degree permanent, necessarily have to struggle with the other inhabitants of the country, the species which are already dominant will be the most likely to yield offspring, which, though in some slight degree modified, still inherit those advantages that enabled their parents to become dominant over their compatriots.

If the plants inhabiting a country and described in any Flora be divided into two equal masses, all those in the larger genera being placed on one side, and all those in the smaller genera on the other side, a somewhat larger number of the very common and much diffused or dominant species will be found on the side of the larger genera. This, again, might have been anticipated; for the mere fact of many species of the same genus inhabiting any country, shows that there is something in the organic or inorganic conditions of that country favourable to the genus; and, consequently, we might have expected to have found in the larger genera, or those including many species, a large proportional number of dominant species. But so many causes tend to obscure this result, that I am surprised that my tables show even a small majority on the side of the larger genera. I will here allude to only two causes of obscurity. Fresh-water and salt-loving plants have generally very wide ranges and are much diffused, but this seems to be connected with the nature of the stations inhabited by them, and has little or no relation to the size of the genera to which the species belong. Again, plants low in the scale of organisation are generally much more widely diffused than plants higher in the scale; and here again there is no close relation to the size of the genera. The cause of lowly-organised plants ranging widely will be discussed in our chapter on geographical distribution.

From looking at species as only strongly-marked and well-defined varieties, I was led to anticipate that the species of the larger genera in each country would oftener present varieties, than the species of the smaller genera; for wherever many closely-related species (*i.e.* species of the same genus) have been formed, many varieties or incipient species ought, as a general rule, to be now forming. Where many large trees grow, we expect to find saplings. Where many species of a genus have been formed through variation, circumstances have been favourable for variation; and hence we might expect that the circumstances would generally be still favourable to variation. On the other hand, if we look at each species as a special act of creation, there is no apparent reason why more varieties should occur in a group having many species, than in one having few.

To test the truth of this anticipation I have arranged the plants of twelve countries, and the coleopterous insects of two districts, into two nearly equal masses, the species of the larger genera on one side, and those of the smaller genera on the other side, and it has invariably proved to be the case that a larger proportion of the species on the side of the larger genera present varieties, than on the side of the smaller genera. Moreover, the species of the large genera which present any varieties, invariably present a larger average number of varieties than do the species of the small genera. Both these results follow when another division is made, and when all the smallest genera, with from only one to four species, are absolutely excluded from the tables. These facts are of plain signification on the view that species are only strongly-marked and permanent varieties; for wherever many species of the same genus have been formed, or where, if we may use the expression, the manufactory of species has been active, we ought generally to find the manufactory still in action, more especially as we have every reason to believe the process of manufacturing new species to be a slow one. And this certainly is the case, if varieties be looked at as incipient species; for my tables clearly show as a general rule that, wherever many species of a genus have been formed, the species of that genus present a number of varieties, that is of incipient species beyond the average. It is not that all large genera are now varying much, and are thus increasing in the number of their species, or that no small genera are now varying and increasing; for if this had been so, it would have been fatal to my theory; inasmuch as geology plainly tells us that small genera have in the lapse of time often increased greatly in size; and that large genera have often come to their maxima, declined, and disappeared. All that we want to show is, that where many species of a genus have been formed, on an average many are still forming; and this holds good.

There are other relations between the species of large genera and their recorded varieties which deserve notice. We have seen that there is no infallible criterion by which to distinguish species and well-marked varieties; and in those cases in which intermediate links have not been found between doubtful forms, naturalists are compelled to come to a determination by the amount of difference between them, judging by analogy whether or not the amount suffices to raise one or both to the rank of species. Hence the amount of difference is one very important criterion in settling whether two forms should be ranked as species or varieties. Now Fries has remarked in regard to plants, and Westwood in regard to insects, that in large genera the amount of difference between the species is often exceedingly small. I have endeavoured to test this numerically by averages, and, as far as my imperfect results go, they confirm the view. I have also consulted some sagacious and experienced observers, and, after deliberation, they concur in this view. In this respect, therefore, the species of the larger genera resemble varieties, more than do the species of the smaller genera. Or the case may be put in another way, and it may be said, that in the larger genera, in which a number of varieties or incipient species greater than the average are now manufacturing, many of the species already manufactured still to a certain extent resemble varieties, for they differ from each other by a less than usual amount of difference.

Moreover, the species of the large genera are related to each other, in the same manner as the varieties of any one species are related to each other. No naturalist pretends that all the species of a genus are equally distinct from each other; they may generally be divided into sub-genera, or sections, or lesser groups. As Fries has well remarked, little groups of species are generally clustered like satellites around certain other species. And what are varieties but groups of forms, unequally related to each other, and clustered round certain forms—that is, round their parent-species? Undoubtedly there is one most important point of difference between varieties and species; namely, that the amount of difference between varieties, when compared with each other or with their parent-species, is much less than that between the species of the same genus. But when we come to discuss the principle, as I call it, of Divergence of Character, we shall see how this may be explained, and how the lesser differences between varieties will tend to increase into the greater differences between species.

There is one other point which seems to me worth notice. Varieties generally have much restricted ranges: this statement is indeed scarcely more than a truism, for if a variety were found to have a wider range than that of its supposed parent-species, their denominations ought to be reversed. But there is also reason to believe, that those species which are very closely allied to other species, and in so far resemble varieties, often have much restricted ranges. For instance, Mr H. C. Watson has marked for me in the well-sifted London Catalogue of plants (4th edition) 63 plants which are therein ranked as species, but which he considers as so closely allied to other species as to be of doubtful value: these 63 reputed species range on an average over 69 of the provinces into which Mr Watson has divided Great Britain. Now, in this same catalogue, 53 acknowledged varieties are recorded, and these range over 77 provinces; whereas, the species to which these varieties belong range over 143 provinces. So that the acknowledged varieties have very nearly the same restricted average range, as have those very closely allied forms, marked for me by Mr Watson as doubtful species, but which are almost universally ranked by British botanists as good and true species.

Finally, then, varieties have the same general characters as species, for they cannot be distinguished from species,—except, firstly, by the discovery of intermediate linking forms, and the occurrence of such links cannot affect the actual characters of the forms which they connect; and except, secondly, by a certain amount of difference, for two forms, if differing very little, are generally ranked as varieties, notwithstanding that intermediate linking forms have not been discovered; but the amount of difference considered necessary to give to two forms the rank of species is quite indefinite. In genera having more than the average number of species in any country, the species of these genera have more than the average number of varieties. In large genera the species are apt to be closely, but unequally allied together, forming little clusters round certain species. Species very closely allied to other species apparently have restricted ranges. In all these several respects the species of large genera present a strong analogy with varieties. And we can clearly understand these analogies, if species have once existed as varieties, and have thus originated: whereas, these analogies are utterly inexplicable if each species has been independently created.

We have, also, seen that it is the most flourishing or dominant species of the larger genera which on an average vary most; and varieties, as we shall hereafter see, tend to become converted into new and distinct species. The larger genera thus tend to become larger; and throughout nature the forms of life which are now dominant tend to become still more dominant by leaving many modified and dominant descendants. But by steps hereafter to be explained, the larger genera also tend to break up into smaller genera. And thus, the forms of life throughout the universe become divided into groups subordinate to groups.

CHAPTER III

STRUGGLE FOR EXISTENCE

Bears on natural selection—The term used in a wide sense—Geometrical powers of increase—Rapid increase of naturalised animals and plants—Nature of the checks to increase—Competition universal—Effects of climate—Protection from the number of individuals—Complex relations of all animals and plants throughout nature—Struggle for life most severe between individuals and varieties of the same species; often severe between species of the same genus—The relation of organism to organism the most important of all relations.

BEFORE entering on the subject of this chapter, I must make a few preliminary remarks, to show how the struggle for existence bears on Natural Selection. It has been seen in the last chapter that amongst organic beings in a state of nature there is some individual variability: indeed I am not aware that this has ever been disputed. It is immaterial for us whether a multitude of doubtful forms be called species or sub-species or varieties; what rank, for instance, the two or three hundred doubtful forms of British plants are entitled to hold, if the existence of any well-marked varieties be admitted. But the mere existence of individual variability and of some few well-marked varieties, though necessary as the foundation for the work, helps us but little in understanding how species arise in nature. How have all those exquisite adaptations of one part of the organisation to another part, and to the conditions of life, and of one distinct organic being to another being, been perfected? We see these beautiful co-adaptations most plainly in the woodpecker and mistletoe; and only a little less plainly in the humblest parasite which clings to the hairs of a quadruped or feathers of a bird; in the structure of the beetle which dives through the water; in the plumed seed which is wafted by the gentlest breeze; in short, we see beautiful adaptations everywhere and in every part of the organic world.

Again, it may be asked, how is it that varieties, which I have called incipient species, become ultimately converted into good and distinct species, which in most cases obviously differ from each other far more than do the varieties of the same species? How do those groups of species, which constitute what are called distinct genera, and which differ from each other more than do the species of the same genus, arise? All these results, as we shall more fully see in the next chapter, follow from the struggle for life. Owing to this struggle for life, any variation, however slight, and from whatever cause proceeding, if it be in any degree profitable to an individual of any species, in its infinitely complex relations to other organic beings and to external nature, will tend to the preservation of that individual, and will generally be inherited by its offspring. The offspring, also, will thus have a better chance of surviving, for, of the many individuals of any species which are periodically born, but a small number can survive. I have called this principle, by which each slight variation, if useful, is preserved, by the term of Natural Selection, in order to mark its relation to man's power of selection. We have seen that man by selection can certainly produce great results, and can adapt organic beings to his own uses, through the accumulation of slight but useful variations, given to him by the hand of Nature. But Natural Selection, as we shall hereafter see, is a power incessantly ready for action, and is as immeasurably superior to man's feeble efforts, as the works of Nature are to those of Art.

We will now discuss in a little more detail the struggle for existence. In my future work this subject shall be treated, as it well deserves, at much greater length. The elder de Candolle and Lyell have largely and philosophically shown that all organic beings are exposed to severe competition. In regard to plants, no one has treated this subject with more spirit and ability than W. Herbert, Dean of Manchester, evidently the result of his great horticultural knowledge. Nothing is easier than to admit in words the truth of the universal struggle for life, or more difficult—at least I have found it so—than constantly to bear this conclusion in mind. Yet unless it be thoroughly engrained in the mind, I am convinced that the whole economy of nature, with every fact on distribution, rarity, abundance, extinction, and variation, will be dimly seen or quite misunderstood. We behold the face of nature bright with gladness, we often see superabundance of food; we do not see, or we forget that the birds which are idly singing round us mostly live on insects or seeds, and are thus constantly destroying life; or we forget how largely these songsters, or their eggs, or their nestlings, are destroyed by birds and beasts of prey; we do not always bear in mind, that though food may be now superabundant, it is not so at all seasons of each recurring year.

I should premise that I use the term Struggle for Existence in a large and metaphorical sense, including dependence of one being on another, and including (which is more important) not only the life of the individual, but success in leaving progeny. Two canine animals in a time of dearth, may be truly said to struggle with each other which shall get food and live. But a plant on the edge of a desert is said to struggle for life against the drought, though more properly it should be said to be dependent on the moisture. A plant which annually produces a thousand seeds, of which on an average only one comes to maturity, may be more truly said to struggle with the plants of the same and other kinds which already clothe the ground. The mistletoe is dependent on the apple and a few other trees, but can only in a far-fetched sense be said to struggle with these trees, for if too many of these parasites grow on the same tree, it will languish and die. But several seedling mistletoes, growing close together on the same branch, may more truly be said to struggle with each other. As the mistletoe is disseminated by birds, its existence depends on birds; and it may metaphorically be said to struggle with other fruit-bearing plants, in order to tempt birds to devour and thus disseminate its seeds rather than those of other plants. In these several senses, which pass into each other, I use for convenience' sake the general term of struggle for existence.

A struggle for existence inevitably follows from the high rate at which all organic beings tend to increase. Every being, which during its natural lifetime produces several eggs or seeds, must suffer destruction during some period of its life, and during some season or occasional year, otherwise, on the principle of geometrical increase, its numbers would quickly become so inordinately great that no country could support the product. Hence, as more individuals are produced than can possibly survive, there must in every case be a struggle for existence, either one individual with another of the same species, or with the individuals of distinct species, or with the physical conditions of life. It is the doctrine of Malthus applied with manifold force to the whole animal and vegetable kingdoms; for in this case there can be no artificial increase of food, and no prudential restraint from marriage. Although some species may be now increasing, more or less rapidly, in numbers, all cannot do so, for the world would not hold them.

There is no exception to the rule that every organic being naturally increases at so high a rate, that if not destroyed, the earth would soon be covered by the progeny of a single pair. Even slow-breeding man has doubled in twenty-five years, and at this rate, in a few thousand years, there would literally not be standing room for his progeny. Linnæus has calculated that if an annual plant produced only two seeds—and there is no plant so unproductive as this—and their seedlings next year produced two, and so on, then in twenty years there would be a million plants. The elephant is reckoned the slowest breeder of all known animals, and I have taken some pains to estimate its probable minimum rate of natural increase: it will be under the mark to assume that it breeds when thirty years old, and goes on breeding till ninety years old, bringing forth three pair of young in this interval; if this be so, at the end of the fifth century there would be alive fifteen million elephants, descended from the first pair.

But we have better evidence on this subject than mere theoretical calculations, namely, the numerous recorded cases of the astonishingly rapid increase of various animals in a state of nature, when circumstances have been favourable to them during two or three following seasons. Still more striking is the evidence from our domestic animals of many kinds which have run wild in several parts of the world: if the statements of the rate of increase of slow-breeding cattle and horses in South America, and latterly in Australia, had not been well authenticated, they would have been incredible. So it is with plants: cases could be given of introduced plants which have become common throughout whole islands in a period of less than ten years. Several of the plants, such as the cardoon and a tall thistle, now most numerous over the wide plains of La Plata, clothing square leagues of surface almost to the exclusion of all other plants, have been introduced from Europe; and there are plants which now range in India, as I hear from Dr Falconer, from Cape Comorin to the Himalaya, which have been imported from America since its discovery. In such cases, and endless instances could be given, no one supposes that the fertility of these animals or plants has been suddenly and temporarily increased in any sensible degree. The obvious explanation is that the conditions of life have been very favourable, and that there has consequently been less destruction of the old and young, and that nearly all the young have been enabled to breed. In such cases the geometrical ratio of increase, the result of which never fails to be surprising, simply explains the extraordinarily rapid increase and wide diffusion of naturalised productions in their new homes.

In a state of nature almost every plant produces seed, and amongst animals there are very few which do not annually pair. Hence we may confidently assert, that all plants and animals are tending to increase at a geometrical ratio, that all would most rapidly stock every station in which they could anyhow exist, and that the geometrical tendency to increase must be checked by destruction at some period of life. Our familiarity with the larger domestic animals tends, I think, to mislead us: we see no great destruction falling on them, and we forget that thousands are annually slaughtered for food, and that in a state of nature an equal number would have somehow to be disposed of.

The only difference between organisms which annually produce eggs or seeds by the thousand, and those which produce extremely few, is, that the slow-breeders would require a few more years to people, under favourable conditions, a whole district, let it be ever so large. The condor lays a couple of eggs and the ostrich a score, and yet in the same country the condor may be the more numerous of the two: the Fulmar petrel lays but one egg, yet it is believed to be the most numerous bird in the world. One fly deposits hundreds of eggs, and another, like the hippobosca, a single one; but this difference does not determine how many individuals of the two species can be supported in a district. A large number of eggs is of some importance to those species which depend on a rapidly fluctuating amount of food, for it allows them rapidly to increase in number. But the real importance of a large number of eggs or seeds is to make up for much destruction at some period of life; and this period in the great majority of cases is an early one. If an animal can in any way protect its own eggs or young, a small number may be produced, and yet the average stock be fully kept up; but if many eggs or young are destroyed, many must be produced, or the species will become extinct. It would suffice to keep up the full number of a tree, which lived on an average for a thousand years, if a single seed were produced once in a thousand years, supposing that this seed were never destroyed, and could be ensured to germinate in a fitting place. So that in all cases, the average number of any animal or plant depends only indirectly on the number of its eggs or seeds.

In looking at Nature, it is most necessary to keep the foregoing considerations always in mind—never to forget that every single organic being around us may be said to be striving to the utmost to increase in numbers; that each lives by a struggle at some period of its life; that heavy destruction inevitably falls either on the young or old, during each generation or at recurrent intervals. Lighten any check, mitigate the destruction ever so little, and the number of the species will almost instantaneously increase to any amount.

The causes which check the natural tendency of each species to increase in number are most obscure. Look at the most vigorous species; by as much as it swarms in numbers, by so much will its tendency to increase be still further increased. We know not exactly what the checks are in even one single instance. Nor will this surprise any one who reflects how ignorant we are on this head, even in regard to mankind, so incomparably better known than any other animal. This subject has been ably treated by several authors, and I shall, in my future work, discuss some of the checks at considerable length, more especially in regard to the feral animals of South America. Here I will make only a few remarks, just to recall to the reader's mind some of the chief points. Eggs or very young animals seem generally to suffer most, but this is not invariably the case. With plants there is a vast destruction of seeds, but, from some observations which I have made, I believe that it is the seedlings which suffer most from germinating in ground already thickly stocked with other plants. Seedlings, also, are destroyed in vast numbers by various enemies; for instance, on a piece of ground three feet long and two wide, dug and cleared, and where there could be no choking from other plants, I marked all the seedlings of our native weeds as they came up, and out of the 357 no less than 295 were destroyed, chiefly by slugs and insects. If turf which has long been mown, and the case would be the same with turf closely browsed by quadrupeds, be let to grow, the more vigorous plants gradually kill the less vigorous, though fully grown, plants: thus out of twenty species growing on a little plot of turf (three feet by four) nine species perished from the other species being allowed to grow up freely.

The amount of food for each species of course gives the extreme limit to which each can increase; but very frequently it is not the obtaining food, but the serving as prey to other animals, which determines the average numbers of a species. Thus, there seems to be little doubt that the stock of partridges, grouse, and hares on any large estate depends chiefly on the destruction of vermin. If not one head of game were shot during the next twenty years in England, and, at the same time, if no vermin were destroyed, there would, in all probability, be less game than at present, although hundreds of thousands of game animals are now annually killed. On the other hand, in some cases, as with the elephant and rhinoceros, none are destroyed by beasts of prey: even the tiger in India most rarely dares to attack a young elephant protected by its dam.

Climate plays an important part in determining the average numbers of a species, and periodical seasons of extreme cold or drought, I believe to be the most effective of all checks. I estimated that the winter of 1854–55 destroyed four-fifths of the birds in my own grounds; and this is a tremendous destruction, when we remember that ten per cent is an extraordinarily severe mortality from epidemics with man. The action of climate seems at first sight to be quite independent of the struggle for existence; but in so far as climate chiefly acts in reducing food, it brings on the most severe struggle between the individuals, whether of the same or of distinct species, which subsist on the same kind of food. Even when climate, for instance extreme cold, acts directly, it will be the least vigorous, or those which have got least food through the advancing winter, which will suffer most. When we travel from south to north, or from a damp region to a dry, we invariably see some species gradually getting rarer and rarer, and finally disappearing; and the change of climate being conspicuous, we are tempted to attribute the whole effect to its direct action. But this is a false view: we forget that each species, even where it most abounds, is constantly suffering enormous destruction at some period of its life, from enemies or from competitors for the same place and food; and if these enemies or competitors be in the least degree favoured by any slight change of climate, they will increase in numbers, and, as each area is already fully stocked with inhabitants, the other species will decrease. When we travel southward and see a species decreasing in numbers, we may feel sure that the cause lies quite as much in other species being favoured, as in this one being hurt. So it is when we travel northward, but in a somewhat lesser degree, for the number of species of all kinds, and therefore of competitors, decreases northwards; hence in going northward, or in ascending a mountain, we far oftener meet with stunted forms, due to the *directly* injurious action of climate, than we do in proceeding southwards or in descending a mountain. When we reach the Arctic regions, or snow-capped summits, or absolute deserts, the struggle for life is almost exclusively with the elements.

That climate acts in main part indirectly by favouring other species, we may clearly see in the prodigious number of plants in our gardens which can perfectly well endure our climate, but which never become naturalised, for they cannot compete with our native plants nor resist destruction by our native animals.

When a species, owing to highly favourable circumstances, increases inordinately in numbers in a small tract, epidemics—

at least, this seems generally to occur with our game animals—often ensue: and here we have a limiting check independent of the struggle for life. But even some of these so-called epidemics appear to be due to parasitic worms, which have from some cause, possibly in part through facility of diffusion amongst the crowded animals, been disproportionably favoured: and here comes in a sort of struggle between the parasite and its prey.

On the other hand, in many cases, a large stock of individuals of the same species, relatively to the numbers of its enemies, is absolutely necessary for its preservation. Thus we can easily raise plenty of corn and rape-seed, etc., in our fields, because the seeds are in great excess compared with the number of birds which feed on them; nor can the birds, though having a superabundance of food at this one season, increase in number proportionally to the supply of seed, as their numbers are checked during winter: but any one who has tried, knows how troublesome it is to get seed from a few wheat or other such plants in a garden: I have in this case lost every single seed. This view of the necessity of a large stock of the same species for its preservation, explains, I believe, some singular facts in nature, such as that of very rare plants being sometimes extremely abundant in the few spots where they do occur; and that of some social plants being social, that is, abounding in individuals, even on the extreme confines of their range. For in such cases, we may believe, that a plant could exist only where the conditions of its life were so favourable that many could exist together, and thus save the species from utter destruction. I should add that the good effects of frequent intercrossing, and the ill effects of close interbreeding, probably come into play in some of these cases; but on this intricate subject I will not here enlarge.

Many cases are on record showing how complex and unexpected are the checks and relations between organic beings which have to struggle together in the same country. I will give only a single instance, which, though a simple one, has interested me. In Staffordshire, on the estate of a relation, where I had ample means of investigation, there was a large and extremely barren heath, which had never been touched by the hand of man; but several hundred acres of exactly the same nature had been enclosed twenty-five years previously and planted with Scotch fir. The change in the native vegetation of the planted part of the heath was most remarkable, more than is generally seen in passing from one quite different soil to another: not only the proportional numbers of the heath-plants were wholly changed, but twelve species of plants (not counting grasses and carices) flourished in the plantations, which could not be found on the heath. The effect on the insects must have been still greater, for six insectivorous birds were very common in the plantations, which were not to be seen on the heath; and the heath was frequented by two or three distinct insectivorous birds. Here we see how potent has been the effect of the introduction of a single tree, nothing whatever else having been done, with the exception that the land had been enclosed, so that cattle could not enter. But how important an element enclosure is, I plainly saw near Farnham, in Surrey. Here there are extensive heaths, with a few clumps of old Scotch firs on the distant hill-tops: within the last ten years large spaces have been enclosed, and self-sown firs are now springing up in multitudes, so close together that all cannot live. When I ascertained that these young trees had not been sown or planted, I was so much surprised at their numbers that I went to several points of view, whence I could examine hundreds of acres of the unenclosed heath, and literally I could not see a single Scotch fir, except the old planted clumps. But on looking closely between the stems of the heath, I found a multitude of seedlings and little trees, which had been perpetually browsed down by the cattle. In one square yard, at a point some hundred yards distant from one of the old clumps, I counted thirty-two little trees; and one of them, with twenty-six rings of growth, had during many years tried to raise its head above the stems of the heath, and had failed. No wonder that, as soon as the land was enclosed, it became thickly clothed with vigorously growing young firs. Yet the heath was so extremely barren and so extensive that no one would ever have imagined that cattle would have so closely and effectually searched it for food.

Here we see that cattle absolutely determine the existence of the Scotch fir; but in several parts of the world insects determine the existence of cattle. Perhaps Paraguay offers the most curious instance of this; for here neither cattle nor horses nor dogs have ever run wild, though they swarm southward and northward in a feral state; and Azara and Rengger have shown that this is caused by the greater number in Paraguay of a certain fly, which lays its eggs in the navels of these animals when first born. The increase of these flies, numerous as they are, must be habitually checked by some means, probably by birds. Hence, if certain insectivorous birds (whose numbers are probably regulated by hawks or beasts of prey) were to increase in Paraguay, the flies would decrease—then cattle and horses would become feral, and this would certainly greatly alter (as indeed I have observed in parts of South America) the vegetation: this again would largely affect the insects; and this, as we just have seen in Staffordshire, the insectivorous birds, and so onwards in ever-increasing circles of complexity. We began this series by insectivorous birds, and we have ended with them. Not that in nature the relations can ever be as simple as this. Battle within battle must ever be recurring with varying success; and yet in the long-run the forces are so nicely balanced, that the face of nature remains uniform for long periods of time, though assuredly the merest trifle would often give the victory to one organic being over another. Nevertheless so profound is our ignorance, and so high our presumption, that we marvel when we hear of the extinction of an organic being; and as we do not see the cause, we invoke cataclysms to desolate the world, or invent laws on the duration of the forms of life!

I am tempted to give one more instance showing how plants and animals, most remote in the scale of nature, are bound together by a web of complex relations. I shall hereafter have occasion to show that the exotic Lobelia fulgens, in this part of England, is never visited by insects, and consequently, from its peculiar structure, never can set a seed. Many of our orchidaceous plants absolutely require the visits of moths to remove their pollen-masses and thus to fertilise them. I have, also, reason to believe that humble-bees are indispensable to the fertilisation of the heartsease (Viola tricolor), for other bees do not visit this flower. From experiments which I have lately tried, I have found that the visits of bees are necessary for the fertilisation of some kinds of clover; but humble-bees alone visit the red clover (Trifolium pratense), as other bees cannot reach the nectar. Hence I have very little doubt, that if the whole genus of humble-bees became extinct or very rare in England, the heartsease and red clover would become very rare, or wholly disappear. The number of humble-bees in any district depends in a great degree on the number of field-mice, which destroy their combs and nests; and Mr H. Newman, who has long attended to the habits of humble-bees, believes that 'more than two-thirds of them are thus destroyed all over England.' Now the number of mice is largely dependent, as every one knows, on the number of cats; and Mr Newman says, 'Near villages and small towns I have found the nests of humble-bees more numerous than elsewhere, which I attribute to the number of cats that destroy the mice.' Hence it is quite credible that the presence of a feline animal in large numbers in a district might determine, through the intervention first of mice and then of bees, the frequency of certain flowers in that district!

In the case of every species, many different checks, acting at different periods of life, and during different seasons or years, probably come into play; some one check or some few being generally the most potent, but all concur in determining the average number or even the existence of the species. In some cases it can be shown that widely-different checks act on the same species in different districts. When we look at the plants and bushes clothing an entangled bank, we are tempted to attribute their proportional numbers and kinds to what we call chance. But how false a view is this! Every one has heard that when an American forest is cut down, a very different vegetation springs up; but it has been observed that ancient Indian ruins in the Southern United States, which must formerly have been cleared of trees, now display the same beautiful diversity and proportion of kinds as in the surrounding virgin forests. What a struggle between the several kinds of trees must here have gone on during long centuries, each annually scattering its seeds by the thousand; what war between insect and insect—between insects, snails, and other animals with birds and beasts of prey—all striving to increase, and all feeding on each other or on the trees or their seeds and seedlings, or on the other plants which first clothed the ground and thus checked the growth of the trees! Throw up a handful of feathers, and all must fall to the ground according to definite laws; but how simple is this problem compared to the action and reaction of the innumerable plants and animals which have determined, in the course of centuries, the proportional numbers and kinds of trees now growing on the old Indian ruins!

The dependency of one organic being on another, as of a parasite on its prey, lies generally between beings remote in the scale of nature. This is often the case with those which may strictly be said to struggle with each other for existence, as in the case of locusts and grass-feeding quadrupeds. But the struggle almost invariably will be most severe between the individuals of the same species, for they frequent the same districts, require the same food, and are exposed to the same dangers. In the case of varieties of the same species, the struggle will generally be almost equally severe, and we sometimes see the contest soon decided: for instance, if several varieties of wheat be sown together, and the mixed seed be resown, some of the varieties which best suit the soil or climate, or are naturally the most fertile, will beat the others and so yield more seed, and consequently in a few years quite supplant the other varieties. To keep up a mixed stock of even such extremely close varieties as the variously coloured sweet-peas, they must be each year harvested separately, and the seed then mixed in due proportion, otherwise the weaker kinds will steadily decrease in numbers and disappear. So again with the varieties of sheep: it has been asserted that certain mountain-varieties will starve out other mountain-varieties, so that they cannot be kept together. The same result has followed from keeping together different varieties of the medicinal leech. It may even be doubted whether the varieties of any one of our domestic plants or animals have so exactly the same strength, habits, and constitution, that the original proportions of a mixed stock could be kept up for half-a-dozen generations, if they were allowed to struggle together, like beings in a state of nature, and if the seed or young were not annually sorted.

As species of the same genus have usually, though by no means invariably, some similarity in habits and constitution, and always in structure, the struggle will generally be more severe between species of the same genus, when they come into competition with each other, than between species of distinct genera. We see this in the recent extension over parts of the United States of one species of swallow having caused the decrease of another species. The recent increase of the missel-thrush in parts of Scotland has caused the decrease of the song-thrush. How frequently we hear of one species of rat taking the place of another species under the most different climates! In Russia the small Asiatic cockroach has everywhere driven before it its great congener. One species of charlock will supplant another, and so in other cases. We can dimly see why the competition should be most severe between allied forms, which fill nearly the same place in the economy of nature; but probably in no one case could we precisely say why one species has been victorious over another in the great battle of life.

A corollary of the highest importance may be deduced from the foregoing remarks, namely, that the structure of every organic being is related, in the most essential yet often hidden manner, to that of all other organic beings, with which it comes into competition for food or residence, or from which it has to escape, or on which it preys. This is obvious in the structure of the teeth and talons of the tiger; and in that of the legs and claws of the parasite which clings to the hair on the tiger's body. But in the beautifully plumed seed of the dandelion, and in the flattened and fringed legs of the water-beetle, the relation seems at first confined to the elements of air and water. Yet the advantage of plumed seeds no doubt stands in the closest relation to the land being already thickly clothed by other plants; so that the seeds may be widely distributed and fall on unoccupied ground. In the water-beetle, the structure of its legs, so well adapted for diving, allows it to compete with other aquatic insects, to hunt for its own prey, and to escape serving as prey to other animals.

The store of nutriment laid up within the seeds of many plants seems at first sight to have no sort of relation to other plants. But from the strong growth of young plants produced from such seeds (as peas and beans), when sown in the midst of long grass, I suspect that the chief use of the nutriment in the seed is to favour the growth of the young seedling, whilst struggling with other plants growing vigorously all around.

Look at a plant in the midst of its range, why does it not double or quadruple its numbers? We know that it can perfectly well withstand a little more heat or cold, dampness or dryness, for elsewhere it ranges into slightly hotter or colder, damper or drier districts. In this case we can clearly see that if we wished in imagination to give the plant the power of increasing in number, we should have to give it some advantage over its competitors, or over the animals which preyed on it. On the confines of its geographical range, a change of constitution with respect to climate would clearly be an advantage to our plant; but we have reason to believe that only a few plants or animals range so far, that they are destroyed by the rigour of the climate alone. Not until we reach the extreme confines of life, in the Arctic regions or on the borders of an utter desert, will competition cease. The land may be extremely cold or dry, yet there will be competition between some few species, or between the individuals of the same species, for the warmest or dampest spots.

Hence, also, we can see that when a plant or animal is placed in a new country amongst new competitors, though the climate may be exactly the same as in its former home, yet the conditions of its life will generally be changed in an essential manner. If we wished to increase its average numbers in its new home, we should have to modify it in a different way to what we should have done in its native country; for we should have to give it some advantage over a different set of competitors or enemies.

It is good thus to try in our imagination to give any form some advantage over another. Probably in no single instance should we know what to do, so as to succeed. It will convince us of our ignorance on the mutual relations of all organic beings; a conviction as necessary, as it seems to be difficult to acquire. All that we can do, is to keep steadily in mind that each organic being is striving to increase at a geometrical ratio; that each at some period of its life, during some season of the year, during each generation or at intervals, has to struggle for life, and to suffer great destruction. When we reflect on this struggle, we may console ourselves with the full belief, that the war of nature is not incessant, that no fear is felt, that death is generally prompt, and that the vigorous, the healthy, and the happy survive and multiply.

CHAPTER IV
NATURAL SELECTION

Natural Selection—its power compared with man's selection—its power on characters of trifling importance—its power at all ages and on both sexes—Sexual Selection—On the generality of intercrosses between individuals of the same species—Circumstances favourable and unfavourable to Natural Selection, namely, intercrossing, isolation, number of individuals—Slow action—Extinction caused by Natural Selection—Divergence of Character, related to the diversity of inhabitants of any small area, and to naturalisation—Action of Natural Selection, through Divergence of Character and Extinction, on the descendants from a common parent—Explains the Grouping of all organic beings.

How will the struggle for existence, discussed too briefly in the last chapter, act in regard to variation? Can the principle of selection, which we have seen is so potent in the hands of man, apply in nature? I think we shall see that it can act most effectually. Let it be borne in mind in what an endless number of strange peculiarities our domestic productions, and, in a lesser degree, those under nature, vary; and how strong the hereditary tendency is. Under domestication, it may be truly said that the whole organisation becomes in some degree plastic. Let it be borne in mind how infinitely complex and close-fitting are the mutual relations of all organic beings to each other and to their physical conditions of life. Can it, then, be thought improbable, seeing that variations useful to man have undoubtedly occurred, that other variations useful in some way to each being in the great and complex battle of life, should sometimes occur in the course of thousands of generations? If such do occur, can we doubt (remembering that many more individuals are born than can possibly survive) that individuals having any advantage, however slight, over others, would have the best chance of surviving and of procreating their kind? On the other hand, we may feel sure that any variation in the least degree injurious would be rigidly destroyed. This preservation of favourable variations and the rejection of injurious variations, I call Natural Selection. Variations neither useful nor injurious would not be affected by natural selection, and would be left a fluctuating element, as perhaps we see in the species called polymorphic.

We shall best understand the probable course of natural selection by taking the case of a country undergoing some physical change, for instance, of climate. The proportional numbers of its inhabitants would almost immediately undergo a change, and some species might become extinct. We may conclude, from what we have seen of the intimate and complex manner in which the inhabitants of each country are bound together, that any change in the numerical proportions of some of the inhabitants, independently of the change of climate itself, would seriously affect many of the others. If the country were open on its borders, new forms would certainly immigrate, and this also would seriously disturb the relations of some of the former inhabitants. Let it be remembered how powerful the influence of a single introduced tree or mammal has been shown to be. But in the case of an island, or of a country partly surrounded by barriers, into which new and better adapted forms could not freely enter, we should then have places in the economy of nature which would assuredly be better filled up, if some of the original inhabitants were in some manner modified; for, had the area been open to immigration, these same places would have been seized on by intruders. In such case, every slight modification, which in the course of ages chanced to arise, and which in any way favoured the individuals of any of the species, by better adapting them to their altered conditions, would tend to be preserved; and natural selection would thus have free scope for the work of improvement.

We have reason to believe, as stated in the first chapter, that a change in the conditions of life, by specially acting on the reproductive system, causes or increases variability; and in the foregoing case the conditions of life are supposed to have undergone a change, and this would manifestly be favourable to natural selection, by giving a better chance of profitable variations occurring; and unless profitable variations do occur, natural selection can do nothing. Not that, as I believe, any extreme amount of variability is necessary; as man can certainly produce great results by adding up in any given direction mere individual differences, so could Nature, but far more easily, from having incomparably longer time at her disposal. Nor do I believe that any great physical change, as of climate, or any unusual degree of isolation to check immigration, is actually necessary to produce new and unoccupied places for natural selection to fill up by modifying and improving some of the varying inhabitants. For as all the inhabitants of each country are struggling together with nicely balanced forces, extremely slight modifications in the structure or habits of one inhabitant would often give it an advantage over others; and still further modifications of the same kind would often still further increase the advantage. No country can be named in which all the native inhabitants are now so perfectly adapted to each other and to the physical conditions under which they live, that none of them could anyhow be improved; for in all countries, the native have been so far conquered by naturalised productions, that they have allowed foreigners to take firm possession of the land. And as foreigners have thus everywhere beaten some of the natives, we may safely conclude that the natives might have been modified with advantage, so as to have better resisted such intruders.

As man can produce and certainly has produced a great result by his methodical and unconscious means of selection, what may not Nature effect? Man can act only on external and visible characters: Nature cares nothing for appearances, except in so far as they may be useful to any being. She can act on every internal organ, on every shade of constitutional difference, on the whole machinery of life. Man selects only for his own good; Nature only for that of the being which she tends. Every selected character is fully exercised by her; and the being is placed under well-suited conditions of life. Man keeps the natives of many climates in the same country; he seldom exercises each selected character in some peculiar and fitting manner; he feeds a long and a short beaked pigeon on the same food; he does not exercise a long-backed or long-legged quadruped in any peculiar manner; he exposes sheep with long and short wool to the same climate. He does not allow the most vigorous males to struggle for the females. He does not rigidly destroy all inferior animals, but protects during each varying season, as far as lies in his power, all his productions. He often begins his selection by some half-monstrous form; or at least by some modification prominent enough to catch his eye, or to be plainly useful to him. Under nature, the slightest difference of structure or constitution may well turn the nicely-balanced scale in the struggle for life, and so be preserved. How fleeting are the wishes and efforts of man! how short his time! and consequently how poor will his products be, compared with those accumulated by Nature during whole geological periods. Can we wonder, then, that Nature's productions should be far 'truer' in character than man's productions; that they should be infinitely better adapted to the most complex conditions of life, and should plainly bear the stamp of far higher workmanship?

It may metaphorically be said that natural selection is daily and hourly scrutinising, throughout the world, every variation, even the slightest; rejecting that which is bad, preserving and adding up all that is good; silently and insensibly working, whenever and wherever opportunity offers, at the improvement of each organic being in relation to its organic and inorganic conditions of life. We see nothing of these slow changes in progress, until the hand of time has marked the long lapse of ages, and then so imperfect is our view into long past geological ages, that we only see that the forms of life are now different from what they formerly were.

Although natural selection can act only through and for the good of each being, yet characters and structures, which we are apt to consider as of very trifling importance, may thus be acted on. When we see leaf-eating insects green, and bark-feeders mottled-grey; the alpine ptarmigan white in winter, the red-grouse the colour of heather, and the black-grouse that of peaty earth, we must believe that these tints are of service to these birds and insects in preserving them from danger. Grouse, if not destroyed at some period of their lives, would increase in countless numbers; they are known to suffer largely from birds of prey; and hawks are guided by eyesight to their prey—so much so, that on parts of the Continent persons are warned not to keep white pigeons, as being the most liable to destruction. Hence I can see no reason to doubt that natural selection might be most effective in giving the proper colour to each kind of grouse, and in keeping that colour, when once acquired, true and constant. Nor ought we to think that the occasional destruction of an animal of any particular colour would produce little effect: we should remember how essential it is in a flock of white sheep to destroy every lamb with the faintest trace of black. In plants the down on the fruit and the colour of the flesh are considered by botanists as characters of the most trifling importance: yet we hear from an excellent horticulturist, Downing, that in the United States smooth-skinned fruits suffer far more from a beetle, a curculio, than those with down; that purple plums suffer far more from a certain disease than yellow plums; whereas another disease attacks yellow-fleshed peaches far more than those with other coloured flesh. If, with all the aids of art, these slight differences make a great difference in cultivating the several varieties, assuredly, in a state of nature, where the trees would have to struggle with other trees and with a host of enemies, such differences would effectually settle which variety, whether a smooth or downy, a yellow or purple fleshed fruit, should succeed.

In looking at many small points of difference between species, which, as far as our ignorance permits us to judge, seem quite unimportant, we must not forget that climate, food, etc., probably produce some slight and direct effect. It is, however, far more necessary to bear in mind that there are many unknown laws of correlation of growth, which, when one part of the organisation is modified through variation, and the modifications are accumulated by natural selection for the good of the being, will cause other modifications, often of the most unexpected nature.

As we see that those variations which under domestication appear at any particular period of life, tend to reappear in the offspring at the same period;—for instance, in the seeds of the many varieties of our culinary and agricultural plants; in the caterpillar and cocoon stages of the varieties of the silkworm;

in the eggs of poultry, and in the colour of the down of their chickens; in the horns of our sheep and cattle when nearly adult;—so in a state of nature, natural selection will be enabled to act on and modify organic beings at any age, by the accumulation of variations profitable at that age, and by their inheritance at a corresponding age. If it profit a plant to have its seeds more and more widely disseminated by the wind, I can see no greater difficulty in this being effected through natural selection, than in the cotton-planter increasing and improving by selection the down in the pods on his cotton-trees. Natural selection may modify and adapt the larva of an insect to a score of contingencies, wholly different from those which concern the mature insect. These modifications will no doubt affect, through the laws of correlation, the structure of the adult; and probably in the case of those insects which live only for a few hours, and which never feed, a large part of their structure is merely the correlated result of successive changes in the structure of their larvæ. So, conversely, modifications in the adult will probably often affect the structure of the larva; but in all cases natural selection will ensure that modifications consequent on other modifications at a different period of life, shall not be in the least degree injurious: for if they became so, they would cause the extinction of the species.

Natural selection will modify the structure of the young in relation to the parent, and of the parent in relation to the young. In social animals it will adapt the structure of each individual for the benefit of the community; if each in consequence profits by the selected change. What natural selection cannot do, is to modify the structure of one species, without giving it any advantage, for the good of another species; and though statements to this effect may be found in works of natural history, I cannot find one case which will bear investigation. A structure used only once in an animal's whole life, if of high importance to it, might be modified to any extent by natural selection; for instance, the great jaws possessed by certain insects, used exclusively for opening the cocoon—or the hard tip to the beak of nestling birds, used for breaking the egg. It has been asserted, that of the best short-beaked tumbler-pigeons more perish in the egg than are able to get out of it; so that fanciers assist in the act of hatching. Now, if nature had to make the beak of a full-grown pigeon very short for the bird's own advantage, the process of modification would be very slow, and there would be simultaneously the most rigorous selection of the young birds within the egg, which had the most powerful and hardest beaks, for all with weak beaks would inevitably perish: or, more delicate and more easily broken shells might be selected, the thickness of the shell being known to vary like every other structure.

Sexual Selection.—Inasmuch as peculiarities often appear under domestication in one sex and become hereditarily attached to that sex, the same fact probably occurs under nature, and if so, natural selection will be able to modify one sex in its functional relations to the other sex, or in relation to wholly different habits of life in the two sexes, as is sometimes the case with insects. And this leads me to say a few words on what I call Sexual Selection. This depends, not on a struggle for existence, but on a struggle between the males for possession of the females; the result is not death to the unsuccessful competitor, but few or no offspring. Sexual selection is, therefore, less rigorous than natural selection. Generally, the most vigorous males, those which are best fitted for their places in nature, will leave most progeny. But in many cases, victory depends not on general vigour, but on having special weapons, confined to the male sex. A hornless stag or spurless cock would have a poor chance of leaving offspring. Sexual selection by always allowing the victor to breed might surely give indomitable courage, length to the spur, and strength to the wing to strike in the spurred leg, as well as the brutal cock-fighter, who knows well that he can improve his breed by careful selection of the best cocks. How low in the scale of nature the law of battle descends, I know not; male alligators have been described as fighting, bellowing, and whirling round, like Indians in a war-dance, for the possession of the females; male salmons have been seen fighting all day long; male stag-beetles often bear wounds from the huge mandibles of other males. The war is, perhaps, severest between the males of polygamous animals, and these seem oftenest provided with special weapons. The males of carnivorous animals are already well armed; though to them and to others, special means of defence may be given through means of sexual selection, as the mane to the lion, the shoulder-pad to the boar, and the hooked jaw to the male salmon; for the shield may be as important for victory, as the sword or spear.

Amongst birds, the contest is often of a more peaceful character. All those who have attended to the subject, believe that there is the severest rivalry between the males of many species to attract by singing the females. The rock-thrush of Guiana, birds of Paradise, and some others, congregate; and successive males display their gorgeous plumage and perform strange antics before the females, which, standing by as spectators, at last choose the most attractive partner. Those who have closely attended to birds in confinement well know that they often take individual preferences and dislikes: thus Sir R. Heron has described how one pied peacock was eminently attractive to all his hen birds. It may appear childish to attribute any effect to such apparently weak means: I cannot here enter on the details necessary to support this view; but if man can in a short time give elegant carriage and beauty to his bantams, according to his standard of beauty, I can see no good reason to doubt that female birds, by selecting, during thousands of generations, the most melodious or beautiful males, according to their standard of beauty, might produce a marked effect. I strongly suspect that some well-known laws, with respect to the plumage of male and female birds, in comparison with the plumage of the young, can be explained on the view of plumage having been chiefly modified by sexual selection, acting when the birds have come to the breeding age or during the breeding season; the modifications thus produced being inherited at corresponding ages or seasons, either by the males alone, or by the males and females; but I have not space here to enter on this subject.

Thus it is, as I believe, that when the males and females of any animal have the same general habits of life, but differ in structure, colour, or ornament, such differences have been mainly caused by sexual selection; that is, individual males have had, in successive generations, some slight advantage over other males, in their weapons, means of defence, or charms; and have transmitted these advantages to their male offspring. Yet, I would not wish to attribute all such sexual differences to this agency: for we see peculiarities arising and becoming attached to the male sex in our domestic animals (as the wattle in male carriers, horn-like protuberances in the cocks of certain fowls, etc.), which we cannot believe to be either useful to the males in battle, or attractive to the females. We see analogous cases under nature, for instance, the tuft of hair on the breast of the turkey-cock, which can hardly be either useful or ornamental to this bird;—indeed, had the tuft appeared under domestication, it would have been called a monstrosity.

Illustrations of the action of Natural Selection.—In order to make it clear how, as I believe, natural selection acts, I must beg permission to give one or two imaginary illustrations. Let us take the case of a wolf, which preys on various animals, securing some by craft, some by strength, and some by fleetness; and let us suppose that the fleetest prey, a deer for instance, had from any change in the country increased in numbers, or that other prey had decreased in numbers, during that season of the year when the wolf is hardest pressed for food. I can under such circumstances see no reason to doubt that the swiftest and slimmest wolves would have the best chance of surviving, and so be preserved or selected,—provided always that they retained strength to master their prey at this or at some other period of the year, when they might be compelled to prey on other animals. I can see no more reason to doubt this, than that man can improve the fleetness of his greyhounds by careful and methodical selection, or by that unconscious selection which results from each man trying to keep the best dogs without any thought of modifying the breed.

Even without any change in the proportional numbers of the animals on which our wolf preyed, a cub might be born with an innate tendency to pursue certain kinds of prey. Nor can this be thought very improbable; for we often observe great differences in the natural tendencies of our domestic animals; one cat, for instance, taking to catch rats, another mice; one cat, according to Mr St John, bringing home winged game, another hares or rabbits, and another hunting on marshy ground and almost nightly catching woodcocks or snipes. The tendency to catch rats rather than mice is known to be inherited. Now, if any slight innate change of habit or of structure benefited an individual wolf, it would have the best chance of surviving and of leaving offspring. Some of its young would probably inherit the same habits or structure, and by the repetition of this process, a new variety might be formed which would either supplant or coexist with the parent form of wolf. Or, again, the wolves inhabiting a mountainous district, and those frequenting the lowlands, would naturally be forced to hunt different prey; and from the continued preservation of the individuals best fitted for the two sites, two varieties might slowly be formed. These varieties would cross and blend where they met; but to this subject of intercrossing we shall soon have to return. I may add, that, according to Mr Pierce, there are two varieties of the wolf inhabiting the Catskill Mountains in the United States, one with a light greyhound-like form, which pursues deer, and the other more bulky, with shorter legs, which more frequently attacks the shepherd's flocks.

Let us now take a more complex case. Certain plants excrete a sweet juice, apparently for the sake of eliminating something injurious from their sap: this is effected by glands at the base of the stipules in some Leguminosæ, and at the back of the leaf of the common laurel. This juice, though small in quantity, is greedily sought by insects. Let us now suppose a little sweet juice or nectar to be excreted by the inner bases of the petals of a flower. In this case insects in seeking the nectar would get dusted with pollen, and would certainly often transport the pollen from one flower to the stigma of another flower. The flowers of two distinct individuals of the same species would thus get crossed; and the act of crossing, we have good reason to believe (as will hereafter be more fully alluded to), would produce very vigorous seedlings, which consequently would have the best chance of flourishing and surviving. Some of these seedlings would probably inherit the nectar-excreting power. Those individual flowers which had the largest glands or nectaries, and which excreted most nectar, would be oftenest visited by insects, and would be oftenest crossed; and so in the long-run would gain the upper hand. Those flowers, also, which had their stamens and pistils placed, in relation to the size and habits of the particular insects which visited them, so as to favour in any degree the transportal of their pollen from flower to flower, would likewise be favoured or selected. We might have taken the case of insects visiting flowers for the sake of collecting pollen instead of nectar; and as pollen is formed for the sole object of fertilisation, its destruction appears a simple loss to the plant; yet if a little pollen were carried, at first occasionally and then habitually, by the pollen-devouring insects from flower to flower, and a cross thus effected, although nine-tenths of the pollen were destroyed, it might still be a great gain to the plant; and those individuals which produced more and more pollen, and had larger and larger anthers, would be selected.

When our plant, by this process of the continued preservation or natural selection of more and more attractive flowers, had been rendered highly attractive to insects, they would, unintentionally on their part, regularly carry pollen from flower to flower; and that they can most effectually do this, I could easily show by many striking instances. I will give only one—not as a very striking case, but as likewise illustrating one step in the separation of the sexes of plants, presently to be alluded to. Some holly-trees bear only male flowers, which have four stamens producing a rather small quantity of pollen, and a rudimentary pistil; other holly-trees bear only female flowers; these have a full-sized pistil, and four stamens with shrivelled anthers, in which not a grain of pollen can be detected. Having found a female tree exactly sixty yards from a male tree, I put the stigmas of twenty flowers, taken from different branches, under the microscope, and on all, without exception, there were pollen-grains, and on some a profusion of pollen. As the wind had set for several days from the female to the male tree, the pollen could not thus have been carried. The weather had been cold and boisterous, and therefore not favourable to bees, nevertheless every female flower which I examined had been effectually fertilised by the bees, accidentally dusted with pollen, having flown from tree to tree in search of nectar. But to return to our imaginary case: as soon as the plant had been rendered so highly attractive to insects that pollen was regularly carried from flower to flower, another process might commence. No naturalist doubts the advantage of what has been called the 'physiological division of labour'; hence we may believe that it would be advantageous to a plant to produce stamens alone in one flower or on one whole plant, and pistils alone in another flower or on another plant. In plants under culture and placed under new conditions of life, sometimes the male organs and sometimes the female organs become more or less impotent; now if we suppose this to occur in ever so slight a degree under nature, then as pollen is already carried regularly from flower to flower, and as a more complete separation of the sexes of our plant would be advantageous on the principle of the division of labour, individuals with this tendency more and more increased, would be continually favoured or selected, until at last a complete separation of the sexes would be effected.

Let us now turn to the nectar-feeding insects in our imaginary case: we may suppose the plant of which we have been slowly increasing the nectar by continued selection, to be a common plant; and that certain insects depended in main part on its nectar for food. I could give many facts, showing how anxious bees are to save time; for instance, their habit of cutting holes and sucking the nectar at the bases of certain flowers, which they can, with a very little more trouble, enter by the mouth. Bearing such facts in mind, I can see no reason to doubt that an accidental deviation in the size and form of the body, or in the curvature and length of the proboscis, etc., far too slight to be appreciated by us, might profit a bee or other insect, so that an individual so characterised would be able to obtain its food more quickly, and so have a better chance of living and leaving descendants. Its descendants would probably inherit a tendency to a similar slight deviation of structure. The tubes of the corollas of the common red and incarnate clovers (Trifolium pratense and incarnatum) do not on a hasty glance appear to differ in length; yet the hive-bee can easily suck the nectar out of the incarnate clover, but not out of the common red clover, which is visited by humble-bees alone; so that whole fields of the red clover offer in vain an abundant supply of precious nectar to the hive-bee. Thus it might be a great advantage to the hive-bee to have a slightly longer or differently constructed proboscis. On the other hand, I have found by experiment that the fertility of clover depends on bees visiting and moving parts of the corolla, so as to push the pollen on to the stigmatic surface. Hence, again, if humble-bees were to become rare in any country, it might be a great advantage to the red clover to have a shorter or more deeply divided tube to its corolla, so that the hive-bee could visit its flowers. Thus I can understand how a flower and a bee might slowly become, either simultaneously or one after the other, modified and adapted in the most perfect manner to each other, by the continued preservation of individuals presenting mutual and slightly favourable deviations of structure.

I am well aware that this doctrine of natural selection, exemplified in the above imaginary instances, is open to the same objections which were at first urged against Sir Charles Lyell's noble views on 'the modern changes of the earth, as illustrative of geology'; but we now seldom hear the action, for instance, of the coast-waves, called a trifling and insignificant cause, when applied to the excavation of gigantic valleys or to the formation of the longest lines of inland cliffs. Natural selection can act only by the preservation and accumulation of infinitesimally small inherited modifications, each profitable to the preserved being; and as modern geology has almost banished such views as the excavation of a great valley by a single diluvial wave, so will natural selection, if it be a true principle, banish the belief of the continued creation of new organic beings, or of any great and sudden modification in their structure.

On the Intercrossing of Individuals.—I must here introduce a short digression. In the case of animals and plants with separated sexes, it is of course obvious that two individuals must always (with the exception of the curious and not well-understood cases of parthenogenesis) unite for each birth; but in the case of hermaphrodites this is far from obvious. Nevertheless I am strongly inclined to believe that with all hermaphrodites two individuals, either occasionally or habitually, concur for the reproduction of their kind. This view was first suggested by Andrew Knight. We shall presently see its importance; but I must here treat the subject with extreme brevity, though I have the materials prepared for an ample discussion. All vertebrate animals, all insects, and some other large groups of animals, pair for each birth. Modern research has much diminished the number of supposed hermaphrodites, and of real hermaphrodites a large number pair; that is, two individuals regularly unite for reproduction, which is all that concerns us. But still there are many hermaphrodite animals which certainly do not habitually pair, and a vast majority of plants are hermaphrodites. What reason, it may be asked, is there for supposing in these cases that two individuals ever concur in reproduction? As it is impossible here to enter on details, I must trust to some general considerations alone.

In the first place, I have collected so large a body of facts, showing, in accordance with the almost universal belief of breeders, that with animals and plants a cross between different varieties, or between individuals of the same variety but of another strain, gives vigour and fertility to the offspring; and on the other hand, that *close* interbreeding diminishes vigour and fertility; that these facts alone incline me to believe that it is a general law of nature (utterly ignorant though we be of the meaning of the law) that no organic being self-fertilises itself for an eternity of generations; but that a cross with another individual is occasionally—perhaps at very long intervals—indispensable.

On the belief that this is a law of nature, we can, I think, understand several large classes of facts, such as the following, which on any other view are inexplicable. Every hybridizer knows how unfavourable exposure to wet is to the fertilisation of a flower, yet what a multitude of flowers have their anthers and stigmas fully exposed to the weather! but if an occasional cross be indispensable, the fullest freedom for the entrance of pollen from another individual will explain this state of exposure, more especially as the plant's own anthers and pistil generally stand so close together that self-fertilisation seems almost inevitable. Many flowers, on the other hand, have their organs of fructification closely enclosed, as in the great papilionaceous or pea-family; but in several, perhaps in all, such flowers, there is a very curious adaptation between the structure of the flower and the manner in which bees suck the nectar; for, in doing this, they either push the flower's own pollen on the stigma, or bring pollen from another flower. So necessary are the visits of bees to papilionaceous flowers, that I have found, by experiments published elsewhere, that their fertility is greatly diminished if these visits be prevented. Now, it is scarcely possible that bees should fly from flower to flower, and not carry pollen from one to the other, to the great good, as I believe, of the plant. Bees will act like a camel-hair pencil, and it is quite sufficient just to touch the anthers of one flower and then the stigma of another with the same brush to ensure fertilisation; but it must not be supposed that bees would thus produce a multitude of hybrids between distinct species; for if you bring on the same brush a plant's own pollen and pollen from another species, the former will have such a prepotent effect, that it will invariably and completely destroy, as has been shown by Gärtner, any influence from the foreign pollen.

When the stamens of a flower suddenly spring towards the pistil, or slowly move one after the other towards it, the contrivance seems adapted solely to ensure self-fertilisation; and no doubt it is useful for this end: but, the agency of insects is often required to cause the stamens to spring forward, as Kölreuter has shown to be the case with the barberry; and in this very genus, which seems to have a special contrivance for self-fertilisation, it is well known that if closely-allied forms or varieties are planted near each other, it is hardly possible to raise pure seedlings, so largely do they naturally cross. In many other cases, far from there being any aids for self-fertilisation, there are special contrivances, as I could show from the writings of C. C. Sprengel and from my own observations, which effectually prevent the stigma receiving pollen from its own flower: for instance, in Lobelia fulgens, there is a really beautiful and elaborate contrivance by which every one of the infinitely numerous pollen-granules are swept out of the conjoined anthers of each flower, before the stigma of that individual flower is ready to receive them; and as this flower is never visited, at least in my garden, by insects, it never sets a seed, though by placing pollen from one flower on the stigma of another, I raised plenty of seedlings; and whilst another species of Lobelia growing close by, which is visited by bees, seeds freely. In very many other cases, though there be no special mechanical contrivance to prevent the stigma of a flower receiving its own pollen, yet, as C. C. Sprengel has shown, and as I can confirm, either the anthers burst before the stigma is ready for fertilisation, or the stigma is ready before the pollen of that flower is ready, so that these plants have in fact separated sexes, and must habitually be crossed. How strange are these facts! How strange that the pollen and stigmatic surface of the same flower, though placed so close together, as if for the very purpose of self-fertilisation, should in so many cases be mutually useless to each other! How simply are these facts explained on the view of an occasional cross with a distinct individual being advantageous or indispensable!

If several varieties of the cabbage, radish, onion, and of some other plants, be allowed to seed near each other, a large majority, as I have found, of the seedlings thus raised will turn out mongrels: for instance, I raised 233 seedling cabbages from some plants of different varieties growing near each other, and of these only 78 were true to their kind, and some even of these were not perfectly true. Yet the pistil of each cabbage-flower is surrounded not only by its own six stamens, but by those of the many other flowers on the same plant. How, then, comes it that such a vast number of the seedlings are mongrelized? I suspect that it must arise from the pollen of a distinct *variety* having a prepotent effect over a flower's own pollen; and that this is part of the general law of good being derived from the intercrossing of distinct individuals of the same species. When distinct *species* are crossed the case is directly the reverse, for a plant's own pollen is always prepotent over foreign pollen; but to this subject we shall return in a future chapter.

In the case of a gigantic tree covered with innumerable flowers, it may be objected that pollen could seldom be carried from tree to tree, and at most only from flower to flower on the same tree, and that flowers on the same tree can be considered as distinct individuals only in a limited sense. I believe this objection to be valid, but that nature has largely provided against it by giving to trees a strong tendency to bear flowers with separated sexes. When the sexes are separated, although the male and female flowers may be produced on the same tree, we can see that pollen must be regularly carried from flower to flower; and this will give a better chance of pollen being occasionally carried from tree to tree. That trees belonging to all Orders have their sexes more often separated than other plants, I find to be the case in this country; and at my request Dr Hooker tabulated the trees of New Zealand, and Dr Asa Gray those of the United States, and the result was as I anticipated. On the other hand, Dr Hooker has recently informed me that he finds that the rule does not hold in Australia; and I have made these few remarks on the sexes of trees simply to call attention to the subject.

Turning for a very brief space to animals: on the land there are some hermaphrodites, as land-mollusca and earth-worms; but these all pair. As yet I have not found a single case of a terrestrial animal which fertilises itself. We can understand this remarkable fact, which offers so strong a contrast with terrestrial plants, on the view of an occasional cross being indispensable, by considering the medium in which terrestrial animals live, and the nature of the fertilising element; for we know of no means, analogous to the action of insects and of the wind in the case of plants, by which an occasional cross could be effected with terrestrial animals without the concurrence of two individuals. Of aquatic animals, there are many self-fertilising hermaphrodites; but here currents in the water offer an obvious means for an occasional cross. And, as in the case of flowers, I have as yet failed, after consultation with one of the highest authorities, namely, Professor Huxley, to discover a single case of an hermaphodrite animal with the organs of reproduction so perfectly enclosed within the body, that access from without and the occasional influence of a distinct individual can be shown to be physically impossible. Cirripedes long appeared to me to present a case of very great difficulty under this point of view; but I have been enabled, by a fortunate chance, elsewhere to prove that two individuals, though both are self-fertilising hermaphrodites, do sometimes cross.

It must have struck most naturalists as a strange anomaly that, in the case of both animals and plants, species of the same family and even of the same genus, though agreeing closely with each other in almost their whole organisation, yet are not rarely, some of them hermaphrodites, and some of them unisexual. But if, in fact, all hermaphrodites do occasionally intercross with other individuals, the difference between hermaphrodites and unisexual species, as far as function is concerned, becomes very small.

From these several considerations and from the many special facts which I have collected, but which I am not here able to give, I am strongly inclined to suspect that, both in the vegetable and animal kingdoms, an occasional intercross with a distinct individual is a law of nature. I am well aware that there are, on this view, many cases of difficulty, some of which I am trying to investigate. Finally then, we may conclude that in many organic beings, a cross between two individuals is an obvious necessity for each birth; in many others it occurs perhaps only at long intervals; but in none, as I suspect, can self-fertilisation go on for perpetuity.

Circumstances favourable to Natural Selection.—This is an extremely intricate subject. A large amount of inheritable and diversified variability is favourable, but I believe mere individual differences suffice for the work. A large number of individuals, by giving a better chance for the appearance within any given period of profitable variations, will compensate for a lesser amount of variability in each individual, and is, I believe, an extremely important element of success. Though nature grants vast periods of time for the work of natural selection, she does not grant an indefinite period; for as all organic beings are striving, it may be said, to seize on each place in the economy of nature, if any one species does not become modified and improved in a corresponding degree with its competitors, it will soon be exterminated.

In man's methodical selection, a breeder selects for some definite object, and free intercrossing will wholly stop his work. But when many men, without intending to alter the breed, have a nearly common standard of perfection, and all try to get and breed from the best animals, much improvement and modification surely but slowly follow from this unconscious process of selection, notwithstanding a large amount of crossing with inferior animals. Thus it will be in nature; for within a confined area, with some place in its polity not so perfectly occupied as might be, natural selection will always tend to preserve all the individuals varying in the right direction, though in different degrees, so as better to fill up the unoccupied place. But if the area be large, its several districts will almost certainly present different conditions of life; and then if natural selection be modifying and improving a species in the several districts, there will be intercrossing with the other individuals of the same species on the confines of each. And in this case the effects of intercrossing can hardly be counterbalanced by natural selection always tending to modify all the individuals in each district in exactly the same manner to the conditions of each; for in a continuous area, the physical conditions at least will generally graduate away insensibly from one district to another. The intercrossing will most affect those animals which unite for each birth, which wander much, and which do not breed at a very quick rate. Hence in animals of this nature, for instance in birds, varieties will generally be confined to separated countries; and this I believe to be the case. In hermaphrodite organisms which cross only occasionally, and likewise in animals which unite for each birth, but which wander little and which can increase at a very rapid rate, a new and improved variety might be quickly formed on any one spot, and might there maintain itself in a body, so that whatever intercrossing took place would be chiefly between the individuals of the same new variety. A local variety when once thus formed might subsequently slowly spread to other districts. On the above principle, nurserymen always prefer getting seed from a large body of plants of the same variety, as the chance of intercrossing with other varieties is thus lessened.

Even in the case of slow-breeding animals, which unite for each birth, we must not overrate the effects of intercrosses in retarding natural selection; for I can bring a considerable catalogue of facts, showing that within the same area, varieties of the same animal can long remain distinct, from haunting different stations, from breeding at slightly different seasons, or from varieties of the same kind preferring to pair together.

Intercrossing plays a very important part in nature in keeping the individuals of the same species, or of the same variety, true and uniform in character. It will obviously thus act far more efficiently with those animals which unite for each birth; but I have already attempted to show that we have reason to believe that occasional intercrosses take place with all animals and with all plants. Even if these take place only at long intervals, I am convinced that the young thus produced will gain so much in vigour and fertility over the offspring from long-continued self-fertilisation, that they will have a better chance of surviving and propagating their kind; and thus, in the long run, the influence of intercrosses, even at rare intervals, will be great. If there exist organic beings which never intercross, uniformity of character can be retained amongst them, as long as their conditions of life remain the same, only through the principle of inheritance, and through natural selection destroying any which depart from the proper type; but if their conditions of life change and they undergo modification, uniformity of character can be given to their modified offspring, solely by natural selection preserving the same favourable variations.

Isolation, also, is an important element in the process of natural selection. In a confined or isolated area, if not very large, the organic and inorganic conditions of life will generally be in a great degree uniform; so that natural selection will tend to modify all the individuals of a varying species throughout the area in the same manner in relation to the same conditions. Intercrosses, also, with the individuals of the same species, which otherwise would have inhabited the surrounding and differently circumstanced districts, will be prevented. But isolation probably acts more efficiently in checking the immigration of better adapted organisms, after any physical change, such as of climate or elevation of the land, etc.; and thus new places in the natural economy of the country are left open for the old inhabitants to struggle for, and become adapted to, through modifications in their structure and constitution. Lastly, isolation, by checking immigration and consequently competition, will give time for any new variety to be slowly improved; and this may sometimes be of importance in the production of new species. If, however, an isolated area be very small, either from being surrounded by barriers, or from having very peculiar physical conditions, the total number of the individuals supported on it will necessarily be very small; and fewness of individuals will greatly retard the production of new species through natural selection, by decreasing the chance of the appearance of favourable variations.

If we turn to nature to test the truth of these remarks, and look at any small isolated area, such as an oceanic island, although the total number of the species inhabiting it, will be found to be small, as we shall see in our chapter on geographical distribution; yet of these species a very large proportion are endemic,—that is, have been produced there, and nowhere else. Hence an oceanic island at first sight seems to have been highly favourable for the production of new species. But we may thus greatly deceive ourselves, for to ascertain whether a small isolated area, or a large open area like a continent, has been most favourable for the production of new organic forms, we ought to make the comparison within equal times; and this we are incapable of doing.

Although I do not doubt that isolation is of considerable importance in the production of new species, on the whole I am inclined to believe that largeness of area is of more importance, more especially in the production of species, which will prove capable of enduring for a long period, and of spreading widely. Throughout a great and open area, not only will there be a better chance of favourable variations arising from the large number of individuals of the same species there supported, but the conditions of life are infinitely complex from the large number of already existing species; and if some of these many species become modified and improved, others will have to be improved in a corresponding degree or they will be exterminated. Each new form, also, as soon as it has been much improved, will be able to spread over the open and continuous area, and will thus come into competition with many others. Hence more new places will be formed, and the competition to fill them will be more severe, on a large than on a small and isolated area. Moreover, great areas, though now continuous, owing to oscillations of level, will often have recently existed in a broken condition, so that the good effects of isolation will generally, to a certain extent, have concurred. Finally, I conclude that, although small isolated areas probably have been in some respects highly favourable for the production of new species, yet that the course of modification will generally have been more rapid on large areas; and what is more important, that the new forms produced on large areas, which already have been victorious over many competitors, will be those that will spread most widely, will give rise to most new varieties and species, and will thus play an important part in the changing history of the organic world.

We can, perhaps, on these views, understand some facts which will be again alluded to in our chapter on geographical distribution; for instance, that the productions of the smaller continent of Australia have formerly yielded, and apparently are now yielding, before those of the larger Europæo-Asiatic area. Thus, also, it is that continental productions have everywhere become so largely naturalised on islands. On a small island, the race for life will have been less severe, and there will have been less modification and less extermination. Hence, perhaps, it comes that the flora of Madeira, according to Oswald Heer, resembles the extinct tertiary flora of Europe. All fresh-water basins, taken together, make a small area compared with that of the sea or of the land; and, consequently, the competition between fresh-water productions will have been less severe than elsewhere; new forms will have been more slowly formed, and old forms more slowly exterminated. And it is in fresh water that we find seven genera of Ganoid fishes, remnants of a once preponderant order: and in fresh water we find some of the most anomalous forms now known in the world, as the Ornithorhynchus and Lepidosiren, which, like fossils, connect to a certain extent orders now widely separated in the natural scale. These anomalous forms may almost be called living fossils; they have endured to the present day, from having inhabited a confined area, and from having thus been exposed to less severe competition.

To sum up the circumstances favourable and unfavourable to natural selection, as far as the extreme intricacy of the subject permits. I conclude, looking to the future, that for terrestrial productions a large continental area, which will probably undergo many oscillations of level, and which consequently will exist for long periods in a broken condition, is the most favourable for the production of many new forms of life, likely to endure long and to spread widely. For the area first existed as a continent, and the inhabitants, at this period numerous in individuals and kinds, will have been subjected to very severe competition. When converted by subsidence into large separate islands, there will still exist many individuals of the same species on each island: intercrossing on the confines of the range of each species will thus be checked: after physical changes of any kind, immigration will be prevented, so that new places in the polity of each island will have to be filled up by modifications of the old inhabitants; and time will be allowed for the varieties in each to become well modified and perfected. When, by renewed elevation, the islands shall be reconverted into a continental area, there will again be severe competition: the most favoured or improved varieties will be enabled to spread: there will be much extinction of the less improved forms, and the relative proportional numbers of the various inhabitants of the renewed continent will again be changed; and again there will be a fair field for natural selection to improve still further the inhabitants, and thus produce new species.

That natural selection will always act with extreme slowness, I fully admit. Its action depends on there being places in the polity of nature, which can be better occupied by some of the inhabitants of the country undergoing modification of some kind. The existence of such places will often depend on physical changes, which are generally very slow, and on the immigration of better adapted forms having been checked. But the action of natural selection will probably still oftener depend on some of the inhabitants becoming slowly modified; the mutual relations of many of the other inhabitants being thus disturbed. Nothing can be effected, unless favourable variations occur, and variation itself is apparently always a very slow process. The process will often be greatly retarded by free intercrossing. Many will exclaim that these several causes are amply sufficient wholly to stop the action of natural selection. I do not believe so. On the other hand, I do believe that natural selection always acts very slowly, often only at long intervals of time, and generally on only a very few of the inhabitants of the same region at the same time. I further believe, that this very slow, intermittent action of natural selection accords perfectly well with what geology tells us of the rate and manner at which the inhabitants of this world have changed.

Slow though the process of selection may be, if feeble man can do much by his powers of artificial selection, I can see no limit to the amount of change, to the beauty and infinite complexity of the coadaptations between all organic beings, one with another and with their physical conditions of life, which may be effected in the long course of time by nature's power of selection.

Extinction.—This subject will be more fully discussed in our chapter on Geology; but it must be here alluded to from being intimately connected with natural selection. Natural selection acts solely through the preservation of variations in some way advantageous, which consequently endure. But as from the high geometrical ratio of increase of all organic beings, each area is already fully stocked with inhabitants, it follows that as each selected and favoured form increases in number, so will the less favoured forms decrease and become rare. Rarity, as geology tells us, is the precursor to extinction. We can, also, see that any form represented by few individuals will, during fluctuations in the seasons or in the number of its enemies, run a good chance of utter extinction. But we may go further than this; for as new forms are continually and slowly being produced, unless we believe that the number of specific forms goes on perpetually and almost indefinitely increasing, numbers inevitably must become extinct. That the number of specific forms has not indefinitely increased, geology shows us plainly; and indeed we can see reason why they should not have thus increased, for the number of places in the polity of nature is not indefinitely great,—not that we have any means of knowing that any one region has as yet got its maximum of species. Probably no region is as yet fully stocked, for at the Cape of Good Hope, where more species of plants are crowded together than in any other quarter of the world, some foreign plants have become naturalised, without causing, as far as we know, the extinction of any natives.

Furthermore, the species which are most numerous in individuals will have the best chance of producing within any given period favourable variations. We have evidence of this, in the facts given in the second chapter, showing that it is the common species which afford the greatest number of recorded varieties, or incipient species. Hence, rare species will be less quickly modified or improved within any given period, and they will consequently be beaten in the race for life by the modified descendants of the commoner species.

From these several considerations I think it inevitably follows, that as new species in the course of time are formed through natural selection, others will become rarer and rarer, and finally extinct. The forms which stand in closest competition with those undergoing modification and improvement, will naturally suffer most. And we have seen in the chapter on the Struggle for Existence that it is the most closely-allied forms,—varieties of the same species, and species of the same genus or of related genera,—which, from having nearly the same structure, constitution, and habits, generally come into the severest competition with each other. Consequently, each new variety or species, during the progress of its formation, will generally press hardest on its nearest kindred, and tend to exterminate them. We see the same process of extermination amongst our domesticated productions, through the selection of improved forms by man. Many curious instances could be given showing how quickly new breeds of cattle, sheep, and other animals, and varieties of flowers, take the place of older and inferior kinds. In Yorkshire, it is historically known that the ancient black cattle were displaced by the long-horns, and that these 'were swept away by the short-horns' (I quote the words of an agricultural writer) 'as if by some murderous pestilence.'

Divergence of Character.—The principle, which I have designated by this term, is of high importance on my theory, and explains, as I believe, several important facts. In the first place, varieties, even strongly-marked ones, though having somewhat of the character of species—as is shown by the hopeless doubts in many cases how to rank them—yet certainly differ from each other far less than do good and distinct species. Nevertheless, according to my view, varieties are species in the process of formation, or are, as I have called them, incipient species. How, then, does the lesser difference between varieties become augmented into the greater difference between species? That this does habitually happen, we must infer from most of the innumerable species throughout nature presenting well-marked differences; whereas varieties, the supposed prototypes and parents of future well-marked species, present slight and ill-defined differences. Mere chance, as we may call it, might cause one variety to differ in some character from its parents, and the offspring of this variety again to differ from its parent in the very same character and in a greater degree; but this alone would never account for so habitual and large an amount of difference as that between varieties of the same species and species of the same genus.

As has always been my practice, let us seek light on this head from our domestic productions. We shall here find something analogous. A fancier is struck by a pigeon having a slightly shorter beak; another fancier is struck by a pigeon having a rather longer beak; and on the acknowledged principle that 'fanciers do not and will not admire a medium standard, but like extremes,' they both go on (as has actually occurred with tumbler-pigeons) choosing and breeding from birds with longer and longer beaks, or with shorter and shorter beaks. Again, we may suppose that at an early period one man preferred swifter horses; another stronger and more bulky horses. The early differences would be very slight; in the course of time, from the continued selection of swifter horses by some breeders, and of stronger ones by others, the differences would become greater, and would be noted as forming two sub-breeds; finally, after the lapse of centuries, the sub-breeds would become converted into two well-established and distinct breeds. As the differences slowly become greater, the inferior animals with intermediate characters, being neither very swift nor very strong, will have been neglected, and will have tended to disappear. Here, then, we see in man's productions the action of what may be called the principle of divergence, causing differences, at first barely appreciable, steadily to increase, and the breeds to diverge in character both from each other and from their common parent.

But how, it may be asked, can any analogous principle apply in nature? I believe it can and does apply most efficiently, from the simple circumstance that the more diversified the descendants from any one species become in structure, constitution, and habits, by so much will they be better enabled to seize on many and widely diversified places in the polity of nature, and so be enabled to increase in numbers.

We can clearly see this in the case of animals with simple habits. Take the case of a carnivorous quadruped, of which the number that can be supported in any country has long ago arrived at its full average. If its natural powers of increase be allowed to act, it can succeed in increasing (the country not undergoing any change in its conditions) only by its varying descendants seizing on places at present occupied by other animals: some of them, for instance, being enabled to feed on new kinds of prey, either dead or alive; some inhabiting new stations, climbing trees, frequenting water, and some perhaps becoming less carnivorous. The more diversified in habits and structure the descendants of our carnivorous animal became, the more places they would be enabled to occupy. What applies to one animal will apply throughout all time to all animals—that is, if they vary—for otherwise natural selection can do nothing. So it will be with plants. It has been experimentally proved, that if a plot of ground be sown with one species of grass, and a similar plot be sown with several distinct genera of grasses, a greater number of plants and a greater weight of dry herbage can thus be raised. The same has been found to hold good when first one variety and then several mixed varieties of wheat have been sown on equal spaces of ground. Hence, if any one species of grass were to go on varying, and those varieties were continually selected which differed from each other in at all the same manner as distinct species and genera of grasses differ from each other, a greater number of individual plants of this species of grass, including its modified descendants, would succeed in living on the same piece of ground. And we well know that each species and each variety of grass is annually sowing almost countless seeds; and thus, as it may be said, is striving its utmost to increase its numbers. Consequently, I cannot doubt that in the course of many thousands of generations, the most distinct varieties of any one species of grass would always have the best chance of succeeding and of increasing in numbers, and thus of supplanting the less distinct varieties; and varieties, when rendered very distinct from each other, take the rank of species.

The truth of the principle, that the greatest amount of life can be supported by great diversification of structure, is seen under many natural circumstances. In an extremely small area, especially if freely open to immigration, and where the contest between individual and individual must be severe, we always find great diversity in its inhabitants. For instance, I found that a piece of turf, three feet by four in size, which had been exposed for many years to exactly the same conditions, supported twenty species of plants, and these belonged to eighteen genera and to eight orders, which shows how much these plants differed from each other. So it is with the plants and insects on small and uniform islets; and so in small ponds of fresh water. Farmers find that they can raise most food by a rotation of plants belonging to the most different orders: nature follows what may be called a simultaneous rotation. Most of the animals and plants which live close round any small piece of ground, could live on it (supposing it not to be in any way peculiar in its nature), and may be said to be striving to the utmost to live there; but, it is seen, that where they come into the closest competition with each other, the advantages of diversification of structure, with the accompanying differences of habit and constitution, determine that the inhabitants, which thus jostle each other most closely, shall, as a general rule, belong to what we call different genera and orders.

The same principle is seen in the naturalisation of plants through man's agency in foreign lands. It might have been expected that the plants which have succeeded in becoming naturalised in any land would generally have been closely allied to the indigenes; for these are commonly looked at as specially created and adapted for their own country. It might, also, perhaps have been expected that naturalised plants would have belonged to a few groups more especially adapted to certain stations in their new homes. But the case is very different; and Alph. De Candolle has well remarked in his great and admirable work, that floras gain by naturalisation, proportionally with the number of the native genera and species, far more in new genera than in new species. To give a single instance: in the last edition of Dr Asa Gray's *Manual of the Flora of the Northern United States*, 260 naturalised plants are enumerated, and these belong to 162 genera. We thus see that these naturalised plants are of a highly diversified nature. They differ, moreover, to a large extent from the indigenes, for out of the 162 genera, no less than 100 genera are not there indigenous, and thus a large proportional addition is made to the genera of these States.

By considering the nature of the plants or animals which have struggled successfully with the indigenes of any country, and have there become naturalised, we may gain some crude idea in what manner some of the natives would have to be modified, in order to gain an advantage over the other natives; and we may at least safely infer that diversification of structure, amounting to new generic differences, would be profitable to them.

The advantage of diversification in the inhabitants of the same region is, in fact, the same as that of the physiological division of labour in the organs of the same individual body—a subject so well elucidated by Milne Edwards. No physiologist doubts that a stomach adapted to digest vegetable matter alone, or flesh alone, draws most nutriment from these substances. So in the general economy of any land, the more widely and perfectly the animals and plants are diversified for different habits of life, so will a greater number of individuals be capable of there supporting themselves. A set of animals, with their organisation but little diversified, could hardly compete with a set more perfectly diversified in structure. It may be doubted, for instance, whether the Australian marsupials, which are divided into groups differing but little from each other, and feebly representing, as Mr Waterhouse and others have remarked, our carnivorous, ruminant, and rodent mammals, could successfully compete with these well-pronounced orders. In the Australian mammals, we see the process of diversification in an early and incomplete stage of development.

After the foregoing discussion, which ought to have been much amplified, we may, I think, assume that the modified descendants of any one species will succeed by so much the better as they become more diversified in structure, and are thus enabled to encroach on places occupied by other beings. Now let us see how this principle of benefit being derived from divergence of character, combined with the principles of natural selection and of extinction, will tend to act.

The accompanying diagram* will aid us in understanding this rather perplexing subject. Let A to L represent the species of a genus large in its own country; these species are supposed to resemble each other in unequal degrees, as is so generally the case in nature, and as is represented in the diagram by the letters standing at unequal distances. I have said a large genus, because we have seen in the second chapter, that on an average more of the species of large genera vary than of small genera; and the varying species of the large genera present a greater number of varieties. We have, also, seen that the species, which are the commonest and the most widely-diffused, vary more than rare species with restricted ranges. Let (A) be a common, widely-diffused, and varying species, belonging to a genus large in its own country. The little fan of diverging dotted lines of unequal lengths proceeding from (A), may represent its varying offspring. The variations are supposed to be extremely slight, but of the most diversified nature; they are not supposed all to appear simultaneously, but often after long intervals of time; nor are they all supposed to endure for equal periods. Only those variations which are in some way profitable will be preserved or naturally selected. And here the

* See diagram at the commencement of volume.

importance of the principle of benefit being derived from divergence of character comes in; for this will generally lead to the most different or divergent variations (represented by the outer dotted lines) being preserved and accumulated by natural selection. When a dotted line reaches one of the horizontal lines, and is there marked by a small numbered letter, a sufficient amount of variation is supposed to have been accumulated to have formed a fairly well-marked variety, such as would be thought worthy of record in a systematic work.

The intervals between the horizontal lines in the diagram, may represent each a thousand generations; but it would have been better if each had represented ten thousand generations. After a thousand generations, species (A) is supposed to have produced two fairly well-marked varieties, namely, a^1 and m^1. These two varieties will generally continue to be exposed to the same conditions which made their parents variable, and the tendency to variability is in itself hereditary, consequently they will tend to vary, and generally to vary in nearly the same manner as their parents varied. Moreover, these two varieties, being only slightly modified forms, will tend to inherit those advantages which made their parent (A) more numerous than most of the other inhabitants of the same country; they will likewise partake of those more general advantages which made the genus to which the parent-species belonged, a large genus in its own country. And these circumstances we know to be favourable to the production of new varieties.

If, then, these two varieties be variable, the most divergent of their variations will generally be preserved during the next thousand generations. And after this interval, variety a^1 is supposed in the diagram to have produced variety a^2, which will, owing to the principle of divergence, differ more from (A) than did variety a^1. Variety m^1 is supposed to have produced two varieties, namely m^2 and s^2, differing from each other, and more considerably from their common parent (A). We may continue the process by similar steps for any length of time; some of the varieties, after each thousand generations, producing only a single variety, but in a more and more modified condition, some producing two or three varieties, and some failing to produce any. Thus the varieties or modified descendants, proceeding from the common parent (A), will generally go on increasing in number and diverging in character. In the diagram the process is represented up to the ten-thousandth generation, and under a condensed and simplified form up to the fourteen-thousandth generation.

But I must here remark that I do not suppose that the process ever goes on so regularly as is represented in the diagram, though in itself made somewhat irregular. I am far from thinking that the most divergent varieties will invariably prevail and multiply: a medium form may often long endure, and may or may not produce more than one modified descendant; for natural selection will always act according to the nature of the places which are either unoccupied or not perfectly occupied by other beings; and this will depend on infinitely complex relations. But as a general rule, the more diversified in structure the descendants from any one species can be rendered, the more places they will be enabled to seize on, and the more their modified progeny will be increased. In our diagram the line of succession is broken at regular intervals by small numbered letters marking the successive forms which have become sufficiently distinct to be recorded as varieties. But these breaks are imaginary, and might have been inserted anywhere, after intervals long enough to have allowed the accumulation of a considerable amount of divergent variation.

As all the modified descendants from a common and widely-diffused species, belonging to a large genus, will tend to partake of the same advantages which made their parent successful in life, they will generally go on multiplying in number as well as diverging in character: this is represented in the diagram by the several divergent branches proceeding from (A). The modified offspring from the later and more highly improved branches in the lines of descent, will, it is probable, often take the place of, and so destroy, the earlier and less improved branches: this is represented in the diagram by some of the lower branches not reaching to the upper horizontal lines. In some cases I do not doubt that the process of modification will be confined to a single line of descent, and the number of the descendants will not be increased; although the amount of divergent modification may have been increased in the successive generations. This case would be represented in the diagram, if all the lines proceeding from (A) were removed, excepting that from a^1 to a^{10}. In the same way, for instance, the English race-horse and English pointer have apparently both gone on slowly diverging in character from their original stocks, without either having given off any fresh branches or races.

After ten thousand generations, species (A) is supposed to have produced three forms, a^{10}, f^{10}, and m^{10}, which, from having diverged in character during the successive generations, will have come to differ largely, but perhaps unequally, from each other and from their common parent. If we suppose the amount of change between each horizontal line in our diagram to be excessively small, these three forms may still be only well-marked varieties; or they may have arrived at the doubtful category of sub-species; but we have only to suppose the steps in the process of modification to be more numerous or greater in amount, to convert these three forms into well-defined species: thus the diagram illustrates the steps by which the small differences distinguishing varieties are increased into the larger differences distinguishing species. By continuing the same process for a greater number of generations (as shown in the diagram in a condensed and simplified manner), we get eight species, marked by the letters between a^{14} and m^{14}, all descended from (A). Thus, as I believe, species are multiplied and genera are formed.

In a large genus it is probable that more than one species would vary. In the diagram I have assumed that a second species (I) has produced, by analogous steps, after ten thousand generations, either two well-marked varieties (w^{10} and z^{10}) or two species, according to the amount of change supposed to be represented between the horizontal lines. After fourteen thousand generations, six new species, marked by the letters n^{14} to z^{14}, are supposed to have been produced. In each genus, the species, which are already extremely different in character, will generally tend to produce the greatest number of modified descendants; for these will have the best chance of filling new and widely different places in the polity of nature: hence in the diagram I have chosen the extreme species (A), and the nearly extreme species (I), as those which have largely varied, and have given rise to new varieties and species. The other nine species (marked by capital letters) of our original genus, may for a long period continue to transmit unaltered descendants; and this is shown in the diagram by the dotted lines not prolonged far upwards from want of space.

But during the process of modification, represented in the diagram, another of our principles, namely that of extinction, will have played an important part. As in each fully stocked country natural selection necessarily acts by the selected form having some advantage in the struggle for life over other forms, there will be a constant tendency in the improved descendants of any one species to supplant and exterminate in each stage of descent their predecessors and their original parent. For it should be remembered that the competition will generally be most severe between those forms which are most nearly related to each other in habits, constitution, and structure. Hence all the intermediate forms between the earlier and later states, that is between the less and more improved state of a species, as well as the original parent-species itself, will generally tend to become extinct. So it probably will be with many whole collateral lines of descent, which will be conquered by later and improved lines of descent. If, however, the modified offspring of a species get into some distinct country, or become quickly adapted to some quite new station, in which child and parent do not come into competition, both may continue to exist.

If then our diagram be assumed to represent a considerable amount of modification, species (A) and all the earlier varieties will have become extinct, having been replaced by eight new species (a^{14} to m^{14}); and (I) will have been replaced by six (n^{14} to z^{14}) new species.

But we may go further than this. The original species of our genus were supposed to resemble each other in unequal degrees, as is so generally the case in nature; species (A) being more nearly related to B, C, and D, than to the other species; and species (I) more to G, H, K, L, than to the others. These two species (A) and (I), were also supposed to be very common and widely diffused species, so that they must originally have had some advantage over most of the other species of the genus. Their modified descendants, fourteen in number at the fourteen-thousandth generation, will probably have inherited some of the same advantages: they have also been modified and improved in a diversified manner at each stage of descent, so as to have become adapted to many related places in the natural economy of their country. It seems, therefore, to me extremely probable that they will have taken the places of, and thus exterminated, not only their parents (A) and (I), but likewise some of the original species which were most nearly related to their parents. Hence very few of the original species will have transmitted offspring to the fourteen-thousandth generation. We may suppose that only one (F), of the two species which were least closely related to the other nine original species, has transmitted descendants to this late stage of descent.

The new species in our diagram descended from the original eleven species, will now be fifteen in number. Owing to the divergent tendency of natural selection, the extreme amount of difference in character between species a^{14} and z^{14} will be much greater than that between the most different of the original eleven species. The new species, moreover, will be allied to each other in a widely different manner. Of the eight descendants from (A) the three marked a^{14}, q^{14}, p^{14}, will be nearly related from having recently branched off from a^{10}; b^{14} and f^{14}, from having diverged at an earlier period from a^5, will be in some degree distinct from the three first-named species; and lastly, o^{14}, e^{14}, and m^{14}, will be nearly related one to the other, but from having diverged at the first commencement of the process of modification, will be widely different from the other five species, and may constitute a sub-genus or even a distinct genus.

The six descendants from (I) will form two subgenera or even genera. But as the original species (I) differed largely from (A), standing nearly at the extreme points of the original genus, the six descendants from (I) will, owing to inheritance alone, differ considerably from the eight descendants from (A); the two groups, moreover, are supposed to have gone on diverging in different directions. The intermediate species, also (and this is a very important consideration), which connected the original species (A) and (I), have all become, excepting (F), extinct, and have left no descendants. Hence the six new species descended from (I), and the eight descended from (A), will have to be ranked as very distinct genera, or even as distinct sub-families.

Thus it is, as I believe, that two or more genera are produced by descent with modification, from two or more species of the same genus. And the two or more parent-species are supposed to have descended from some one species of an earlier genus. In our diagram, this is indicated by the broken lines, beneath the capital letters, converging in sub-branches downwards towards a single point; this point representing a single species, the supposed single parent of our several new sub-genera and genera.

It is worth while to reflect for a moment on the character of the new species f^{14}, which is supposed not to have diverged much in character, but to have retained the form of (F), either unaltered or altered only in a slight degree. In this case, its affinities to the other fourteen new species will be of a curious and circuitous nature. Having descended from a form which stood between the two parent-species (A) and (I), now supposed to be extinct and unknown, it will be in some degree intermediate in character between the two groups descended from these species. But as these two groups have gone on diverging in character from the type of their parents, the new species (f^{14}) will not be directly intermediate between them, but rather between types of the two groups; and every naturalist will be able to bring some such case before his mind.

In the diagram, each horizontal line has hitherto been supposed to represent a thousand generations, but each may represent a million or hundred million generations, and likewise a section of the successive strata of the earth's crust including extinct remains. We shall, when we come to our chapter on Geology, have to refer again to this subject, and I think we shall then see that the diagram throws light on the affinities of extinct beings, which, though generally belonging to the same orders, or families, or genera, with those now living, yet are often, in some degree, intermediate in character between existing groups; and we can understand this fact, for the extinct species lived at very ancient epochs when the branching lines of descent had diverged less.

I see no reason to limit the process of modification, as now explained, to the formation of genera alone. If, in our diagram, we suppose the amount of change represented by each successive group of diverging dotted lines to be very great, the forms marked a^{14} to p^{14}, those marked b^{14} and f^{14}, and those marked o^{14} to m^{14}, will form three very distinct genera. We shall also have two very distinct genera descended from (I); and as these latter two genera, both from continued divergence of character and from inheritance from a different parent, will differ widely from the three genera descended from (A), the two little groups of genera will form two distinct families, or even orders, according to the amount of divergent modification supposed to be represented in the diagram. And the two new families, or orders, will have descended from two species of the original genus; and these two species are supposed to have descended from one species of a still more ancient and unknown genus.

We have seen that in each country it is the species of the larger genera which oftenest present varieties or incipient species. This, indeed, might have been expected; for as natural selection acts through one form having some advantage over other forms in the struggle for existence, it will chiefly act on those which already have some advantage; and the largeness of any group shows that its species have inherited from a common ancestor some advantage in common. Hence, the struggle for the production of new and modified descendants, will mainly lie between the larger groups, which are all trying to increase in number. One large group will slowly conquer another large group, reduce its numbers, and thus lessen its chance of further variation and improvement. Within the same large group, the later and more highly perfected sub-groups from branching out and seizing on many new places in the polity of Nature, will constantly tend to supplant and destroy the earlier and less improved sub-groups. Small and broken groups and sub-groups will finally disappear. Looking to the future, we can predict that the groups of organic beings which are now large and triumphant, and which are least broken up, that is, which as yet have suffered least extinction, will for a long period continue to increase. But which groups will ultimately prevail, no man can predict; for we well know that many groups, formerly most extensively developed, have now become extinct. Looking still more remotely to the future, we may predict that, owing to the continued and steady increase of the larger groups, a multitude of smaller groups will become utterly extinct, and leave no modified descendants; and consequently that of the species living at any one period, extremely few will transmit descendants to a remote futurity. I shall have to return to this subject in the chapter on Classification, but I may add that on this view of extremely few of the more ancient species having transmitted descendants, and on the view of all the descendants of the same species making a class, we can understand how it is that there exist but very few classes in each main division of the animal and vegetable kingdoms. Although extremely few of the most ancient species may now have living and modified descendants, yet at the most remote geological period, the earth may have been as well peopled with many species of many genera, families, orders, and classes, as at the present day.

Summary of Chapter.—If during the long course of ages and under varying conditions of life, organic beings vary at all in the several parts of their organisation, and I think this cannot be disputed; if there be, owing to the high geometrical ratio of increase of each species, a severe struggle for life at some age, season, or year, and this certainly cannot be disputed; then, considering the infinite complexity of the relations of all organic beings to each other and to their conditions of existence, causing an infinite diversity in structure, constitution, and habits, to be advantageous to them, I think it would be a most extraordinary fact if no variation ever had occurred useful to each being's own welfare, in the same manner as so many variations have occurred useful to man. But if variations useful to any organic being do occur, assuredly individuals thus characterised will have the best chance of being preserved in the struggle for life; and from the strong principle of inheritance they will tend to produce offspring similarly characterised. This principle of preservation, I have called, for the sake of brevity, Natural Selection; and it leads to the improvement of each creature in relation to its organic and inorganic conditions of life.

Natural selection, on the principle of qualities being inherited at corresponding ages, can modify the egg, seed, or young, as easily as the adult. Amongst many animals, sexual selection will give its aid to ordinary selection, by assuring to the most vigorous and best adapted males the greatest number of offspring. Sexual selection will also give characters useful to the males alone, in their struggles with other males.

Whether natural selection has really thus acted in nature, in modifying and adapting the various forms of life to their several conditions and stations, must be judged of by the general tenor and balance of evidence given in the following chapters. But we already see how it entails extinction; and how largely extinction has acted in the world's history, geology plainly declares. Natural selection, also, leads to divergence of character; for more living beings can be supported on the same area the more they diverge in structure, habits, and constitution, of which we see proof by looking to the inhabitants of any small spot or to naturalised productions. Therefore during the modification of the descendants of any one species, and during the incessant struggle of all species to increase in numbers, the more diversified these descendants become, the better will be their chance of succeeding in the battle for life. Thus the small differences distinguishing varieties of the same species, steadily tend to increase till they come to equal the greater differences between species of the same genus, or even of distinct genera.

We have seen that it is the common, the widely-diffused, and widely-ranging species, belonging to the larger genera, which vary most; and these tend to transmit to their modified offspring that superiority which now makes them dominant in their own countries. Natural selection, as has just been remarked, leads to divergence of character and to much extinction of the less improved and intermediate forms of life. On these principles, I believe, the nature of the affinities of all organic beings may be explained. It is a truly wonderful fact—the wonder of which we are apt to overlook from familiarity—that all animals and all plants throughout all time and space should be related to each other in group subordinate to group, in the manner which we everywhere behold—namely, varieties of the same species most closely related together, species of the same genus less closely and unequally related together, forming sections and sub-genera, species of distinct genera much less closely related, and genera related in different degrees, forming sub-families, families, orders, sub-classes, and classes. The several subordinate groups in any class cannot be ranked in a single file, but seem rather to be clustered round points, and these round other points, and so on in almost endless cycles. On the view that each species has been independently created, I can see no explanation of this great fact in the classification of all organic beings; but, to the best of my judgment, it is explained through inheritance and the complex action of natural selection, entailing extinction and divergence of character, as we have seen illustrated in the diagram.

The affinities of all the beings of the same class have sometimes been represented by a great tree. I believe this simile largely speaks the truth. The green and budding twigs may represent existing species; and those produced during each former year may represent the long succession of extinct species. At each period of growth all the growing twigs have tried to branch out on all sides, and to overtop and kill the surrounding twigs and branches, in the same manner as species and groups of species have tried to overmaster other species in the great battle for life. The limbs divided into great branches, and these into lesser and lesser branches, were themselves once, when the tree was small, budding twigs; and this connection of the former and present buds by ramifying branches may well represent the classification of all extinct and living species in groups subordinate to groups. Of the many twigs which flourished when the tree was a mere bush, only two or three, now grown into great branches, yet survive and bear all the other branches; so with the species which lived during long-past geological periods, very few now have living and modified descendants. From the first growth of the tree, many a limb and branch have decayed and dropped off; and these lost branches of various sizes may represent those whole orders, families, and genera which have now no living representatives, and which are known to us only from having been found in a fossil state. As we here and there see a thin straggling branch springing from a fork low down in a tree, and which by some chance has been favoured and is still alive on its summit, so we occasionally see an animal like the Ornithorhynchus or Lepidosiren, which in some small degree connects by its affinities two large branches of life, and which has apparently been saved from fatal competition by having inhabited a protected station. As buds give rise by growth to fresh buds, and these, if vigorous, branch out and overtop on all sides many a feebler branch, so by generation I believe it has been with the great Tree of Life, which fills with its dead and broken branches the crust of the earth, and covers the surface with its ever branching and beautiful ramifications.

CHAPTER V

LAWS OF VARIATION

Effects of external conditions—Use and disuse, combined with natural selection; organs of flight and of vision—Acclimatisation—Correlation of growth—Compensation and economy of growth—False correlation—Multiple, rudimentary, and lowly organised structures variable—Parts developed in an unusual manner are highly variable: specific characters more variable than generic: secondary sexual characters variable—Species of the same genus vary in an analogous manner—Reversions to long-lost characters—Summary.

I HAVE hitherto sometimes spoken as if the variations—so common and multiform in organic beings under domestication, and in a lesser degree in those in a state of nature—had been due to chance. This, of course, is a wholly incorrect expression, but it serves to acknowledge plainly our ignorance of the cause of each particular variation. Some authors believe it to be as much the function of the reproductive system to produce individual differences, or very slight deviations of structure, as to make the child like its parents. But the much greater variability, as well as the greater frequency of monstrosities, under domestication or cultivation, than under nature, leads me to believe that deviations of structure are in some way due to the nature of the conditions of life, to which the parents and their more remote ancestors have been exposed during several generations. I have remarked in the first chapter—but a long catalogue of facts which cannot be here given would be necessary to show the truth of the remark—that the reproductive system is eminently susceptible to changes in the conditions of life; and to this system being functionally disturbed in the parents, I chiefly attribute the varying or plastic condition of the offspring. The male and female sexual elements seem to be affected before that union takes place which is to form a new being. In the case of 'sporting' plants, the bud, which in its earliest condition does not apparently differ essentially from an ovule, is alone affected. But why, because the reproductive system is disturbed, this or that part should vary more or less, we are profoundly ignorant. Nevertheless, we can here and there dimly catch a faint ray of light, and we may feel sure that there must be some cause for each deviation of structure, however slight.

How much direct effect difference of climate, food, etc., produces on any being is extremely doubtful. My impression is, that the effect is extremely small in the case of animals, but perhaps rather more in that of plants. We may, at least, safely conclude that such influences cannot have produced the many striking and complex co-adaptations of structure between one organic being and another, which we see everywhere throughout nature. Some little influence may be attributed to climate, food, etc.: thus, E. Forbes speaks confidently that shells at their southern limit, and when living in shallow water, are more brightly coloured than those of the same species further north or from greater depths. Gould believes that birds of the same species are more brightly coloured under a clear atmosphere, than when living on islands or near the coast. So with insects, Wollaston is convinced that residence near the sea affects their colours. Moquin-Tandon gives a list of plants which when growing near the sea-shore have their leaves in some degree fleshy, though not elsewhere fleshy. Several other such cases could be given.

The fact of varieties of one species, when they range into the zone of habitation of other species, often acquiring in a very slight degree some of the characters of such species, accords with our view that species of all kinds are only well-marked and permanent varieties. Thus the species of shells which are confined to tropical and shallow seas are generally brighter-coloured than those confined to cold and deeper seas. The birds which are confined to continents are, according to Mr Gould, brighter-coloured than those of islands. The insect-species confined to sea-coasts, as every collector knows, are often brassy or lurid. Plants which live exclusively on the sea-side are very apt to have fleshy leaves. He who believes in the creation of each species, will have to say that this shell, for instance, was created with bright colours for a warm sea; but that this other shell became bright-coloured by variation when it ranged into warmer or shallower waters.

When a variation is of the slightest use to a being, we cannot tell how much of it to attribute to the accumulative action of natural selection, and how much to the conditions of life. Thus, it is well known to furriers that animals of the same species have thicker and better fur the more severe the climate is under which they have lived; but who can tell how much of this difference may be due to the warmest-clad individuals having been favoured and preserved during many generations, and how much to the direct action of the severe climate? for it would appear that climate has some direct action on the hair of our domestic quadrupeds.

Instances could be given of the same variety being produced under conditions of life as different as can well be conceived; and, on the other hand, of different varieties being produced from the same species under the same conditions. Such facts show how indirectly the conditions of life act. Again, innumerable instances are known to every naturalist of species keeping true, or not varying at all, although living under the most opposite climates. Such considerations as these incline me to lay very little weight on the direct action of the conditions of life. Indirectly, as already remarked, they seem to play an important part in affecting the reproductive system, and in thus inducing variability; and natural selection will then accumulate all profitable variations, however slight, until they become plainly developed and appreciable by us.

Effects of Use and Disuse.—From the facts alluded to in the first chapter, I think there can be little doubt that use in our domestic animals strengthens and enlarges certain parts, and disuse diminishes them; and that such modifications are inherited. Under free nature, we can have no standard of comparison, by which to judge of the effects of long-continued use or disuse, for we know not the parent-forms; but many animals have structures which can be explained by the effects of disuse. As Professor Owen has remarked, there is no greater anomaly in nature than a bird that cannot fly; yet there are several in this state. The logger-headed duck of South America can only flap along the surface of the water, and has its wings in nearly the same condition as the domestic Aylesbury duck. As the larger ground-feeding birds seldom take flight except to escape danger, I believe that the nearly wingless condition of several birds, which now inhabit or have lately inhabited several oceanic islands, tenanted by no beast of prey, has been caused by disuse. The ostrich indeed inhabits continents and is exposed to danger from which it cannot escape by flight, but by kicking it can defend itself from enemies, as well as any of the smaller quadrupeds. We may imagine that the early progenitor of the ostrich had habits like those of a bustard, and that as natural selection increased in successive generations the size and weight of its body, its legs were used more, and its wings less, until they became incapable of flight.

Kirby has remarked (and I have observed the same fact) that the anterior tarsi, or feet, of many male dung-feeding beetles are very often broken off; he examined seventeen specimens in his own collection, and not one had even a relic left. In the Onites apelles the tarsi are so habitually lost, that the insect has been described as not having them. In some other genera they are present, but in a rudimentary condition. In the Ateuchus or sacred beetle of the Egyptians, they are totally deficient. There is not sufficient evidence to induce me to believe that mutilations are ever inherited; and I should prefer explaining the entire absence of the anterior tarsi in Ateuchus, and their rudimentary condition in some other genera, by the long-continued effects of disuse in their progenitors; for as the tarsi are almost always lost in many dung-feeding beetles, they must be lost early in life, and therefore cannot be much used by these insects.

In some cases we might easily put down to disuse modifications of structure which are wholly, or mainly, due to natural selection. Mr Wollaston has discovered the remarkable fact that 200 beetles, out of the 550 species inhabiting Madeira, are so far deficient in wings that they cannot fly; and that of the twenty-nine endemic genera, no less than twenty-three genera have all their species in this condition! Several facts, namely, that beetles in many parts of the world are frequently blown to sea and perish; that the beetles in Madeira, as observed by Mr

Wollaston, lie much concealed, until the wind lulls and the sun shines; that the proportion of wingless beetles is larger on the exposed Desertas than in Madeira itself; and especially the extraordinary fact, so strongly insisted on by Mr Wollaston, of the almost entire absence of certain large groups of beetles, elsewhere excessively numerous, and which groups have habits of life almost necessitating frequent flight;—these several considerations have made me believe that the wingless condition of so many Madeira beetles is mainly due to the action of natural selection, but combined probably with disuse. For during thousands of successive generations each individual beetle which flew least, either from its wings having been ever so little less perfectly developed or from indolent habit, will have had the best chance of surviving from not being blown out to sea; and, on the other hand, those beetles which most readily took to flight would oftenest have been blown to sea and thus have been destroyed.

The insects in Madeira which are not ground-feeders, and which, as the flower-feeding coleoptera and lepidoptera, must habitually use their wings to gain their subsistence, have, as Mr Wollaston suspects, their wings not at all reduced, but even enlarged. This is quite compatible with the action of natural selection. For when a new insect first arrived on the island, the tendency of natural selection to enlarge or to reduce the wings, would depend on whether a greater number of individuals were saved by successfully battling with the winds, or by giving up the attempt and rarely or never flying. As with mariners shipwrecked near a coast, it would have been better for the good swimmers if they had been able to swim still further, whereas it would have been better for the bad swimmers if they had not been able to swim at all and had stuck to the wreck.

The eyes of moles and of some burrowing rodents are rudimentary in size, and in some cases are quite covered up by skin and fur. This state of the eyes is probably due to gradual reduction from disuse, but aided perhaps by natural selection. In South America, a burrowing rodent, the tuco-tuco, or Ctenomys, is even more subterranean in its habits than the mole; and I was assured by a Spaniard, who had often caught them, that they were frequently blind; one which I kept alive was certainly in this condition, the cause, as appeared on dissection, having been inflammation of the nictitating membrane. As frequent inflammation of the eyes must be injurious to any animal, and as eyes are certainly not indispensable to animals with subterranean habits, a reduction in their size with the adhesion of the eye-lids and growth of fur over them, might in such case be an advantage; and if so, natural selection would constantly aid the effects of disuse.

It is well known that several animals, belonging to the most different classes, which inhabit the caves of Styria and of Kentucky, are blind. In some of the crabs the foot-stalk for the eye remains, though the eye is gone; the stand for the telescope is there, though the telescope with its glasses has been lost. As it is difficult to imagine that eyes, though useless, could be in any way injurious to animals living in darkness, I attribute their loss wholly to disuse. In one of the blind animals, namely, the cave-rat, the eyes are of immense size; and Professor Silliman thought that it regained, after living some days in the light, some slight power of vision. In the same manner as in Madeira the wings of some of the insects have been enlarged, and the wings of others have been reduced by natural selection aided by use and disuse, so in the case of the cave-rat natural selection seems to have struggled with the loss of light and to have increased the size of the eyes; whereas with all the other inhabitants of the caves, disuse by itself seems to have done its work.

It is difficult to imagine conditions of life more similar than deep limestone caverns under a nearly similar climate; so that on the common view of the blind animals having been separately created for the American and European caverns, close similarity in their organisation and affinities might have been expected; but, as Schiödte and others have remarked, this is not the case, and the cave-insects of the two continents are not more closely allied than might have been anticipated from the general resemblance of the other inhabitants of North America and Europe. On my view we must suppose that American animals, having ordinary powers of vision, slowly migrated by successive generations from the outer world into the deeper and deeper recesses of the Kentucky caves, as did

European animals into the caves of Europe. We have some evidence of this gradation of habit; for, as Schiödte remarks, 'animals not far remote from ordinary forms, prepare the transition from light to darkness. Next follow those that are constructed for twilight; and, last of all, those destined for total darkness.' By the time that an animal had reached, after numberless generations, the deepest recesses, disuse will on this view have more or less perfectly obliterated its eyes, and natural selection will often have effected other changes, such as an increase in the length of the antennæ or palpi, as a compensation for blindness. Notwithstanding such modifications, we might expect still to see in the cave-animals of America, affinities to the other inhabitants of that continent, and in those of Europe, to the inhabitants of the European continent. And this is the case with some of the American cave-animals, as I hear from Professor Dana; and some of the European cave-insects are very closely allied to those of the surrounding country. It would be most difficult to give any rational explanation of the affinities of the blind cave-animals to the other inhabitants of the two continents on the ordinary view of their independent creation. That several of the inhabitants of the caves of the Old and New Worlds should be closely related, we might expect from the well-known relationship of most of their other productions. Far from feeling any surprise that some of the cave-animals should be very anomalous, as Agassiz has remarked in regard to the blind fish, the Amblyopsis, and as is the case with the blind Proteus with reference to the reptiles of Europe, I am only surprised that more wrecks of ancient life have not been preserved, owing to the less severe competition to which the inhabitants of these dark abodes will probably have been exposed.

Acclimatisation.—Habit is hereditary with plants, as in the period of flowering, in the amount of rain requisite for seeds to germinate, in the time of sleep, etc., and this leads me to say a few words on acclimatisation. As it is extremely common for species of the same genus to inhabit very hot and very cold countries, and as I believe that all the species of the same genus have descended from a single parent, if this view be correct, acclimatisation must be readily effected during long-continued descent. It is notorious that each species is adapted to the climate of its own home: species from an arctic or even from a temperate region cannot endure a tropical climate, or conversely. So again, many succulent plants cannot endure a damp climate. But the degree of adaptation of species to the climates under which they live is often overrated. We may infer this from our frequent inability to predict whether or not an imported plant will endure our climate, and from the number of plants and animals brought from warmer countries which here enjoy good health. We have reason to believe that species in a state of nature are limited in their ranges by the competition of other organic beings quite as much as, or more than, by adaptation to particular climates. But whether or not the adaptation be generally very close, we have evidence, in the case of some few plants, of their becoming, to a certain extent, naturally habituated to different temperatures, or becoming acclimatised: thus the pines and rhododendrons, raised from seed collected by Dr Hooker from trees growing at different heights on the Himalaya, were found in this country to possess different constitutional powers of resisting cold. Mr Thwaites informs me that he has observed similar facts in Ceylon, and analogous observations have been made by Mr H. C. Watson on European species of plants brought from the Azores to England. In regard to animals, several authentic cases could be given of species within historical times having largely extended their range from warmer to cooler latitudes, and conversely; but we do not positively know that these animals were strictly adapted to their native climate, but in all ordinary cases we assume such to be the case; nor do we know that they have subsequently become acclimatised to their new homes.

As I believe that our domestic animals were originally chosen by uncivilised man because they were useful and bred readily under confinement, and not because they were subsequently found capable of far-extended transportation, I think the common and extraordinary capacity in our domestic animals of not only withstanding the most different climates but of being perfectly fertile (a far severer test) under them, may be used as an argument that a large proportion of other animals, now in a state of nature, could easily be brought to bear widely different climates. We must not, however, push the foregoing argument too far, on account of the probable origin of some of our domestic animals from several wild stocks: the blood, for instance, of a tropical and arctic wolf or wild dog may perhaps be mingled in our domestic breeds. The rat and mouse cannot be considered as domestic animals, but they have been transported by man to many parts of the world, and now have a far wider range than any other rodent, living free under the cold climate of Faroe in the north and of the Falklands in the south, and on many islands in the torrid zones. Hence I am inclined to look at adaptation to any special climate as a quality readily grafted on an innate wide flexibility of constitution, which is common to most animals. On this view, the capacity of enduring the most different climates by man himself and by his domestic animals, and such facts as that former species of the elephant and rhinoceros were capable of enduring a glacial climate, whereas the living species are now all tropical or sub-tropical in their habits, ought not to be looked at as anomalies, but merely as examples of a very common flexibility of constitution, brought, under peculiar circumstances, into play.

How much of the acclimatisation of species to any peculiar climate is due to mere habit, and how much to the natural selection of varieties having different innate constitutions, and how much to both means combined, is a very obscure question. That habit or custom has some influence I must believe, both from analogy, and from the incessant advice given in agricultural works, even in the ancient Encyclopædias of China, to be very cautious in transposing animals from one district to another; for it is not likely that man should have succeeded in selecting so many breeds and sub-breeds with constitutions specially fitted for their own districts: the result must, I think, be due to habit. On the other hand, I can see no reason to doubt that natural selection will continually tend to preserve those individuals which are born with constitutions best adapted to their native countries. In treatises on many kinds of cultivated plants, certain varieties are said to withstand certain climates better than others: this is very strikingly shown in works on fruit trees published in the United States, in which certain varieties are habitually recommended for the northern, and others for the southern States; and as most of these varieties are of recent origin, they cannot owe their constitutional differences to habit. The case of the Jerusalem artichoke, which is never propagated by seed, and of which consequently new varieties have not been produced, has even been advanced—for it is now as tender as ever it was—as proving that acclimatisation cannot be effected! The case, also, of the kidney-bean has been often cited for a similar purpose, and with much greater weight; but until some one will sow, during a score of generations, his kidney-beans so early that a very large proportion are destroyed by frost, and then collect seed from the few survivors, with care to prevent accidental crosses, and then again get seed from these seedlings, with the same precautions, the experiment cannot be said to have been even tried. Nor let it be supposed that no differences in the constitution of seedling kidney-beans ever appear, for an account has been published how much more hardy some seedlings appeared to be than others.

On the whole, I think we may conclude that habit, use, and disuse, have, in some cases, played a considerable part in the modification of the constitution, and of the structure of various organs; but that the effects of use and disuse have often been largely combined with, and sometimes overmastered by the natural selection of innate variations.

Correlation of Growth.—I mean by this expression that the whole organisation is so tied together during its growth and development, that when slight variations in any one part occur, and are accumulated through natural selection, other parts become modified. This is a very important subject, most imperfectly understood. The most obvious case is, that modifications accumulated solely for the good of the young or larva, will, it may safely be concluded, affect the structure of the adult; in the same manner as any malconformation affecting the early embryo, seriously affects the whole organisation of the adult. The several parts of the body which are homologous, and which, at an early embryonic period, are alike, seem liable to vary in an allied manner: we see this in the right and left sides of the body varying in the same manner; in the front and hind legs, and even in the jaws and limbs, varying together, for the lower jaw is believed to be homologous with the limbs. These tendencies, I do not doubt, may be mastered more or less completely by natural selection: thus a family of stags once existed with an antler only on one side; and if this had been of any great use to the breed it might probably have been rendered permanent by natural selection.

Homologous parts, as has been remarked by some authors, tend to cohere; this is often seen in monstrous plants; and nothing is more common than the union of homologous parts in normal structures, as the union of the petals of the corolla into a tube. Hard parts seem to affect the form of adjoining soft parts; it is believed by some authors that the diversity in the shape of the pelvis in birds causes the remarkable diversity in the shape of their kidneys. Others believe that the shape of the pelvis in the human mother influences by pressure the shape of the head of the child. In snakes, according to Schlegel, the shape of the body and the manner of swallowing determine the position of several of the most important viscera.

The nature of the bond of correlation is very frequently quite obscure. M. Is. Geoffroy St Hilaire has forcibly remarked, that certain malconformations very frequently, and that others rarely coexist, without our being able to assign any reason. What can be more singular than the relation between blue eyes and deafness in cats, and the tortoise-shell colour with the female sex; the feathered feet and skin between the outer toes in pigeons, and the presence of more or less down on the young birds when first hatched, with the future colour of their plumage; or, again, the relation between the hair and teeth in the naked Turkish dog, though here probably homology comes into play? With respect to this latter case of correlation, I think it can hardly be accidental, that if we pick out the two orders of mammalia which are most abnormal in their dermal covering, viz. Cetacea (whales) and Edentata (armadilloes, scaly ant-eaters, etc.), that these are likewise the most abnormal in their teeth.

I know of no case better adapted to show the importance of the laws of correlation in modifying important structures, independently of utility and, therefore, of natural selection, than that of the difference between the outer and inner flowers in some Compositous and Umbelliferous plants. Every one knows the difference in the ray and central florets of, for instance, the daisy, and this difference is often accompanied with the abortion of parts of the flower. But, in some Compositous plants, the seeds also differ in shape and sculpture; and even the ovary itself, with its accessory parts, differs, as has been described by Cassini. These differences have been attributed by some authors to pressure, and the shape of the seeds in the ray-florets in some Compositæ countenances this idea; but, in the case of the corolla of the Umbelliferæ, it is by no means, as Dr Hooker informs me, in species with the densest heads that the inner and outer flowers most frequently differ. It might have been thought that the development of the ray-petals by drawing nourishment from certain other parts of the flower had caused their abortion; but in some Compositæ there is a difference in the seeds of the outer and inner florets without any difference in the corolla. Possibly, these several differences may be connected with some difference in the flow of nutriment towards the central and external flowers: we know, at least, that in irregular flowers, those nearest to the axis are oftenest subject to peloria, and become regular. I may add, as an instance of this, and of a striking case of correlation, that I have recently observed in some garden pelargoniums, that the central flower of the truss often loses the patches of darker colour in the two upper petals; and that when this occurs, the adherent nectary is quite aborted; when the colour is absent from only one of the two upper petals, the nectary is only much shortened.

With respect to the difference in the corolla of the central and exterior flowers of a head or umbel, I do not feel at all sure that C. C. Sprengel's idea that the ray-florets serve to attract insects, whose agency is highly advantageous in the fertilisation of plants of these two orders, is so far-fetched, as it may at first appear: and if it be advantageous, natural selection may have come into play. But in regard to the differences both in the internal and external structure of the seeds, which are not always correlated with any differences in the flowers, it seems impossible that they can be in any way advantageous to the plant: yet in the Umbelliferæ these differences are of such apparent importance—the seeds being in some cases, according to Tausch, orthospermous in the exterior flowers and cœlospermous in the central flowers,—that the elder De Candolle founded his main divisions of the order on analogous differences. Hence we see that modifications of structure, viewed by systematists as of high value, may be wholly due to unknown laws of correlated growth, and without being, as far as we can see, of the slightest service to the species.

We may often falsely attribute to correlation of growth, structures which are common to whole groups of species, and which in truth are simply due to inheritance; for an ancient progenitor may have acquired through natural selection some one modification in structure, and, after thousands of generations, some other and independent modification; and these two modifications, having been transmitted to a whole group of descendants with diverse habits, would naturally be thought to be correlated in some necessary manner. So, again, I do not doubt that some apparent correlations, occurring throughout whole orders, are entirely due to the manner alone in which natural selection can act. For instance, Alph. De Candolle has remarked that winged seeds are never found in fruits which do not open: I should explain the rule by the fact that seeds could not gradually become winged through natural selection, except in fruits which opened; so that the individual plants producing seeds which were a little better fitted to be wafted further, might get an advantage over those producing seed less fitted for dispersal; and this process could not possibly go on in fruit which did not open.

The elder Geoffroy and Goethe propounded, at about the same period, their law of compensation or balancement of growth; or, as Goethe expressed it, 'in order to spend on one side, nature is forced to economise on the other side.' I think this holds true to a certain extent with our domestic productions: if nourishment flows to one part or organ in excess, it rarely flows, at least in excess, to another part; thus it is difficult to get a cow to give much milk and to fatten readily. The same varieties of the cabbage do not yield abundant and nutritious foliage and a copious supply of oil-bearing seeds. When the seeds in our fruits become atrophied, the fruit itself gains largely in size and quality. In our poultry, a large tuft of feathers on the head is generally accompanied by a diminished comb, and a large beard by diminished wattles. With species in a state of nature it can hardly be maintained that the law is of universal application; but many good observers, more especially botanists, believe in its truth. I will not, however, here give any instances, for I see hardly any way of distinguishing between the effects, on the one hand, of a part being largely developed through natural selection and another and adjoining part being reduced by this same process or by disuse, and, on the other hand, the actual withdrawal of nutriment from one part owing to the excess of growth in another and adjoining part.

I suspect, also, that some of the cases of compensation which have been advanced, and likewise some other facts, may be merged under a more general principle, namely, that natural selection is continually trying to economise in every part of the organisation. If under changed conditions of life a structure before useful becomes less useful, any diminution, however slight, in its development, will be seized on by natural selection, for it will profit the individual not to have its nutriment wasted in building up an useless structure. I can thus only understand a fact with which I was much struck when examining cirripedes, and of which many other instances could be given: namely, that when a cirripede is parasitic within another and is thus protected, it loses more or less completely its own shell or carapace. This is the case with the male Ibla, and in a truly extraordinary manner with the Proteolepas: for the carapace in all other cirripedes consists of the three highly-important anterior segments of the head enormously developed, and furnished with great nerves and muscles; but in the parasitic and protected Proteolepas, the whole anterior part of the head is reduced to the merest rudiment attached to the basis of the prehensile antennæ. Now the saving of a large and complex structure, when rendered superfluous by the parasitic habits of the Proteolepas, though effected by slow steps, would be a decided advantage to each successive individual of the species; for in the struggle for life to which every animal is exposed, each individual Proteolepas would have a better chance of supporting itself, by less nutriment being wasted in developing a structure now become useless.

Thus, as I believe, natural selection will always succeed in the long run in reducing and saving every part of the organisation, as soon as it is rendered superfluous, without by any means causing some other part to be largely developed in a corresponding degree. And, conversely, that natural selection may perfectly well succeed in largely developing any organ, without requiring as a necessary compensation the reduction of some adjoining part.

It seems to be a rule, as remarked by Is. Geoffroy St Hilaire, both in varieties and in species, that when any part or organ is repeated many times in the structure of the same individual (as the vertebræ in snakes, and the stamens in polyandrous flowers) the number is variable; whereas the number of the same part or organ, when it occurs in lesser numbers, is constant. The same author and some botanists have further remarked that multiple parts are also very liable to variation in structure. Inasmuch as this 'vegetative repetition,' to use Professor Owen's expression, seems to be a sign of low organisation, the foregoing remark seems connected with the very general opinion of naturalists, that beings low in the scale of nature are more variable than those which are higher. I presume that lowness in this case means that the several parts of the organisation have been but little specialised for particular functions; and as long as the same part has to perform diversified work, we can perhaps see why it should remain variable, that is, why natural selection should have preserved or rejected each little deviation of form less carefully than when the part has to serve for one special purpose alone. In the same way that a knife which has to cut all sorts of things may be of almost any shape; whilst a tool for some particular object had better be of some particular shape. Natural selection, it should never be forgotten, can act on each part of each being, solely through and for its advantage.

Rudimentary parts, it has been stated by some authors, and I believe with truth, are apt to be highly variable. We shall have to recur to the general subject of rudimentary and aborted organs; and I will here only add that their variability seems to be owing to their uselessness, and therefore to natural selection having no power to check deviations in their structure. Thus rudimentary parts are left to the free play of the various laws of growth, to the effects of long-continued disuse, and to the tendency to reversion.

A part developed in any species in an extraordinary degree or manner, in comparison with the same part in allied species, tends to be highly variable.—Several years ago I was much struck with a remark, nearly to the above effect, published by Mr Waterhouse. I infer also from an observation made by Professor Owen, with respect to the length of the arms of the ourang-outang, that he has come to a nearly similar conclusion. It is hopeless to attempt to convince any one of the truth of this proposition without giving the long array of facts which I have collected, and which cannot possibly be here introduced. I can only state my conviction that it is a rule of high generality. I am aware of several causes of error, but I hope that I have made due allowance for them. It should be understood that the rule by no means applies to any part, however unusually developed, unless it be unusually developed in comparison with the same part in closely allied species. Thus, the bat's wing is a most abnormal structure in the class mammalia; but the rule would not here apply, because there is a whole group of bats having wings; it would apply only if some one species of bat had its wings developed in some remarkable manner in comparison with the other species of the same genus. The rule applies very strongly in the case of secondary sexual characters, when displayed in any unusual manner. The term, secondary sexual characters, used by Hunter, applies to characters which are attached to one sex, but are not directly connected with the act of reproduction. The rule applies to males and females; but as females more rarely offer remarkable secondary sexual characters, it applies more rarely to them. The rule being so plainly applicable in the case of secondary sexual characters, may be due to the great variability of these characters, whether or not displayed in any unusual manner—of which fact I think there can be little doubt. But that our rule is not confined to secondary sexual characters is clearly shown in the case of hermaphrodite cirripedes; and I may here add, that I particularly attended to Mr Waterhouse's remark, whilst investigating this Order, and I am fully convinced that the rule almost invariably holds good with cirripedes. I shall, in my future work, give a list of the more remarkable cases; I will here only briefly give one, as it illustrates the rule in its largest application. The opercular valves of sessile cirripedes (rock barnacles) are, in every sense of the word, very important structures, and they differ extremely little even in different genera; but in the several species of one genus, Pyrgoma, these valves present a marvellous amount of diversification: the homologous valves in the different species being sometimes wholly unlike in shape; and the amount of variation in the individuals of several of the species is so great, that it is no exaggeration to state that the varieties differ more from each other in the characters of these important valves than do other species of distinct genera.

As birds within the same country vary in a remarkably small degree, I have particularly attended to them, and the rule seems to me certainly to hold good in this class. I cannot make out that it applies to plants, and this would seriously have shaken my belief in its truth, had not the great variability in plants made it particularly difficult to compare their relative degrees of variability.

When we see any part or organ developed in a remarkable degree or manner in any species, the fair presumption is that it is of high importance to that species; nevertheless the part in this case is eminently liable to variation. Why should this be so? On the view that each species has been independently created, with all its parts as we now see them, I can see no explanation. But on the view that groups of species have descended from other species, and have been modified through natural selection, I think we can obtain some light. In our domestic animals, if any part, or the whole animal, be neglected and no selection be applied, that part (for instance, the comb in the Dorking fowl) or the whole breed will cease to have a nearly uniform character. The breed will then be said to have degenerated. In rudimentary organs, and in those which have been but little specialised for any particular purpose, and perhaps in polymorphic groups, we see a nearly parallel natural case; for in such cases natural selection either has not or cannot come into full play, and thus the organisation is left in a fluctuating condition. But what here more especially concerns us is, that in our domestic animals those points, which at the present time are undergoing rapid change by continued selection, are also eminently liable to variation. Look at the breeds of the pigeon; see what a prodigious amount of difference there is in the beak of the different tumblers, in the beak and wattle of the different carriers, in the carriage and tail of our fantails, etc., these being the points now mainly attended to by English fanciers. Even in the sub-breeds, as in the short-faced tumbler, it is notoriously difficult to breed them nearly to perfection, and frequently individuals are born which depart widely from the standard. There may be truly said to be a constant struggle going on between, on the one hand, the tendency to reversion to a less modified state, as well as an innate tendency to further variability of all kinds, and, on the other hand, the power of steady selection to keep the breed true. In the long run selection gains the day, and we do not expect to fail so far as to breed a bird as coarse as a common tumbler from a good short-faced strain. But as long as selection is rapidly going on, there may always be expected to be much variability in the structure undergoing modification. It further deserves notice that these variable characters, produced by man's selection, sometimes become attached, from causes quite unknown to us, more to one sex than to the other, generally to the male sex, as with the wattle of carriers and the enlarged crop of pouters.

Now let us turn to nature. When a part has been developed in an extraordinary manner in any one species, compared with the other species of the same genus, we may conclude that this part has undergone an extraordinary amount of modification since the period when the species branched off from the common progenitor of the genus. This period will seldom be remote in any extreme degree, as species very rarely endure for more than one geological period. An extraordinary amount of modification implies an unusually large and long-continued amount of variability, which has continually been accumulated by natural selection for the benefit of the species. But as the variability of the extraordinarily-developed part or organ has been so great and long-continued within a period not excessively remote, we might, as a general rule, expect still to find more variability in such parts than in other parts of the organisation which have remained for a much longer period nearly constant. And this, I am convinced, is the case. That the struggle between natural selection on the one hand, and the tendency to reversion and variability on the other hand, will in the course of time cease; and that the most abnormally developed organs may be made constant, I can see no reason to doubt. Hence when an organ, however abnormal it may be, has been transmitted in approximately the same condition to many modified descendants, as in the case of the wing of the bat, it must have existed, according to my theory, for an immense period in nearly the same state; and thus it comes to be no more variable than any other structure. It is only in those cases in which the modification has been comparatively recent and extraordinarily great that we ought to find the *generative variability*, as it may be called, still present in a high degree. For in this case the variability will seldom as yet have been fixed by the continued selection of the individuals varying in the required manner and degree, and by the continued rejection of those tending to revert to a former and less modified condition.

The principle included in these remarks may be extended. It is notorious that specific characters are more variable than generic. To explain by a simple example what is meant. If some species in a large genus of plants had blue flowers and some had red, the colour would be only a specific character, and no one would be surprised at one of the blue species varying into red, or conversely; but if all the species had blue flowers, the colour would become a generic character, and its variation would be a more unusual circumstance. I have chosen this example because an explanation is not in this case applicable, which most naturalists would advance, namely, that specific characters are more variable than generic, because they are taken from parts of less physiological importance than those commonly used for classing genera. I believe this explanation is partly, yet only indirectly, true; I shall, however, have to return to this subject in our chapter on Classification. It would be almost superfluous to adduce evidence in support of the above statement, that specific characters are more variable than generic; but I have repeatedly noticed in works on natural history, that when an author has remarked with surprise that some *important* organ or part, which is generally very constant throughout large groups of species, has *differed* considerably in closely-allied species, that it has, also, been *variable* in the individuals of some of the species. And this fact shows that a character, which is generally of generic value, when it sinks in value and becomes only of specific value, often becomes variable, though its physiological importance may remain the same. Something of the same kind applies to monstrosities: at least Is. Geoffroy St Hilaire seems to entertain no doubt, that the more an organ normally differs in the different species of the same group, the more subject it is to individual anomalies.

On the ordinary view of each species having been independently created, why should that part of the structure, which differs from the same part in other independently-created species of the same genus, be more variable than those parts which are closely alike in the several species? I do not see that any explanation can be given. But on the view of species being only strongly marked and fixed varieties, we might surely expect to find them still often continuing to vary in those parts of their structure which have varied within a moderately recent period, and which have thus come to differ. Or to state the case in another manner:—the points in which all the species of a genus resemble each other, and in which they differ from the species of some other genus, are called generic characters; and these characters in common I attribute to inheritance from a common progenitor, for it can rarely have happened that natural selection will have modified several species, fitted to more or less widely-different habits, in exactly the same manner: and as these so-called generic characters have been inherited from a remote period, since that period when the species first branched off from their common progenitor, and subsequently have not varied or come to differ in any degree, or only in a slight degree, it is not probable that they should vary at the present day. On the other hand, the points in which species differ from other species of the same genus, are called specific characters; and as these specific characters have varied and come to differ within the period of the branching off of the species from a common progenitor, it is probable that they should still often be in some degree variable,—at least more variable than those parts of the organisation which have for a very long period remained constant.

In connection with the present subject, I will make only two other remarks. I think it will be admitted, without my entering on details, that secondary sexual characters are very variable; I think it also will be admitted that species of the same group differ from each other more widely in their secondary sexual characters, than in other parts of their organisation; compare, for instance, the amount of difference between the males of gallinaceous birds, in which secondary sexual characters are strongly displayed, with the amount of difference between their females; and the truth of this proposition will be granted. The cause of the original variability of secondary sexual characters is not manifest; but we can see why these characters should not have been rendered as constant and uniform as other parts of the organisation; for secondary sexual characters have been accumulated by sexual selection, which is less rigid in its action than ordinary selection, as it does not entail death, but only gives fewer offspring to the less favoured males. Whatever the cause may be of the variability of secondary sexual characters, as they are highly variable, sexual selection will have had a wide scope for action, and may thus readily have succeeded in giving to the species of the same group a greater amount of difference in their sexual characters, than in other parts of their structure.

It is a remarkable fact, that the secondary sexual differences between the two sexes of the same species are generally displayed in the very same parts of the organisation in which the different species of the same genus differ from each other. Of this fact I will give in illustration two instances, the first which happen to stand on my list; and as the differences in these cases are of a very unusual nature, the relation can hardly be accidental. The same number of joints in the tarsi is a character generally common to very large groups of beetles, but in the Engidæ, as Westwood has remarked, the number varies greatly; and the number likewise differs in the two sexes of the same species: again in fossorial hymenoptera, the manner of neuration of the wings is a character of the highest importance, because common to large groups; but in certain genera the neuration differs in the different species, and likewise in the two sexes of the same species. This relation has a clear meaning on my view of the subject: I look at all the species of the same genus as having as certainly descended from the same progenitor, as have the two sexes of any one of the species. Consequently, whatever part of the structure of the common progenitor, or of its early descendants, became variable; variations of this part would, it is highly probable, be taken advantage of by natural and sexual selection, in order to fit the several species to their several places in the economy of nature, and likewise to fit the two sexes of the same species to each other, or to fit the males and females to different habits of life, or the males to struggle with other males for the possession of the females.

Finally, then, I conclude that the greater variability of specific characters, or those which distinguish species from species, than of generic characters, or those which the species possess in common;—that the frequent extreme variability of any part which is developed in a species in an extraordinary manner in comparison with the same part in its congeners; and the slight degree of variability in a part, however extraordinarily it may be developed, if it be common to a whole group of species;—that the great variability of secondary sexual characters, and the great amount of difference in these same characters between closely-allied species;—that secondary sexual and ordinary specific differences are generally displayed in the same parts of the organisation,—are all principles closely connected together. All being mainly due to the species of the same group having descended from a common progenitor, from whom they have inherited much in common,—to parts which have recently and largely varied being more likely still to go on varying than parts which have long been inherited and have not varied,—to natural selection having more or less completely, according to the lapse of time, overmastered the tendency to reversion and to further variability,—to sexual selection being less rigid than ordinary selection,—and to variations in the same parts having been accumulated by natural and sexual selection, and having been thus adapted for secondary sexual, and for ordinary specific purposes.

Distinct species present analogous variations; and a variety of one species often assumes some of the characters of an allied species, or reverts to some of the characters of an early progenitor.—These propositions will be most readily understood by looking to our domestic races. The most distinct breeds of pigeons, in countries most widely apart, present sub-varieties with reversed feathers on the head and feathers on the feet,—characters not possessed by the aboriginal rock-pigeon; these then are analogous variations in two or more distinct races. The frequent presence of fourteen or even sixteen tail-feathers in the pouter, may be considered as a variation representing the normal structure of another race, the fantail. I presume that no one will doubt that all such analogous variations are due to the several races of the pigeon having inherited from a common parent the same constitution and tendency to variation, when acted on by similar unknown influences. In the vegetable kingdom we have a case of analogous variation, in the enlarged stems, or roots as commonly called, of the Swedish turnip and Ruta baga, plants which several botanists rank as varieties produced by cultivation from a common parent: if this be not so, the case will then be one of analogous variation in two so-called distinct species; and to these a third may be added, namely, the common turnip. According to the ordinary view of each species having been independently created, we should have to attribute this similarity in the enlarged stems of these three plants, not to the *vera causa* of community of descent, and a consequent tendency to vary in a like manner, but to three separate yet closely related acts of creation.

With pigeons, however, we have another case, namely, the occasional appearance in all the breeds, of slaty-blue birds with two black bars on the wings, a white rump, a bar at the end of the tail, with the outer feathers externally edged near their bases with white. As all these marks are characteristic of the parent rock-pigeon, I presume that no one will doubt that this is a case of reversion, and not of a new yet analogous variation appearing in the several breeds. We may, I think, confidently come to this conclusion, because, as we have seen, these coloured marks are eminently liable to appear in the crossed offspring of two distinct and differently coloured breeds; and in this case there is nothing in the external conditions of life to cause the reappearance of the slaty-blue, with the several marks, beyond the influence of the mere act of crossing on the laws of inheritance.

No doubt it is a very surprising fact that characters should reappear after having been lost for many, perhaps for hundreds of generations. But when a breed has been crossed only once by some other breed, the offspring occasionally show a tendency to revert in character to the foreign breed for many generations—some say, for a dozen or even a score of generations. After twelve generations, the proportion of blood, to use a common expression, of any one ancestor, is only 1 in 2048; and yet, as we see, it is generally believed that a tendency to reversion is retained by this very small proportion of foreign blood. In a breed which has not been crossed, but in which *both* parents have lost some character which their progenitor possessed, the tendency, whether strong or weak, to reproduce the lost character might be, as was formerly remarked, for all that we can see to the contrary, transmitted for almost any number of generations. When a character which has been lost in a breed, reappears after a great number of generations, the most probable hypothesis is, not that the offspring suddenly takes after an ancestor some hundred generations distant, but that in each successive generation there has been a tendency to reproduce the character in question, which at last, under unknown favourable conditions, gains an ascendancy. For instance, it is probable that in each generation of the barb-pigeon, which produces most rarely a blue and black-barred bird, there has been a tendency in each generation in the plumage to assume this colour. This view is hypothetical, but could be supported by some facts; and I can see no more abstract improbability in a tendency to produce any character being inherited for an endless number of generations, than in quite useless or rudimentary organs being, as we all know them to be, thus inherited. Indeed, we may sometimes observe a mere tendency to produce a rudiment inherited: for instance, in the common snapdragon (Antirrhinum) a rudiment of a fifth stamen so often appears, that this plant must have an inherited tendency to produce it.

As all the species of the same genus are supposed, on my theory, to have descended from a common parent, it might be expected that they would occasionally vary in an analogous manner; so that a variety of one species would resemble in some of its characters another species; this other species being on my view only a well-marked and permanent variety. But characters thus gained would probably be of an unimportant nature, for the presence of all important characters will be governed by natural selection, in accordance with the diverse habits of the species, and will not be left to the mutual action of the conditions of life and of a similar inherited constitution. It might further be expected that the species of the same genus would occasionally exhibit reversions to lost ancestral characters. As, however, we never know the exact character of the common ancestor of a group, we could not distinguish these two cases: if, for instance, we did not know that the rock-pigeon was not feather-footed or turn-crowned, we could not have told, whether these characters in our domestic breeds were reversions or only analogous variations; but we might have inferred that the blueness was a case of reversion, from the number of the markings, which are correlated with the blue tint, and which it does not appear probable would all appear together from simple variation. More especially we might have inferred this, from the blue colour and marks so often appearing when distinct breeds of diverse colours are crossed. Hence, though under nature it must generally be left doubtful, what cases are reversions to an anciently existing character, and what are new but analogous variations, yet we ought, on my theory, sometimes to find the varying offspring of a species assuming characters (either from reversion or from analogous variation) which already occur in some other members of the same group. And this undoubtedly is the case in nature.

A considerable part of the difficulty in recognising a variable species in our systematic works, is due to its varieties mocking, as it were, some of the other species of the same genus. A considerable catalogue, also, could be given of forms intermediate between two other forms, which themselves must be doubtfully ranked as either varieties or species; and this shows, unless all these forms be considered as independently created species, that the one in varying has assumed some of the characters of the other, so as to produce the intermediate form. But the best evidence is afforded by parts or organs of an important and uniform nature occasionally varying so as to acquire, in some degree, the character of the same part or organ in an allied species. I have collected a long list of such cases; but here, as before, I lie under a great disadvantage in not being able to give them. I can only repeat that such cases certainly do occur, and seem to me very remarkable.

I will, however, give one curious and complex case, not indeed as affecting any important character, but from occurring in several species of the same genus, partly under domestication and partly under nature. It is a case apparently of reversion. The ass not rarely has very distinct transverse bars on its legs, like those on the legs of the zebra: it has been asserted that these are plainest in the foal, and from inquiries which I have made, I believe this to be true. It has also been asserted that the stripe on each shoulder is sometimes double. The shoulder-stripe is certainly very variable in length and outline. A white ass, but *not* an albino, has been described without either spinal or shoulder stripe; and these stripes are sometimes very obscure, or actually quite lost, in dark-coloured asses. The koulan of Pallas is said to have been seen with a double shoulder-stripe. The hemionus has no shoulder-stripe; but traces of it, as stated by Mr Blyth and others, occasionally appear: and I have been informed by Colonel Poole that the foals of this species are generally striped on the legs, and faintly on the shoulder. The quagga, though so plainly barred like a zebra over the body, is without bars on the legs; but Dr Gray has figured one specimen with very distinct zebra-like bars on the hocks.

With respect to the horse, I have collected cases in England of the spinal stripe in horses of the most distinct breeds, and of *all* colours; transverse bars on the legs are not rare in duns, mouse-duns, and in one instance in a chestnut: a faint shoulder-stripe may sometimes be seen in duns, and I have seen a trace in a bay horse. My son made a careful examination and sketch for me of a dun Belgian cart-horse with a double stripe on each shoulder and with leg-stripes; and a man, whom I can implicitly trust, has examined for me a small dun Welch pony with *three* short parallel stripes on each shoulder.

In the north-west part of India the Kattywar breed of horses is so generally striped, that, as I hear from Colonel Poole, who examined the breed for the Indian Government, a horse without stripes is not considered as purely-bred. The spine is always striped; the legs are generally barred; and the shoulder-stripe, which is sometimes double and sometimes treble, is common; the side of the face, moreover, is sometimes striped. The stripes are plainest in the foal; and sometimes quite disappear in old horses. Colonel Poole has seen both gray and bay Kattywar horses striped when first foaled. I have, also, reason to suspect, from information given me by Mr W. W. Edwards, that with the English race-horse the spinal stripe is much commoner in the foal than in the full-grown animal. Without here entering on further details, I may state that I have collected cases of leg and shoulder stripes in horses of very different breeds, in various countries from Britain to Eastern China; and from Norway in the north to the Malay Archipelago in the south. In all parts of the world these stripes occur far oftenest in duns and mouse-duns; by the term dun a large range of colour is included, from one between brown and black to a close approach to cream-colour.

I am aware that Colonel Hamilton Smith, who has written on this subject, believes that the several breeds of the horse have descended from several aboriginal species—one of which, the dun, was striped; and that the above-described appearances are all due to ancient crosses with the dun stock. But I am not at all satisfied with this theory, and should be loth to apply it to breeds so distinct as the heavy Belgian cart-horse, Welch ponies, cobs, the lanky Kattywar race, etc., inhabiting the most distant parts of the world.

Now let us turn to the effects of crossing the several species of the horse-genus. Rollin asserts, that the common mule from the ass and horse is particularly apt to have bars on its legs: according to Mr Gosse, in certain parts of the United States about nine out of ten mules have striped legs. I once saw a mule with its legs so much striped that any one would at first have thought that it must have been the product of a zebra; and Mr W. C. Martin, in his excellent treatise on the horse, has given a figure of a similar mule. In four coloured drawings, which I have seen, of hybrids between the ass and zebra, the legs were much more plainly barred than the rest of the body; and in one of them there was a double shoulder-stripe. In Lord Morton's famous hybrid from a chestnut mare and male quagga, the hybrid, and even the pure offspring subsequently produced from the mare by a black Arabian sire, were much more plainly barred across the legs than is even the pure quagga. Lastly, and this is another most remarkable case, a hybrid has been figured by Dr Gray (and he informs me that he knows of a second case) from the ass and the hemionus; and this hybrid, though the ass seldom has stripes on his legs and the hemionus has none and has not even a shoulder-stripe, nevertheless had all four legs barred, and had three short shoulder-stripes, like those on the dun Welch pony, and even had some zebra-like stripes on the sides of its face. With respect to this last fact, I was so convinced that not even a stripe of colour appears from what would commonly be called an accident, that I was led solely from the occurrence of the face-stripes on this hybrid from the ass and hemionus to ask Colonel Poole whether such face-stripes ever occur in the eminently striped Kattywar breed of horses, and was, as we have seen, answered in the affirmative.

What now are we to say to these several facts? We see several very distinct species of the horse-genus becoming, by simple variation, striped on the legs like a zebra, or striped on the shoulders like an ass. In the horse we see this tendency strong whenever a dun tint appears—a tint which approaches to that of the general colouring of the other species of the genus. The appearance of the stripes is not accompanied by any change of form or by any other new character. We see this tendency to become striped most strongly displayed in hybrids from between several of the most distinct species. Now observe the case of the several breeds of pigeons: they are descended from a pigeon (including two or three sub-species or geographical races) of a bluish colour, with certain bars and other marks; and when any breed assumes by simple variation a bluish tint, these bars and other marks invariably reappear; but without any other change of form or character. When the oldest and truest breeds of various colours are crossed, we see a strong tendency for the blue tint and bars and marks to reappear in the mongrels. I have stated that the most probable hypothesis to account for the reappearance of very ancient characters, is—that there is a *tendency* in the young of each successive generation to produce the long-lost character, and that this tendency, from unknown causes, sometimes prevails. And we have just seen that in several species of the horse-genus the stripes are either plainer or appear more commonly in the young than in the old. Call the breeds of pigeons, some of which have bred true for centuries, species; and how exactly parallel is the case with that of the species of the horse-genus! For myself, I venture confidently to look back thousands on thousands of generations, and I see an animal striped like a zebra, but perhaps otherwise very differently constructed, the common parent of our domestic horse, whether or not it be descended from one or more wild stocks, of the ass, the hemionus, quagga, and zebra.

He who believes that each equine species was independently created, will, I presume, assert that each species has been created with a tendency to vary, both under nature and under domestication, in this particular manner, so as often to become striped like other species of the genus; and that each has been created with a strong tendency, when crossed with species inhabiting distant quarters of the world, to produce hybrids resembling in their stripes, not their own parents, but other species of the genus. To admit this view is, as it seems to me, to reject a real for an unreal, or at least for an unknown, cause. It makes the works of God a mere mockery and deception; I would almost as soon believe with the old and ignorant cosmogonists, that fossil shells had never lived, but had been created in stone so as to mock the shells now living on the sea-shore.

Summary.—Our ignorance of the laws of variation is profound. Not in one case out of a hundred can we pretend to assign any reason why this or that part differs, more or less, from the same part in the parents. But whenever we have the means of instituting a comparison, the same laws appear to have acted in producing the lesser differences between varieties of the same species, and the greater differences between species of the same genus. The external conditions of life, as climate and food, etc., seem to have induced some slight modifications. Habit in producing constitutional differences, and use in strengthening and disuse in weakening and diminishing organs, seem to have been more potent in their effects. Homologous parts tend to vary in the same way, and homologous parts tend to cohere. Modifications in hard parts and in external parts sometimes affect softer and internal parts. When one part is largely developed, perhaps it tends to draw nourishment from the adjoining parts; and every part of the structure which can be saved without detriment to the individual, will be saved. Changes of structure at an early age will generally affect parts subsequently developed; and there are very many other correlations of growth, the nature of which we are utterly unable to understand. Multiple parts are variable in number and in structure, perhaps arising from such parts not having been closely specialised to any particular function, so that their modifications have not been closely checked by natural selection. It is probably from this same cause that organic beings low in the scale of nature are more variable than those which have their whole organisation more specialised, and are higher in the scale. Rudimentary organs, from being useless, will be disregarded by natural selection, and hence probably are variable. Specific characters—that is, the characters which have come to differ since the several species of the same genus branched off from a common parent—are more variable than generic characters, or those which have long been inherited, and have not differed within this same period. In these remarks we have referred to special parts or organs being still variable, because they have recently varied and thus come to differ; but we have also seen in the second Chapter that the same principle applies to the whole individual; for in a district where many species of any genus are found—that is, where there has been much former variation and differentiation, or where the manufactory of new specific forms has been actively at work—there, on an average, we now find most varieties or incipient species. Secondary sexual characters are highly variable, and such characters differ much in the species of the same group. Variability in the same parts of the organisation has generally been taken advantage of in giving secondary sexual differences to the sexes of the same species, and specific differences to the several species of the same genus. Any part or organ developed to an extraordinary size or in an extraordinary manner, in comparison with the same part or organ in the allied species, must have gone through an extraordinary amount of modification since the genus arose; and thus we can understand why it should often still be variable in a much higher degree than other parts; for variation is a long-continued and slow process, and natural selection will in such cases not as yet have had time to overcome the tendency to further variability and to reversion to a less modified state. But when a species with any extraordinarily-developed organ has become the parent of many modified descendants—which on my view must be a very slow process, requiring a long lapse of time—in this case, natural selection may readily have succeeded in giving a fixed character to the organ, in however extraordinary a manner it may be developed. Species inheriting nearly the same constitution from a common parent and exposed to similar influences will naturally tend to present analogous variations, and these same species may occasionally revert to some of the characters of their ancient progenitors. Although new and important modifications may not arise from reversion and analogous variation, such modifications will add to the beautiful and harmonious diversity of nature.

Whatever the cause may be of each slight difference in the offspring from their parents—and a cause for each must exist—it is the steady accumulation, through natural selection, of such differences, when beneficial to the individual, that gives rise to all the more important modifications of structure, by which the innumerable beings on the face of this earth are enabled to struggle with each other, and the best adapted to survive.

CHAPTER VI

DIFFICULTIES ON THEORY

Difficulties on the theory of descent with modification—Transitions—Absence or rarity of transitional varieties—Transitions in habits of life—Diversified habits in the same species—Species with habits widely different from those of their allies—Organs of extreme perfection—Means of transition—Cases of difficulty—*Natura non facit saltum*—Organs of small importance—Organs not in all cases absolutely perfect—The law of Unity of Type and of the Conditions of Existence embraced by the theory of Natural Selection.

LONG before having arrived at this part of my work, a crowd of difficulties will have occurred to the reader. Some of them are so grave that to this day I can never reflect on them without being staggered; but, to the best of my judgment, the greater number are only apparent, and those that are real are not, I think, fatal to my theory.

These difficulties and objections may be classed under the following heads:—Firstly, why, if species have descended from other species by insensibly fine gradations, do we not everywhere see innumerable transitional forms? Why is not all nature in confusion instead of the species being, as we see them, well defined?

Secondly, is it possible that an animal having, for instance, the structure and habits of a bat, could have been formed by the modification of some animal with wholly different habits? Can we believe that natural selection could produce, on the one hand, organs of trifling importance, such as the tail of a giraffe, which serves as a fly-flapper, and, on the other hand, organs of such wonderful structure, as the eye, of which we hardly as yet fully understand the inimitable perfection?

Thirdly, can instincts be acquired and modified through natural selection? What shall we say to so marvellous an instinct as that which leads the bee to make cells, which has practically anticipated the discoveries of profound mathematicians?

Fourthly, how can we account for species, when crossed, being sterile and producing sterile offspring, whereas, when varieties are crossed, their fertility is unimpaired?

The two first heads shall be here discussed—Instinct and Hybridism in separate chapters.

On the absence or rarity of transitional varieties.—As natural selection acts solely by the preservation of profitable modifications, each new form will tend in a fully-stocked country to take the place of, and finally to exterminate, its own less improved parent or other less-favoured forms with which it comes into competition. Thus extinction and natural selection will, as we have seen, go hand in hand. Hence, if we look at each species as descended from some other unknown form, both the parent and all the transitional varieties will generally have been exterminated by the very process of formation and perfection of the new form.

But, as by this theory innumerable transitional forms must have existed, why do we not find them embedded in countless numbers in the crust of the earth? It will be much more convenient to discuss this question in the chapter on the Imperfection of the geological record; and I will here only state that I believe the answer mainly lies in the record being incomparably less perfect than is generally supposed; the imperfection of the record being chiefly due to organic beings not inhabiting profound depths of the sea, and to their remains being embedded and preserved to a future age only in masses of sediment sufficiently thick and extensive to withstand an enormous amount of future degradation; and such fossiliferous masses can be accumulated only where much sediment is deposited on the shallow bed of the sea, whilst it slowly subsides. These contingencies will concur only rarely, and after enormously long intervals. Whilst the bed of the sea is stationary or is rising, or when very little sediment is being deposited, there will be blanks in our geological history. The crust of the earth is a vast museum; but the natural collections have been made only at intervals of time immensely remote.

But it may be urged that when several closely-allied species inhabit the same territory we surely ought to find at the present time many transitional forms. Let us take a simple case: in travelling from north to south over a continent, we generally meet at successive intervals with closely allied or representative species, evidently filling nearly the same place in the natural economy of the land. These representative species often meet and interlock; and as the one becomes rarer and rarer, the other becomes more and more frequent, till the one replaces the other. But if we compare these species where they intermingle, they are generally as absolutely distinct from each other in every detail of structure as are specimens taken from the metropolis inhabited by each. By my theory these allied species have descended from a common parent; and during the process of modification, each has become adapted to the conditions of life of its own region, and has supplanted and exterminated its original parent and all the transitional varieties between its past and present states. Hence we ought not to expect at the present time to meet with numerous transitional varieties in each region, though they must have existed there, and may be embedded there in a fossil condition. But in the intermediate region, having intermediate conditions of life, why do we not now find closely-linking intermediate varieties? This difficulty for a long time quite confounded me. But I think it can be in large part explained.

In the first place we should be extremely cautious in inferring, because an area is now continuous, that it has been continuous during a long period. Geology would lead us to believe that almost every continent has been broken up into islands even during the later tertiary periods; and in such islands distinct species might have been separately formed without the possibility of intermediate varieties existing in the intermediate zones. By changes in the form of the land and of climate, marine areas now continuous must often have existed within recent times in a far less continuous and uniform condition than at present. But I will pass over this way of escaping from the difficulty; for I believe that many perfectly defined species have been formed on strictly continuous areas; though I do not doubt that the formerly broken condition of areas now continuous has played an important part in the formation of new species, more especially with freely-crossing and wandering animals.

In looking at species as they are now distributed over a wide area, we generally find them tolerably numerous over a large territory, then becoming somewhat abruptly rarer and rarer on the confines, and finally disappearing. Hence the neutral territory between two representative species is generally narrow in comparison with the territory proper to each. We see the same fact in ascending mountains, and sometimes it is quite remarkable how abruptly, as Alph. De Candolle has observed, a common alpine species disappears. The same fact has been noticed by E. Forbes in sounding the depths of the sea with the dredge. To those who look at climate and the physical conditions of life as the all-important elements of distribution, these facts ought to cause surprise, as climate and height or depth graduate away insensibly. But when we bear in mind that almost every species, even in its metropolis, would increase immensely in numbers, were it not for other competing species; that nearly all either prey on or serve as prey for others; in short, that each organic being is either directly or indirectly related in the most important manner to other organic beings, we must see that the range of the inhabitants of any country by no means exclusively depends on insensibly changing physical conditions, but in large part on the presence of other species, on which it depends, or by which it is destroyed, or with which it comes into competition; and as these species are already defined objects (however they may have become so), not blending one into another by insensible gradations, the range of any one species, depending as it does on the range of others, will tend to be sharply defined. Moreover, each species on the confines of its range, where it exists in lessened numbers, will, during fluctuations in the number of its enemies or of its prey, or in the seasons, be extremely liable to utter extermination; and thus its geographical range will come to be still more sharply defined.

If I am right in believing that allied or representative species, when inhabiting a continuous area, are generally so distributed that each has a wide range, with a comparatively narrow neutral territory between them, in which they become rather suddenly rarer and rarer; then, as varieties do not essentially differ from species, the same rule will probably apply to both; and if we in imagination adapt a varying species to a very large area, we shall have to adapt two varieties to two large areas, and a third variety to a narrow intermediate zone. The intermediate variety, consequently, will exist in lesser numbers from inhabiting a narrow and lesser area; and practically, as far as I can make out, this rule holds good with varieties in a state of nature. I have met with striking instances of the rule in the case of varieties intermediate between well-marked varieties in the genus Balanus. And it would appear from information given me by Mr Watson, Dr Asa Gray, and Mr Wollaston, that generally when varieties intermediate between two other forms occur, they are much rarer numerically than the forms which they connect. Now, if we may trust these facts and inferences, and therefore conclude that varieties linking two other varieties together have generally existed in lesser numbers than the forms which they connect, then, I think, we can understand why intermediate varieties should not endure for very long periods;—why as a general rule they should be exterminated and disappear, sooner than the forms which they originally linked together.

For any form existing in lesser numbers would, as already remarked, run a greater chance of being exterminated than one existing in large numbers; and in this particular case the intermediate form would be eminently liable to the inroads of closely-allied forms existing on both sides of it. But a far more important consideration, as I believe, is that, during the process of further modification, by which two varieties are supposed on my theory to be converted and perfected into two distinct species, the two which exist in larger numbers from inhabiting larger areas, will have a great advantage over the intermediate variety, which exists in smaller numbers in a narrow and intermediate zone. For forms existing in larger numbers will always have a better chance, within any given period, of presenting further favourable variations for natural selection to seize on, than will the rarer forms which exist in lesser numbers. Hence, the more common forms, in the race for life, will tend to beat and supplant the less common forms, for these will be more slowly modified and improved. It is the same principle which, as I believe, accounts for the common species in each country, as shown in the second chapter, presenting on an average a greater number of well-marked varieties than do the rarer species. I may illustrate what I mean by supposing three varieties of sheep to be kept, one adapted to an extensive mountainous region; a second to a comparatively narrow, hilly tract; and a third to wide plains at the base; and that the inhabitants are all trying with equal steadiness and skill to improve their stocks by selection; the chances in this case will be strongly in favour of the great holders on the mountains or on the plains improving their breeds more quickly than the small holders on the intermediate narrow, hilly tract; and consequently the improved mountain or plain breed will soon take the place of the less improved hill breed; and thus the two breeds, which originally existed in greater numbers, will come into close contact with each other, without the interposition of the supplanted, intermediate hill-variety.

To sum up, I believe that species come to be tolerably well-defined objects, and do not at any one period present an inextricable chaos of varying and intermediate links: firstly, because new varieties are very slowly formed, for variation is a very slow process, and natural selection can do nothing until favourable variations chance to occur, and until a place in the natural polity of the country can be better filled by some modification of some one or more of its inhabitants. And such new places will depend on slow changes of climate, or on the occasional immigration of new inhabitants, and, probably, in a still more important degree, on some of the old inhabitants becoming slowly modified, with the new forms thus produced and the old ones acting and reacting on each other. So that, in any one region and at any one time, we ought only to see a few species presenting slight modifications of structure in some degree permanent; and this assuredly we do see.

Secondly, areas now continuous must often have existed within the recent period in isolated portions, in which many forms, more especially amongst the classes which unite for each birth and wander much, may have separately been rendered sufficiently distinct to rank as representative species. In this case, intermediate varieties between the several representative species and their common parent, must formerly have existed in each broken portion of the land, but these links will have been supplanted and exterminated during the process of natural selection, so that they will no longer exist in a living state.

Thirdly, when two or more varieties have been formed in different portions of a strictly continuous area, intermediate varieties will, it is probable, at first have been formed in the intermediate zones, but they will generally have had a short duration. For these intermediate varieties will, from reasons already assigned (namely from what we know of the actual distribution of closely allied or representative species, and likewise of acknowledged varieties), exist in the intermediate zones in lesser numbers than the varieties which they tend to connect. From this cause alone the intermediate varieties will be liable to accidental extermination; and during the process of further modification through natural selection, they will almost certainly be beaten and supplanted by the forms which they connect; for these from existing in greater numbers will, in the aggregate, present more variation, and thus be further improved through natural selection and gain further advantages.

Lastly, looking not to any one time, but to all time, if my theory be true, numberless intermediate varieties, linking most closely all the species of the same group together, must assuredly have existed; but the very process of natural selection constantly tends, as has been so often remarked, to exterminate the parent-forms and the intermediate links. Consequently evidence of their former existence could be found only amongst fossil remains, which are preserved, as we shall in a future chapter attempt to show, in an extremely imperfect and intermittent record.

On the origin and transitions of organic beings with peculiar habits and structure.—It has been asked by the opponents of such views as I hold, how, for instance, a land carnivorous animal could have been converted into one with aquatic habits; for how could the animal in its transitional state have subsisted? It would be easy to show that within the same group carnivorous animals exist having every intermediate grade between truly aquatic and strictly terrestrial habits; and as each exists by a struggle for life, it is clear that each is well adapted in its habits to its place in nature. Look at the Mustela vison of North America, which has webbed feet and which resembles an otter in its fur, short legs, and form of tail; during summer this animal dives for and preys on fish, but during the long winter it leaves the frozen waters, and preys like other pole-cats on mice and land animals. If a different case had been taken, and it had been asked how an insectivorous quadruped could possibly have been converted into a flying bat, the question would have been far more difficult, and I could have given no answer. Yet I think such difficulties have very little weight.

Here, as on other occasions, I lie under a heavy disadvantage, for out of the many striking cases which I have collected, I can give only one or two instances of transitional habits and structures in closely allied species of the same genus; and of diversified habits, either constant or occasional, in the same species. And it seems to me that nothing less than a long list of such cases is sufficient to lessen the difficulty in any particular case like that of the bat.

Look at the family of squirrels; here we have the finest gradation from animals with their tails only slightly flattened, and from others, as Sir J. Richardson has remarked, with the posterior part of their bodies rather wide and with the skin on their flanks rather full, to the so-called flying squirrels; and flying squirrels have their limbs and even the base of the tail united by a broad expanse of skin, which serves as a parachute and allows them to glide through the air to an astonishing distance from tree to tree. We cannot doubt that each structure is of use to each kind of squirrel in its own country, by enabling it to escape birds or beasts of prey, or to collect food more quickly, or, as there is reason to believe, by lessening the danger from occasional falls. But it does not follow from this fact that the structure of each squirrel is the best that it is possible to conceive under all natural conditions. Let the climate and vegetation change, let other competing rodents or new beasts of prey immigrate, or old ones become modified, and all analogy would lead us to believe that some at least of the squirrels would decrease in numbers or become exterminated, unless they also became modified and improved in structure in a corresponding manner. Therefore, I can see no difficulty, more especially under changing conditions of life, in the continued preservation of individuals with fuller and fuller flank-membranes, each modification being useful, each being propagated, until by the accumulated effects of this process of natural selection, a perfect so-called flying squirrel was produced.

Now look at the Galeopithecus or flying lemur, which formerly was falsely ranked amongst bats. It has an extremely wide flank-membrane, stretching from the corners of the jaw to the tail, and including the limbs and the elongated fingers: the flank-membrane is, also, furnished with an extensor muscle. Although no graduated links of structure, fitted for gliding through the air, now connect the Galeopithecus with the other Lemuridæ, yet I see no difficulty in supposing that such links formerly existed, and that each had been formed by the same steps as in the case of the less perfectly gliding squirrels; and that each grade of structure was useful to its possessor. Nor can I see any insuperable difficulty in further believing it possible that the membrane-connected fingers and fore-arm of the Galeopithecus might be greatly lengthened by natural selection; and this, as far as the organs of flight are concerned, would convert it into a bat. In bats which have the wing-membrane extended from the top of the shoulder to the tail, including the hind-legs, we perhaps see traces of an apparatus originally constructed for gliding through the air rather than for flight.

If about a dozen genera of birds had become extinct or were unknown, who would have ventured to have surmised that birds might have existed which used their wings solely as flappers, like the logger-headed duck (Micropterus of Eyton); as fins in the water and front legs on the land, like the penguin; as sails, like the ostrich; and functionally for no purpose, like the Apteryx? Yet the structure of each of these birds is good for it, under the conditions of life to which it is exposed, for each has to live by a struggle; but it is not necessarily the best possible under all possible conditions. It must not be inferred from these remarks that any of the grades of wing-structure here alluded to, which perhaps may all have resulted from disuse, indicate the natural steps by which birds have acquired their perfect power of flight; but they serve, at least, to show what diversified means of transition are possible.

Seeing that a few members of such water-breathing classes as the Crustacea and Mollusca are adapted to live on the land; and seeing that we have flying birds and mammals, flying insects of the most diversified types, and formerly had flying reptiles, it is conceivable that flying-fish, which now glide far through the air, slightly rising and turning by the aid of their fluttering fins, might have been modified into perfectly winged animals. If this had been effected, who would have ever imagined that in an early transitional state they had been inhabitants of the open ocean, and had used their incipient organs of flight exclusively, as far as we know, to escape being devoured by other fish?

When we see any structure highly perfected for any particular habit, as the wings of a bird for flight, we should bear in mind that animals displaying early transitional grades of the structure will seldom continue to exist to the present day, for they will have been supplanted by the very process of perfection through natural selection. Furthermore, we may conclude that transitional grades between structures fitted for very different habits of life will rarely have been developed at an early period in great numbers and under many subordinate forms. Thus, to return to our imaginary illustration of the flying-fish, it does not seem probable that fishes capable of true flight would have been developed under many subordinate forms, for taking prey of many kinds in many ways, on the land and in the water, until their organs of flight had come to a high stage of perfection, so as to have given them a decided advantage over other animals in the battle of life. Hence the chance of discovering species with transitional grades of structure in a fossil condition will always be less, from their having existed in lesser numbers, than in the case of species with fully developed structures.

I will now give two or three instances of diversified and of changed habits in the individuals of the same species. When either case occurs, it would be easy for natural selection to fit the animal, by some modification of its structure, for its changed habits, or exclusively for one of its several different habits. But it is difficult to tell, and immaterial for us, whether habits generally change first and structure afterwards; or whether slight modifications of structure lead to changed habits; both probably often change almost simultaneously. Of cases of changed habits it will suffice merely to allude to that of the many British insects which now feed on exotic plants, or exclusively on artificial substances. Of diversified habits innumerable instances could be given: I have often watched a tyrant flycatcher (Saurophagus sulphuratus) in South America, hovering over one spot and then proceeding to another, like a kestrel, and at other times standing stationary on the margin of water, and then dashing like a kingfisher at a fish. In our own country the larger titmouse (Parus major) may be seen climbing branches, almost like a creeper; it often, like a shrike, kills small birds by blows on the head; and I have many times seen and heard it hammering the seeds of the yew on a branch, and thus breaking them like a nuthatch. In North America the black bear was seen by Hearne swimming for hours with widely open mouth, thus catching, almost like a whale, insects in the water.

As we sometimes see individuals of a species following habits widely different from those of their own species and of the other species of the same genus, we might expect, on my theory, that such individuals would occasionally have given rise to new species, having anomalous habits, and with their structure either slightly or considerably modified from that of their proper type. And such instances do occur in nature. Can a more striking instance of adaptation be given than that of a woodpecker for climbing trees and for seizing insects in the chinks of the bark? Yet in North America there are woodpeckers which feed largely on fruit, and others with elongated wings which chase insects on the wing; and on the plains of La Plata, where not a tree grows, there is a woodpecker, which in every essential part of its organisation, even in its colouring, in the harsh tone of its voice, and undulatory flight, told me plainly of its close blood-relationship to our common species; yet it is a woodpecker which never climbs a tree!

Petrels are the most aërial and oceanic of birds, yet in the quiet Sounds of Tierra del Fuego, the Puffinuria berardi, in its general habits, in its astonishing power of diving, its manner of swimming, and of flying when unwillingly it takes flight, would be mistaken by any one for an auk or grebe; nevertheless, it is essentially a petrel, but with many parts of its organisation profoundly modified. On the other hand, the acutest observer by examining the dead body of the water-ouzel would never have suspected its sub-aquatic habits; yet this anomalous member of the strictly terrestrial thrush family wholly subsists by diving,—grasping the stones with its feet and using its wings under water.

He who believes that each being has been created as we now see it, must occasionally have felt surprise when he has met with an animal having habits and structure not at all in agreement. What can be plainer than that the webbed feet of ducks and geese are formed for swimming? yet there are upland geese with webbed feet which rarely or never go near the water; and no one except Audubon has seen the frigate-bird, which has all its four toes webbed, alight on the surface of the sea. On the other hand grebes and coots are eminently aquatic, although their toes are only bordered by membrane. What seems plainer than that the long toes of grallatores are formed for walking over swamps and floating plants, yet the water-hen is nearly as aquatic as the coot; and the landrail nearly as terrestrial as the quail or partridge. In such cases, and many others could be given, habits have changed without a corresponding change of structure. The webbed feet of the upland goose may be said to have become rudimentary in function, though not in structure. In the frigate-bird, the deeply-scooped membrane between the toes shows that structure has begun to change.

He who believes in separate and innumerable acts of creation will say, that in these cases it has pleased the Creator to cause a being of one type to take the place of one of another type; but this seems to me only re-stating the fact in dignified language. He who believes in the struggle for existence and in the principle of natural selection, will acknowledge that every organic being is constantly endeavouring to increase in numbers; and that if any one being vary ever so little, either in habits or structure, and thus gain an advantage over some other inhabitant of the country, it will seize on the place of that inhabitant, however different it may be from its own place. Hence it will cause him no surprise that there should be geese and frigate-birds with webbed feet, living on the dry land or most rarely alighting on the water; that there should be long-toed corncrakes living in meadows instead of in swamps; that there should be woodpeckers where not a tree grows; that there should be diving thrushes, and petrels with the habits of auks.

Organs of extreme perfection and complication.—To suppose that the eye, with all its inimitable contrivances for adjusting the focus to different distances, for admitting different amounts of light, and for the correction of spherical and chromatic aberration, could have been formed by natural selection, seems, I freely confess, absurd in the highest possible degree. Yet reason tells me, that if numerous gradations from a perfect and complex eye to one very imperfect and simple, each grade being useful to its possessor, can be shown to exist; if further, the eye does vary ever so slightly, and the variations be inherited, which is certainly the case; and if any variation or modification in the organ be ever useful to an animal under changing conditions of life, then the difficulty of believing that a perfect and complex eye could be formed by natural selection, though insuperable by our imagination, can hardly be considered real. How a nerve comes to be sensitive to light, hardly concerns us more than how life itself first originated; but I may remark that several facts make me suspect that any sensitive nerve may be rendered sensitive to light, and likewise to those coarser vibrations of the air which produce sound.

In looking for the gradations by which an organ in any species has been perfected, we ought to look exclusively to its lineal ancestors; but this is scarcely ever possible, and we are forced in each case to look to species of the same group, that is to the collateral descendants from the same original parent-form, in order to see what gradations are possible, and for the chance of some gradations having been transmitted from the earlier stages of descent, in an unaltered or little altered condition. Amongst existing Vertebrata, we find but a small amount of gradation in the structure of the eye, and from fossil species we can learn nothing on this head. In this great class we should probably have to descend far beneath the lowest known fossiliferous stratum to discover the earlier stages, by which the eye has been perfected.

In the Articulata we can commence a series with an optic nerve merely coated with pigment, and without any other mechanism; and from this low stage, numerous gradations of structure, branching off in two fundamentally different lines, can be shown to exist, until we reach a moderately high stage of perfection. In certain crustaceans, for instance, there is a double cornea, the inner one divided into facets, within each of which there is a lens-shaped swelling. In other crustaceans the transparent cones which are coated by pigment, and which properly act only by excluding lateral pencils of light, are convex at their upper ends and must act by convergence; and at their lower ends there seems to be an imperfect vitreous substance. With these facts, here far too briefly and imperfectly given, which show that there is much graduated diversity in the eyes of living crustaceans, and bearing in mind how small the number of living animals is in proportion to those which have become extinct, I can see no very great difficulty (not more than in the case of many other structures) in believing that natural selection has converted the simple apparatus of an optic nerve merely coated with pigment and invested by transparent membrane, into an optical instrument as perfect as is possessed by any member of the great Articulate class.

He who will go thus far, if he find on finishing this treatise that large bodies of facts, otherwise inexplicable, can be explained by the theory of descent, ought not to hesitate to go further, and to admit that a structure even as perfect as the eye of an eagle might be formed by natural selection, although in this case he does not know any of the transitional grades. His reason ought to conquer his imagination; though I have felt the difficulty far too keenly to be surprised at any degree of hesitation in extending the principle of natural selection to such startling lengths.

It is scarcely possible to avoid comparing the eye to a telescope. We know that this instrument has been perfected by the long-continued efforts of the highest human intellects; and we naturally infer that the eye has been formed by a somewhat analogous process. But may not this inference be presumptuous? Have we any right to assume that the Creator works by intellectual powers like those of man? If we must compare the eye to an optical instrument, we ought in imagination to take a thick layer of transparent tissue, with a nerve sensitive to light beneath, and then suppose every part of this layer to be continually changing slowly in density, so as to separate into layers of different densities and thicknesses, placed at different distances from each other, and with the surfaces of each layer slowly changing in form. Further we must suppose that there is a power always intently watching each slight accidental alteration in the transparent layers; and carefully selecting each alteration which, under varied circumstances, may in any way, or in any degree, tend to produce a distincter image. We must suppose each new state of the instrument to be multiplied by the million; and each to be preserved till a better be produced, and then the old ones to be destroyed. In living bodies, variation will cause the slight alterations, generation will multiply them almost infinitely, and natural selection will pick out with unerring skill each improvement. Let this process go on for millions on millions of years; and during each year on millions of individuals of many kinds; and may we not believe that a living optical instrument might thus be formed as superior to one of glass, as the works of the Creator are to those of man?

If it could be demonstrated that any complex organ existed, which could not possibly have been formed by numerous, successive, slight modifications, my theory would absolutely break down. But I can find out no such case. No doubt many organs exist of which we do not know the transitional grades, more especially if we look to much-isolated species, round which, according to my theory, there has been much extinction. Or again, if we look to an organ common to all the members of a large class, for in this latter case the organ must have been first formed at an extremely remote period, since which all the many members of the class have been developed; and in order to discover the early transitional grades through which the organ has passed, we should have to look to very ancient ancestral forms, long since become extinct.

We should be extremely cautious in concluding that an organ could not have been formed by transitional gradations of some kind. Numerous cases could be given amongst the lower animals of the same organ performing at the same time wholly distinct functions; thus the alimentary canal respires, digests, and excretes in the larva of the dragon-fly and in the fish Cobites. In the Hydra, the animal may be turned inside out, and the exterior surface will then digest and the stomach respire. In such cases natural selection might easily specialise, if any advantage were thus gained, a part or organ, which had performed two functions, for one function alone, and thus wholly change its nature by insensible steps. Two distinct organs sometimes perform simultaneously the same function in the same individual; to give one instance, there are fish with gills or branchiæ that breathe the air dissolved in the water, at the same time that they breathe free air in their swimbladders, this latter organ having a ductus pneumaticus for its supply, and being divided by highly vascular partitions. In these cases one of the two organs might with ease be modified and perfected so as to perform all the work by itself, being aided during the process of modification by the other organ; and then this other organ might be modified for some other and quite distinct purpose, or be quite obliterated.

The illustration of the swimbladder in fishes is a good one, because it shows us clearly the highly important fact that an organ originally constructed for one purpose, namely flotation, may be converted into one for a wholly different purpose, namely respiration. The swimbladder has, also, been worked in as an accessory to the auditory organs of certain fish, or, for I do not know which view is now generally held, a part of the auditory apparatus has been worked in as a complement to the swimbladder. All physiologists admit that the swimbladder is homologous, or 'ideally similar' in position and structure with the lungs of the higher vertebrate animals: hence there seems to me to be no great difficulty in believing that natural selection has actually converted a swimbladder into a lung, or organ used exclusively for respiration.

I can, indeed, hardly doubt that all vertebrate animals having true lungs have descended by ordinary generation from an ancient prototype, of which we know nothing, furnished with a floating apparatus or swimbladder. We can thus, as I infer from Professor Owen's interesting description of these parts, understand the strange fact that every particle of food and drink which we swallow has to pass over the orifice of the trachea, with some risk of falling into the lungs, notwithstanding the beautiful contrivance by which the glottis is closed. In the higher Vertebrata the branchiæ have wholly disappeared—the slits on the sides of the neck and the loop-like course of the arteries still marking in the embryo their former position. But it is conceivable that the now utterly lost branchiæ might have been gradually worked in by natural selection for some quite distinct purpose: in the same manner as, on the view entertained by some naturalists that the branchiæ and dorsal scales of Annelids are homologous with the wings and wing-covers of insects, it is probable that organs which at a very ancient period served for respiration have been actually converted into organs of flight.

In considering transitions of organs, it is so important to bear in mind the probability of conversion from one function to another, that I will give one more instance. Pedunculated cirripedes have two minute folds of skin, called by me the ovigerous frena, which serve, through the means of a sticky secretion, to retain the eggs until they are hatched within the sack. These cirripedes have no branchiæ, the whole surface of the body and sack, including the small frena, serving for respiration. The Balanidæ or sessile cirripedes, on the other hand, have no ovigerous frena, the eggs lying loose at the bottom of the sack, in the well-enclosed shell; but they have large folded branchiæ. Now I think no one will dispute that the ovigerous frena in the one family are strictly homologous with the branchiæ of the other family; indeed, they graduate into each other. Therefore I do not doubt that little folds of skin, which originally served as ovigerous frena, but which, likewise, very slightly aided the act of respiration, have been gradually converted by natural selection into branchiæ, simply through an increase in their size and the obliteration of their adhesive glands. If all pedunculated cirripedes had become extinct, and they have already suffered far more extinction than have sessile cirripedes, who would ever have imagined that the branchiæ in this latter family had originally existed as organs for preventing the ova from being washed out of the sack?

Although we must be extremely cautious in concluding that any organ could not possibly have been produced by successive transitional gradations, yet, undoubtedly, grave cases of difficulty occur, some of which will be discussed in my future work.

One of the gravest is that of neuter insects, which are often very differently constructed from either the males or fertile females; but this case will be treated of in the next chapter. The electric organs of fishes offer another case of special difficulty; it is impossible to conceive by what steps these wondrous organs have been produced; but, as Owen and others have remarked, their intimate structure closely resembles that of common muscle; and as it has lately been shown that Rays have an organ closely analogous to the electric apparatus, and yet do not, as Matteucei asserts, discharge any electricity, we must own that we are far too ignorant to argue that no transition of any kind is possible.

The electric organs offer another and even more serious difficulty; for they occur in only about a dozen fishes, of which several are widely remote in their affinities. Generally when the same organ appears in several members of the same class, especially if in members having very different habits of life, we may attribute its presence to inheritance from a common ancestor; and its absence in some of the members to its loss through disuse or natural selection. But if the electric organs had been inherited from one ancient progenitor thus provided, we might have expected that all electric fishes would have been specially related to each other. Nor does geology at all lead to the belief that formerly most fishes had electric organs, which most of their modified descendants have lost. The presence of luminous organs in a few insects, belonging to different families and orders, offers a parallel case of difficulty. Other cases could be given; for instance in plants, the very curious contrivance of a mass of pollen-grains, borne on a foot-stalk with a sticky gland at the end, is the same in Orchis and Asclepias,—genera almost as remote as possible amongst flowering plants. In all these cases of two very distinct species furnished with apparently the same anomalous organ, it should be observed that, although the general appearance and function of the organ may be the same, yet some fundamental difference can generally be detected. I am inclined to believe that in nearly the same way as two men have sometimes independently hit on the very same invention, so natural selection, working for the good of each being and taking advantage of analogous variatious, has sometimes modified in very nearly the same manner two parts in two organic beings, which beings owe but little of their structure in common to inheritance from the same ancestor.

Although in many cases it is most difficult to conjecture by what transitions organs could have arrived at their present state; yet, considering that the proportion of living and known forms to the extinct and unknown is very small, I have been astonished how rarely an organ can be named, towards which no transitional grade is known to lead. The truth of this remark is indeed shown by that old but somewhat exaggerated canon in natural history of 'Natura non facit saltum.' We meet with this admission in the writings of almost every experienced naturalist; or, as Milne Edwards has well expressed it, Nature is prodigal in variety, but niggard in innovation. Why, on the theory of Creation, should this be so? Why should all the parts and organs of many independent beings, each supposed to have been separately created for its proper place in nature, be so commonly linked together by graduated steps? Why should not Nature have taken a leap from structure to structure? On the theory of natural selection, we can clearly understand why she should not; for natural selection can act only by taking advantage of slight successive variations; she can never take a leap, but must advance by the shortest and slowest steps.

Organs of little apparent importance.—As natural selection acts by life and death,—by the preservation of individuals with any favourable variation, and by the destruction of those with any unfavourable deviation of structure,—I have sometimes felt much difficulty in understanding the origin of simple parts, of which the importance does not seem sufficient to cause the preservation of successively varying individuals. I have sometimes felt as much difficulty, though of a very different kind, on this head, as in the case of an organ as perfect and complex as the eye.

In the first place, we are much too ignorant in regard to the whole economy of any one organic being, to say what slight modifications would be of importance or not. In a former chapter I have given instances of most trifling characters, such as the down on fruit and the colour of its flesh, which, from determining the attacks of insects or from being correlated with constitutional differences, might assuredly be acted on by natural selection. The tail of the giraffe looks like an artificially constructed fly-flapper; and it seems at first incredible that this could have been adapted for its present purpose by successive slight modifications, each better and better, for so trifling an object as driving away flies; yet we should pause before being too positive even in this case, for we know that the distribution and existence of cattle and other animals in South America absolutely depends on their power of resisting the attacks of insects: so that individuals which could by any means defend themselves from these small enemies, would be able to range into new pastures and thus gain a great advantage. It is not that the larger quadrupeds are actually destroyed (except in some rare cases) by flies, but they are incessantly harassed and their strength reduced, so that they are more subject to disease, or not so well enabled in a coming dearth to search for food, or to escape from beasts of prey.

Organs now of trifling importance have probably in some cases been of high importance to an early progenitor, and, after having been slowly perfected at a former period, have been transmitted in nearly the same state, although now become of very slight use; and any actually injurious deviations in their structure will always have been checked by natural selection. Seeing how important an organ of locomotion the tail is in most aquatic animals, its general presence and use for many purposes in so many land animals, which in their lungs or modified swimbladders betray their aquatic origin, may perhaps be thus accounted for. A well-developed tail having been formed in an aquatic animal, it might subsequently come to be worked in for all sorts of purposes, as a fly-flapper, an organ of prehension, or as an aid in turning, as with the dog, though the aid must be slight, for the hare, with hardly any tail, can double quickly enough.

In the second place, we may sometimes attribute importance to characters which are really of very little importance, and which have originated from quite secondary causes, independently of natural selection. We should remember that climate, food, etc., probably have some little direct influence on the organisation; that characters reappear from the law of reversion; that correlation of growth will have had a most important influence in modifying various structures; and finally, that sexual selection will often have largely modified the external characters of animals having a will, to give one male an advantage in fighting with another or in charming the females. Moreover when a modification of structure has primarily arisen from the above or other unknown causes, it may at first have been of no advantage to the species, but may subsequently have been taken advantage of by the descendants of the species under new conditions of life and with newly acquired habits.

To give a few instances to illustrate these latter remarks. If green woodpeckers alone had existed, and we did not know that there were many black and pied kinds, I dare say that we should have thought that the green colour was a beautiful adaptation to hide this tree-frequenting bird from its enemies; and consequently that it was a character of importance and might have been acquired through natural selection; as it is, I have no doubt that the colour is due to some quite distinct cause, probably to sexual selection. A trailing bamboo in the Malay Archipelago climbs the loftiest trees by the aid of exquisitely constructed hooks clustered around the ends of the branches, and this contrivance, no doubt, is of the highest service to the plant; but as we see nearly similar hooks on many trees which are not climbers, the hooks on the bamboo may have arisen from unknown laws of growth, and have been subsequently taken advantage of by the plant undergoing further modification and becoming a climber. The naked skin on the head of a vulture is generally looked at as a direct adaptation for wallowing in putridity; and so it may be, or it may possibly be due to the direct action of putrid matter; but we should be very cautious in drawing any such inference, when we see that the skin on the head of the clean-feeding male turkey is likewise naked. The sutures in the skulls of young mammals have been advanced as a beautiful adaptation for aiding parturition, and no doubt they facilitate, or may be indispensable for this act; but as sutures occur in the skulls of young birds and reptiles, which have only to escape from a broken egg, we may infer that this structure has arisen from the laws of growth, and has been taken advantage of in the parturition of the higher animals.

We are profoundly ignorant of the causes producing slight and unimportant variations; and we are immediately made conscious of this by reflecting on the differences in the breeds of our domesticated animals in different countries,—more especially in the less civilised countries where there has been but little artificial selection. Careful observers are convinced that a damp climate affects the growth of the hair, and that with the hair the horns are correlated. Mountain breeds always differ from lowland breeds; and a mountainous country would probably affect the hind limbs from exercising them more, and possibly even the form of the pelvis; and then by the law of homologous variation, the front limbs and even the head would probably be affected. The shape, also, of the pelvis might affect by pressure the shape of the head of the young in the womb. The laborious breathing necessary in high regions would, we have some reason to believe, increase the size of the chest; and again correlation would come into play. Animals kept by savages in different countries often have to struggle for their own subsistence, and would be exposed to a certain extent to natural selection, and individuals with slightly different constitutions would succeed best under different climates;

and there is reason to believe that constitution and colour are correlated. A good observer, also, states that in cattle susceptibility to the attacks of flies is correlated with colour, as is the liability to be poisoned by certain plants; so that colour would be thus subjected to the action of natural selection. But we are far too ignorant to speculate on the relative importance of the several known and unknown laws of variation; and I have here alluded to them only to show that, if we are unable to account for the characteristic differences of our domestic breeds, which nevertheless we generally admit to have arisen through ordinary generation, we ought not to lay too much stress on our ignorance of the precise cause of the slight analogous differences between species. I might have adduced for this same purpose the differences between the races of man, which are so strongly marked; I may add that some little light can apparently be thrown on the origin of these differences, chiefly through sexual selection of a particular kind, but without here entering on copious details my reasoning would appear frivolous.

The foregoing remarks lead me to say a few words on the protest lately made by some naturalists, against the utilitarian doctrine that every detail of structure has been produced for the good of its possessor. They believe that very many structures have been created for beauty in the eyes of man, or for mere variety. This doctrine, if true, would be absolutely fatal to my theory. Yet I fully admit that many structures are of no direct use to their possessors. Physical conditions probably have had some little effect on structure, quite independently of any good thus gained. Correlation of growth has no doubt played a most important part, and a useful modification of one part will often have entailed on other parts diversified changes of no direct use. So again characters which formerly were useful, or which formerly had arisen from correlation of growth, or from other unknown cause, may reappear from the law of reversion, though now of no direct use. The effects of sexual selection, when displayed in beauty to charm the females, can be called useful only in rather a forced sense. But by far the most important consideration is that the chief part of the organisation of every being is simply due to inheritance;

and consequently, though each being assuredly is well fitted for its place in nature, many structures now have no direct relation to the habits of life of each species. Thus, we can hardly believe that the webbed feet of the upland goose or of the frigate-bird are of special use to these birds; we cannot believe that the same bones in the arm of the monkey, in the fore-leg of the horse, in the wing of the bat, and in the flipper of the seal, are of special use to these animals. We may safely attribute these structures to inheritance. But to the progenitor of the upland goose and of the frigate-bird, webbed feet no doubt were as useful as they now are to the most aquatic of existing birds. So we may believe that the progenitor of the seal had not a flipper, but a foot with five toes fitted for walking or grasping; and we may further venture to believe that the several bones in the limbs of the monkey, horse, and bat, which have been inherited from a common progenitor, were formerly of more special use to that progenitor, or its progenitors, than they now are to these animals having such widely diversified habits. Therefore we may infer that these several bones might have been acquired through natural selection, subjected formerly, as now, to the several laws of inheritance, reversion, correlation of growth, etc. Hence every detail of structure in every living creature (making some little allowance for the direct action of physical conditions) may be viewed, either as having been of special use to some ancestral form, or as being now of special use to the descendants of this form—either directly, or indirectly through the complex laws of growth.

Natural selection cannot possibly produce any modification in any one species exclusively for the good of another species; though throughout nature one species incessantly takes advantage of, and profits by, the structure of another. But natural selection can and does often produce structures for the direct injury of other species, as we see in the fang of the adder, and in the ovipositor of the ichneumon, by which its eggs are deposited in the living bodies of other insects. If it could be proved that any part of the structure of any one species had been formed for the exclusive good of another species, it would annihilate my theory, for such could not have been produced through natural selection. Although many statements may be found in works on natural history to this effect, I cannot find even one which seems to me of any weight. It is admitted that the rattlesnake has a poison-fang for its own defence and for the destruction of its prey; but some authors suppose that at the same time this snake is furnished with a rattle for its own injury, namely, to warn its prey to escape. I would almost as soon believe that the cat curls the end of its tail when preparing to spring, in order to warn the doomed mouse. But I have not space here to enter on this and other such cases.

Natural selection will never produce in a being anything injurious to itself, for natural selection acts solely by and for the good of each. No organ will be formed, as Paley has remarked, for the purpose of causing pain or for doing an injury to its possessor. If a fair balance be struck between the good and evil caused by each part, each will be found on the whole advantageous. After the lapse of time, under changing conditions of life, if any part comes to be injurious, it will be modified; or if it be not so, the being will become extinct, as myriads have become extinct.

Natural selection tends only to make each organic being as perfect as, or slightly more perfect than, the other inhabitants of the same country with which it has to struggle for existence. And we see that this is the degree of perfection attained under nature. The endemic productions of New Zealand, for instance, are perfect one compared with another; but they are now rapidly yielding before the advancing legions of plants and animals introduced from Europe. Natural selection will not produce absolute perfection, nor do we always meet, as far as we can judge, with this high standard under nature. The correction for the aberration of light is said, on high authority, not to be perfect even in that most perfect organ, the eye. If our reason leads us to admire with enthusiasm a multitude of inimitable contrivances in nature, this same reason tells us, though we may easily err on both sides, that some other contrivances are less perfect. Can we consider the sting of the wasp or of the bee as perfect, which, when used against many attacking animals, cannot be withdrawn, owing to the backward serratures, and so inevitably causes the death of the insect by tearing out its viscera?

If we look at the sting of the bee, as having originally existed in a remote progenitor as a boring and serrated instrument, like that in so many members of the same great order, and which has been modified but not perfected for its present purpose, with the poison originally adapted to cause galls subsequently intensified, we can perhaps understand how it is that the use of the sting should so often cause the insect's own death: for if on the whole the power of stinging be useful to the community, it will fulfil all the requirements of natural selection, though it may cause the death of some few members. If we admire the truly wonderful power of scent by which the males of many insects find their females, can we admire the production for this single purpose of thousands of drones, which are utterly useless to the community for any other end, and which are ultimately slaughtered by their industrious and sterile sisters? It may be difficult, but we ought to admire the savage instinctive hatred of the queen-bee, which urges her instantly to destroy the young queens her daughters as soon as born, or to perish herself in the combat; for undoubtedly this is for the good of the community; and maternal love or maternal hatred, though the latter fortunately is most rare, is all the same to the inexorable principle of natural selection. If we admire the several ingenious contrivances, by which the flowers of the orchis and of many other plants are fertilised through insect agency, can we consider as equally perfect the elaboration by our fir-trees of dense clouds of pollen, in order that a few granules may be wafted by a chance breeze on to the ovules?

Summary of Chapter.—We have in this chapter discussed some of the difficulties and objections which may be urged against my theory. Many of them are very serious; but I think that in the discussion light has been thrown on several facts, which on the theory of independent acts of creation are utterly obscure. We have seen that species at any one period are not indefinitely variable, and are not linked together by a multitude of intermediate gradations, partly because the process of natural selection will always be very slow, and will act, at any one time, only on a very few forms; and partly because the very process of natural selection almost implies the continual supplanting and extinction of preceding and intermediate gradations. Closely allied species, now living on a continuous area, must often have been formed when the area was not continuous, and when the conditions of life did not insensibly graduate away from one part to another. When two varieties are formed in two districts of a continuous area, an intermediate variety will often be formed, fitted for an intermediate zone; but from reasons assigned, the intermediate variety will usually exist in lesser numbers than the two forms which it connects; consequently the two latter, during the course of further modification, from existing in greater numbers, will have a great advantage over the less numerous intermediate variety, and will thus generally succeed in supplanting and exterminating it.

We have seen in this chapter how cautious we should be in concluding that the most different habits of life could not graduate into each other; that a bat, for instance, could not have been formed by natural selection from an animal which at first could only glide through the air.

We have seen that a species may under new conditions of life change its habits, or have diversified habits, with some habits very unlike those of its nearest congeners. Hence we can understand, bearing in mind that each organic being is trying to live wherever it can live, how it has arisen that there are upland geese with webbed feet, ground woodpeckers, diving thrushes, and petrels with the habits of auks.

Although the belief that an organ so perfect as the eye could have been formed by natural selection, is more than enough to stagger any one; yet in the case of any organ, if we know of a long series of gradations in complexity, each good for its possessor, then, under changing conditions of life there is no logical impossibility in the acquirement of any conceivable degree of perfection through natural selection. In the cases in which we know of no intermediate or transitional states, we should be very cautious in concluding that none could have existed, for the homologies of many organs and their intermediate states show that wonderful metamorphoses in function are at least possible. For instance, a swim-bladder has apparently been converted into an air-breathing lung. The same organ having performed simultaneously very different functions, and then having been specialised for one function; and two very distinct organs having performed at the same time the same function, the one having been perfected whilst aided by the other, must often have largely facilitated transitions.

We are far too ignorant, in almost every case, to be enabled to assert that any part or organ is so unimportant for the welfare of a species, that modifications in its structure could not have been slowly accumulated by means of natural selection. But we may confidently believe that many modifications, wholly due to the laws of growth, and at first in no way advantageous to a species, have been subsequently taken advantage of by the still further modified descendants of this species. We may, also, believe that a part formerly of high importance has often been retained (as the tail of an aquatic animal by its terrestrial descendants), though it has become of such small importance that it could not, in its present state, have been acquired by natural selection,—a power which acts solely by the preservation of profitable variations in the struggle for life.

Natural selection will produce nothing in one species for the exclusive good or injury of another; though it may well produce parts, organs, and excretions highly useful or even indispensable, or highly injurious to another species, but in all cases at the same time useful to the owner. Natural selection in each well-stocked country, must act chiefly through the competition of the inhabitants one with another, and consequently will produce perfection, or strength in the battle for life, only according to the standard of that country. Hence the inhabitants of one country, generally the smaller one, will often yield, as we see they do yield, to the inhabitants of another and generally larger country. For in the larger country there will have existed more individuals, and more diversified forms, and the competition will have been severer, and thus the standard of perfection will have been rendered higher. Natural selection will not necessarily produce absolute perfection; nor, as far as we can judge by our limited faculties, can absolute perfection be everywhere found.

On the theory of natural selection we can clearly understand the full meaning of that old canon in natural history, 'Natura non facit saltum.' This canon, if we look only to the present inhabitants of the world, is not strictly correct, but if we include all those of past times, it must by my theory be strictly true.

It is generally acknowledged that all organic beings have been formed on two great laws—Unity of Type, and the Conditions of Existence. By unity of type is meant that fundamental agreement in structure, which we see in organic beings of the same class, and which is quite independent of their habits of life. On my theory, unity of type is explained by unity of descent. The expression of conditions of existence, so often insisted on by the illustrious Cuvier, is fully embraced by the principle of natural selection. For natural selection acts by either now adapting the varying parts of each being to its organic and inorganic conditions of life; or by having adapted them during long-past periods of time: the adaptations being aided in some cases by use and disuse, being slightly affected by the direct action of the external conditions of life, and being in all cases subjected to the several laws of growth. Hence, in fact, the law of the Conditions of Existence is the higher law; as it includes, through the inheritance of former adaptations, that of Unity of Type.

CHAPTER VII

INSTINCT

Instincts comparable with habits, but different in their origin—Instincts graduated—Aphides and ants—Instincts variable—Domestic instincts, their origin—Natural instincts of the cuckoo, ostrich, and parasitic bees—Slave-making-ants—Hive-bee, its cell-making instinct—Difficulties on the theory of the Natural Selection of instincts—Neuter or sterile insects—Summary.

THE subject of instinct might have been worked into the previous chapters; but I have thought that it would be more convenient to treat the subject separately, especially as so wonderful an instinct as that of the hive-bee making its cells will probably have occurred to many readers, as a difficulty sufficient to overthrow my whole theory. I must premise, that I have nothing to do with the origin of the primary mental powers, any more than I have with that of life itself. We are concerned only with the diversities of instinct and of the other mental qualities of animals within the same class.

I will not attempt any definition of instinct. It would be easy to show that several distinct mental actions are commonly embraced by this term; but every one understands what is meant, when it is said that instinct impels the cuckoo to migrate and to lay her eggs in other birds' nests. An action, which we ourselves should require experience to enable us to perform, when performed by an animal, more especially by a very young one, without any experience, and when performed by many individuals in the same way, without their knowing for what purpose it is performed, is usually said to be instinctive. But I could show that none of these characters of instinct are universal. A little dose, as Pierre Huber expresses it, of judgment or reason, often comes into play, even in animals very low in the scale of nature.

Frederick Cuvier and several of the older metaphysicians have compared instinct with habit. This comparison gives, I think, a remarkably accurate notion of the frame of mind under which an instinctive action is performed, but not of its origin. How unconsciously many habitual actions are performed, indeed not rarely in direct opposition to our conscious will! yet they may be modified by the will or reason. Habits easily become associated with other habits, and with certain periods of time and states of the body. When once acquired, they often remain constant throughout life. Several other points of resemblance between instincts and habits could be pointed out. As in repeating a well-known song, so in instincts, one action follows another by a sort of rhythm; if a person be interrupted in a song, or in repeating anything by rote, he is generally forced to go back to recover the habitual train of thought: so P. Huber found it was with a caterpillar, which makes a very complicated hammock; for if he took a caterpillar which had completed its hammock up to, say, the sixth stage of construction, and put it into a hammock completed up only to the third stage, the caterpillar simply re-performed the fourth, fifth, and sixth stages of construction. If, however, a caterpillar were taken out of a hammock made up, for instance, to the third stage, and were put into one finished up to the sixth stage, so that much of its work was already done for it, far from feeling the benefit of this, it was much embarrassed, and, in order to complete its hammock, seemed forced to start from the third stage, where it had left off, and thus tried to complete the already finished work.

If we suppose any habitual action to become inherited—and I think it can be shown that this does sometimes happen—then the resemblance between what originally was a habit and an instinct becomes so close as not to be distinguished. If Mozart, instead of playing the pianoforte at three years old with wonderfully little practice, had played a tune with no practice at all, he might truly be said to have done so instinctively. But it would be the most serious error to suppose that the greater number of instincts have been acquired by habit in one generation, and then transmitted by inheritance to succeeding generations. It can be clearly shown that the most wonderful instincts with which we are acquainted, namely, those of the hive-bee and of many ants, could not possibly have been thus acquired.

It will be universally admitted that instincts are as important as corporeal structure for the welfare of each species, under its present conditions of life. Under changed conditions of life, it is at least possible that slight modifications of instinct might be profitable to a species; and if it can be shown that instincts do vary ever so little, then I can see no difficulty in natural selection preserving and continually accumulating variations of instinct to any extent that may be profitable. It is thus, as I believe, that all the most complex and wonderful instincts have originated. As modifications of corporeal structure arise from, and are increased by, use or habit, and are diminished or lost by disuse, so I do not doubt it has been with instincts. But I believe that the effects of habit are of quite subordinate importance to the effects of the natural selection of what may be called accidental variations of instincts;—that is of variations produced by the same unknown causes which produce slight deviations of bodily structure.

No complex instinct can possibly be produced through natural selection, except by the slow and gradual accumulation of numerous, slight, yet profitable, variations. Hence, as in the case of corporeal structures, we ought to find in nature, not the actual transitional gradations by which each complex instinct has been acquired—for these could be found only in the lineal ancestors of each species—but we ought to find in the collateral lines of descent some evidence of such gradations; or we ought at least to be able to show that gradations of some kind are possible; and this we certainly can do. I have been surprised to find, making allowance for the instincts of animals having been but little observed except in Europe and North America, and for no instinct being known amongst extinct species, how very generally gradations, leading to the most complex instincts, can be discovered. Changes of instinct may sometimes be facilitated by the same species having different instincts at different periods of life, or at different seasons of the year, or when placed under different circumstances, etc.; in which case either one or the other instinct might be preserved by natural selection. And such instances of diversity of instinct in the same species can be shown to occur in nature.

Again as in the case of corporeal structure, and conformably with my theory, the instinct of each species is good for itself, but has never, as far as we can judge, been produced for the exclusive good of others. One of the strongest instances of an animal apparently performing an action for the sole good of another, with which I am acquainted, is that of aphides voluntarily yielding their sweet excretion to ants: that they do so voluntarily, the following facts show. I removed all the ants from a group of about a dozen aphides on a dock-plant, and prevented their attendance during several hours. After this interval, I felt sure that the aphides would want to excrete. I watched them for some time through a lens, but not one excreted; I then tickled and stroked them with a hair in the same manner, as well as I could, as the ants do with their antennæ; but not one excreted. Afterwards I allowed an ant to visit them, and it immediately seemed, by its eager way of running about, to be well aware what a rich flock it had discovered; it then began to play with its antennæ on the abdomen first of one aphis and then of another; and each aphis, as soon as it felt the antennæ, immediately lifted up its abdomen and excreted a limpid drop of sweet juice, which was eagerly devoured by the ant. Even the quite young aphides behaved in this manner, showing that the action was instinctive, and not the result of experience. But as the excretion is extremely viscid, it is probably a convenience to the aphides to have it removed; and therefore probably the aphides do not instinctively excrete for the sole good of the ants. Although I do not believe that any animal in the world performs an action for the exclusive good of another of a distinct species, yet each species tries to take advantage of the instincts of others, as each takes advantage of the weaker bodily structure of others. So again, in some few cases, certain instincts cannot be considered as absolutely perfect; but as details on this and other such points are not indispensable, they may be here passed over.

As some degree of variation in instincts under a state of nature, and the inheritance of such variations, are indispensable for the action of natural selection, as many instances as possible ought to be here given; but want of space prevents me. I can only assert, that instincts certainly do vary—for instance, the migratory instinct, both in extent and direction, and in its total loss. So it is with the nests of birds, which vary partly in dependence on the situations chosen, and on the nature and temperature of the country inhabited, but often from causes wholly unknown to us: Audubon has given several remarkable cases of differences in the nests of the same species in the northern and southern United States. Fear of any particular enemy is certainly an instinctive quality, as may be seen in nestling birds, though it is strengthened by experience, and by the sight of fear of the same enemy in other animals. But fear of man is slowly acquired, as I have elsewhere shown, by various animals inhabiting desert islands; and we may see an instance of this, even in England, in the greater wildness of all our large birds than of our small birds; for the large birds have been most persecuted by man. We may safely attribute the greater wildness of our large birds to this cause; for in uninhabited islands large birds are not more fearful than small; and the magpie, so wary in England, is tame in Norway, as is the hooded crow in Egypt.

That the general disposition of individuals of the same species, born in a state of nature, is extremely diversified, can be shown by a multitude of facts. Several cases also, could be given, of occasional and strange habits in certain species, which might, if advantageous to the species, give rise, through natural selection, to quite new instincts. But I am well aware that these general statements, without facts given in detail, can produce but a feeble effect on the reader's mind. I can only repeat my assurance, that I do not speak without good evidence.

The possibility, or even probability, or inherited variations of instinct in a state of nature will be strengthened by briefly considering a few cases under domestication. We shall thus also be enabled to see the respective parts which habit and the selection of so-called accidental variations have played in modifying the mental qualities of our domestic animals. A number of curious and authentic instances could be given of the inheritance of all shades of disposition and tastes, and likewise of the oddest tricks, associated with certain frames of mind or periods of time. But let us look to the familiar case of the several breeds of dogs: it cannot be doubted that young pointers (I have myself seen a striking instance) will sometimes point and even back other dogs the very first time that they are taken out; retrieving is certainly in some degree inherited by retrievers; and a tendency to run round, instead of at, a flock of sheep, by shepherd-dogs. I cannot see that these actions, performed without experience by the young, and in nearly the same manner by each individual, performed with eager delight by each breed, and without the end being known,—for the young pointer can no more know that he points to aid his master, than the white butterfly knows why she lays her eggs on the leaf of the cabbage,—I cannot see that these actions differ essentially from true instincts. If we were to see one kind of wolf, when young and without any training, as soon as it scented its prey, stand motionless like a statue, and then slowly crawl forward with a peculiar gait; and another kind of wolf rushing round, instead of at, a herd of deer, and driving them to a distant point, we should assuredly call these actions instinctive. Domestic instincts, as they may be called, are certainly far less fixed or invariable than natural instincts; but they have been acted on by far less rigorous selection, and have been transmitted for an incomparably shorter period, under less fixed conditions of life.

How strongly these domestic instincts, habits, and dispositions are inherited, and how curiously they become mingled, is well shown when different breeds of dogs are crossed. Thus it is known that a cross with a bull-dog has affected for many generations the courage and obstinacy of greyhounds; and a cross with a greyhound has given to a whole family of shepherd-dogs a tendency to hunt hares. These domestic instincts, when thus tested by crossing, resemble natural instincts, which in a like manner become curiously blended together, and for a long period exhibit traces of the instincts of either parent: for example, Le Roy describes a dog, whose great-grandfather was a wolf, and this dog showed a trace of its wild parentage only in one way, by not coming in a straight line to his master when called.

Domestic instincts are sometimes spoken of as actions which have become inherited solely from long-continued and compulsory habit, but this, I think, is not true. No one would ever have thought of teaching, or probably could have taught, the tumbler-pigeon to tumble,—an action which, as I have witnessed, is performed by young birds, that have never seen a pigeon tumble. We may believe that some one pigeon showed a slight tendency to this strange habit, and that the long-continued selection of the best individuals in successive generations made tumblers what they now are; and near Glasgow there are house-tumblers, as I hear from Mr Brent, which cannot fly eighteen inches high without going head over heels. It may be doubted whether any one would have thought of training a dog to point, had not some one dog naturally shown a tendency in this line; and this is known occasionally to happen, as I once saw in a pure terrier: the act of pointing is probably, as many have thought, only the exaggerated pause of an animal preparing to spring on its prey. When the first tendency to point was once displayed, methodical selection and the inherited effects of compulsory training in each successive generation would soon complete the work; and unconscious selection is still at work, as each man tries to procure, without intending to improve the breed, dogs which will stand and hunt best. On the other hand, habit alone in some cases has sufficed; no animal is more difficult to tame than the young of the wild rabbit; scarcely any animal is tamer than the young of the tame rabbit; but I do not suppose that domestic rabbits have ever been selected for tameness; and I presume that we must attribute the whole of the inherited change from extreme wildness to extreme tameness, simply to habit and long-continued close confinement.

Natural instincts are lost under domestication: a remarkable instance of this is seen in those breeds of fowls which very rarely or never become 'broody,' that is, never wish to sit on their eggs. Familiarity alone prevents our seeing how universally and largely the minds of our domestic animals have been modified by domestication. It is scarcely possible to doubt that the love of man has become instinctive in the dog. All wolves, foxes, jackals, and species of the cat genus, when kept tame, are most eager to attack poultry, sheep, and pigs; and this tendency has been found incurable in dogs which have been brought home as puppies from countries, such as Tierra del

Fuego and Australia, where the savages do not keep these domestic animals. How rarely, on the other hand, do our civilised dogs, even when quite young, require to be taught not to attack poultry, sheep, and pigs! No doubt they occasionally do make an attack, and are then beaten; and if not cured, they are destroyed; so that habit, with some degree of selection, has probably concurred in civilising by inheritance our dogs. On the other hand, young chickens have lost, wholly by habit, that fear of the dog and cat which no doubt was originally instinctive in them, in the same way as it is so plainly instinctive in young pheasants, though reared under a hen. It is not that chickens have lost all fear, but fear only of dogs and cats, for if the hen gives the danger-chuckle, they will run (more especially young turkeys) from under her, and conceal themselves in the surrounding grass or thickets; and this is evidently done for the instinctive purpose of allowing, as we see in wild ground-birds, their mother to fly away. But this instinct retained by our chickens has become useless under domestication, for the mother-hen has almost lost by disuse the power of flight.

Hence, we may conclude, that domestic instincts have been acquired and natural instincts have been lost partly by habit, and partly by man selecting and accumulating during successive generations, peculiar mental habits and actions, which at first appeared from what we must in our ignorance call an accident. In some cases compulsory habit alone has sufficed to produce such inherited mental changes; in other cases compulsory habit has done nothing, and all has been the result of selection, pursued both methodically and unconsciously; but in most cases, probably, habit and selection have acted together.

We shall, perhaps, best understand how instincts in a state of nature have become modified by selection, by considering a few cases. I will select only three, out of the several which I shall have to discuss in my future work,—namely, the instinct which leads the cuckoo to lay her eggs in other birds' nests; the slave-making instinct of certain ants; and the comb-making power of the hive-bee; these two latter instincts have generally, and most justly, been ranked by naturalists as the most wonderful of all known instincts.

It is now commonly admitted that the more immediate and final cause of the cuckoo's instinct is, that she lays her eggs, not daily, but at intervals of two or three days; so that, if she were to make her own nest and sit on her own eggs, those first laid would have to be left for some time unincubated, or there would be eggs and young birds of different ages in the same nest. If this were the case, the process of laying and hatching might be inconveniently long, more especially as she has to migrate at a very early period; and the first hatched young would probably have to be fed by the male alone. But the American cuckoo is in this predicament; for she makes her own nest and has eggs and young successively hatched, all at the same time. It has been asserted that the American cuckoo occasionally lays her eggs in other birds' nests; but I hear on the high authority of Dr Brewer, that this is a mistake. Nevertheless, I could give several instances of various birds which have been known occasionally to lay their eggs in other birds' nests. Now let us suppose that the ancient progenitor of our European cuckoo had the habits of the American cuckoo; but that occasionally she laid an egg in another bird's nest. If the old bird profited by this occasional habit, or if the young were made more vigorous by advantage having been taken of the mistaken maternal instinct of another bird, than by their own mother's care, encumbered as she can hardly fail to be by having eggs and young of different ages at the same time; then the old birds or the fostered young would gain an advantage. And analogy would lead me to believe, that the young thus reared would be apt to follow by inheritance the occasional and aberrant habit of their mother, and in their turn would be apt to lay their eggs in other birds' nests, and thus be successful in rearing their young. By a continued process of this nature, I believe that the strange instinct of our cuckoo could be, and has been, generated. I may add that, according to Dr Gray and to some other observers, the European cuckoo has not utterly lost all maternal love and care for her own offspring.

The occasional habit of birds laying their eggs in other birds' nests, either of the same or of a distinct species, is not very uncommon with the Gallinaceæ; and this perhaps explains the origin of a singular instinct in the allied group of ostriches. For several hen ostriches, at least in the case of the American species, unite and lay first a few eggs in one nest and then in another; and these are hatched by the males. This instinct may probably be accounted for by the fact of the hens laying a large number of eggs; but, as in the case of the cuckoo, at intervals of two or three days. This instinct, however, of the American ostrich has not as yet been perfected; for a surprising number of eggs lie strewed over the plains, so that in one day's hunting I picked up no less than twenty lost and wasted eggs.

Many bees are parasitic, and always lay their eggs in the nests of bees of other kinds. This case is more remarkable than that of the cuckoo; for these bees have not only their instincts but their structure modified in accordance with their parasitic habits; for they do not possess the pollen-collecting apparatus which would be necessary if they had to store food for their own young. Some species, likewise, of Sphegidæ (wasp-like insects) are parasitic on other species; and M. Fabre has lately shown good reason for believing that although the Tachytes nigra generally makes its own burrow and stores it with paralysed prey for its own larvæ to feed on, yet that when this insect finds a burrow already made and stored by another sphex, it takes advantage of the prize, and becomes for the occasion parasitic. In this case, as with the supposed case of the cuckoo, I can see no difficulty in natural selection making an occasional habit permanent, if of advantage to the species, and if the insect whose nest and stored food are thus feloniously appropriated, be not thus exterminated.

Slave-making instinct.—This remarkable instinct was first discovered in the Formica (Polyerges) rufescens by Pierre Huber, a better observer even than his celebrated father. This ant is absolutely dependent on its slaves; without their aid, the species would certainly become extinct in a single year. The males and fertile females do no work. The workers or sterile females, though most energetic and courageous in capturing slaves, do no other work. They are incapable of making their own nests, or of feeding their own larvæ. When the old nest is found inconvenient, and they have to migrate, it is the slaves which determine the migration, and actually carry their masters in their jaws. So utterly helpless are the masters, that when Huber shut up thirty of them without a slave, but with plenty of the food which they like best and with their larvæ and pupæ to stimulate them to work, they did nothing; they could not even feed themselves, and many perished of hunger. Huber then introduced a single slave (F. fusca), and she instantly set to work, fed and saved the survivors; made some cells and tended the larvæ, and put all to rights. What can be more extraordinary than these well-ascertained facts? If we had not known of any other slave-making ant, it would have been hopeless to have speculated how so wonderful an instinct could have been perfected.

Another species, Formica sanguinea, was likewise first discovered by P. Huber to be a slave-making ant. This species is found in the southern parts of England, and its habits have been attended to by Mr F. Smith, of the British Museum, to whom I am much indebted for information on this and other subjects. Although fully trusting to the statements of Huber and Mr Smith, I tried to approach the subject in a sceptical frame of mind, as any one may well be excused for doubting the truth of so extraordinary and odious an instinct as that of making slaves. Hence I will give the observations which I have myself made, in some little detail. I opened fourteen nests of F. sanguinea, and found a few slaves in all. Males and fertile females of the slave-species (F. fusca) are found only in their own proper communities, and have never been observed in the nests of F. sanguinea. The slaves are black and not above half the size of their red masters, so that the contrast in their appearance is very great. When the nest is slightly disturbed, the slaves occasionally come out, and like their masters are much agitated and defend the nest: when the nest is much disturbed and the larvæ and pupæ are exposed, the slaves work energetically with their masters in carrying them away to a place of safety. Hence, it is clear, that the slaves feel quite at home. During the months of June and July, in three successive years, I have watched for many hours several nests in Surrey and Sussex, and never saw a slave either leave or enter a nest. As, during these months, the slaves are very few in number, I thought that they might behave differently when more numerous; but Mr Smith informs me that he has watched the nests at various hours during May, June and August, both in Surrey and Hampshire, and has never seen the slaves, though present in large numbers in August, either leave or enter the nest. Hence he considers them as strictly household slaves. The masters, on the other hand, may be constantly seen bringing in materials for the nest, and food of all kinds. During the present year, however, in the month of July, I came across a community with an unusually large stock of slaves, and I observed a few slaves mingled with their masters leaving the nest, and marching along the same road to a tall Scotch fir-tree, twenty-five yards distant, which they ascended together, probably in search of aphides or cocci. According to Huber, who had ample opportunities for observation, in Switzerland the slaves habitually work with their masters in making the nest, and they alone open and close the doors in the morning and evening; and, as Huber expressly states, their principal office is to search for aphides. This difference in the usual habits of the masters and slaves in the two countries, probably depends merely on the slaves being captured in greater numbers in Switzerland than in England.

One day I fortunately witnessed a migration of F. sanguinea from one nest to another, and it was a most interesting spectacle to behold the masters carefully carrying (instead of being carried by, as in the case of F. rufescens) their slaves in their jaws. Another day my attention was struck by about a score of the slave-makers haunting the same spot, and evidently not in search of food; they approached and were vigorously repulsed by an independent community of the slave-species (F. fusca); sometimes as many as three of these ants clinging to the legs of the slave-making F. sanguinea. The latter ruthlessly killed their small opponents, and carried their dead bodies as food to their nest, twenty-nine yards distant; but they were prevented from getting any pupæ to rear as slaves. I then dug up a small parcel of the pupæ of F. fusca from another nest, and put them down on a bare spot near the place of combat; they were eagerly seized, and carried off by the tyrants, who perhaps fancied that, after all, they had been victorious in their late combat.

At the same time I laid on the same place a small parcel of the pupæ of another species, F. flava, with a few of these little yellow ants still clinging to the fragments of the nest. This species is sometimes, though rarely, made into slaves, as has been described by Mr Smith. Although so small a species, it is very courageous, and I have seen it ferociously attack other ants. In one instance I found to my surprise an independent community of F. flava under a stone beneath a nest of the slave-making F. sanguinea; and when I had accidentally disturbed both nests, the little ants attacked their big neighbours with surprising courage. Now I was curious to ascertain whether F. sanguinea could distinguish the pupæ of F. fusca, which they habitually make into slaves, from those of the little and furious F. flava, which they rarely capture, and it was evident that they did at once distinguish them: for we have seen that they eagerly and instantly seized the pupæ of F. fusca, whereas they were much terrified when they came across the pupæ, or even the earth from the nest of F. flava, and quickly ran away; but in about a quarter of an hour, shortly after all the little yellow ants had crawled away, they took heart and carried off the pupæ.

One evening I visited another community of F. sanguinea, and found a number of these ants returning home and entering their nests, carrying the dead bodies of F. fusca (showing that it was not a migration) and numerous pupæ. I traced a long file of ants burthened with booty, for about forty yards, to a very thick clump of heath, whence I saw the last individual of F. sanguinea emerge, carrying a pupa; but I was not able to find the desolated nest in the thick heath. The nest, however, must have been close at hand, for two or three individuals of F. fusca were rushing about in the greatest agitation, and one was perched motionless with its own pupa in its mouth on the top of a spray of heath, an image of despair, over its ravaged home.

Such are the facts, though they did not need confirmation by me, in regard to the wonderful instinct of making slaves. Let it be observed what a contrast the instinctive habits of F. sanguinea present with those of the continental F. rufescens. The latter does not build its own nest, does not determine its own migrations, does not collect food for itself or its young, and cannot even feed itself: it is absolutely dependent on its numerous slaves. Formica sanguinea, on the other hand, possesses much fewer slaves, and in the early part of the summer extremely few: the masters determine when and where a new nest shall be formed, and when they migrate, the masters carry the slaves. Both in Switzerland and England the slaves seem to have the exclusive care of the larvæ, and the masters alone go on slave-making expeditions. In Switzerland the slaves and masters work together, making and bringing materials for the nest: both, but chiefly the slaves, tend, and milk as it may be called, their aphides; and thus both collect food for the community. In England the masters alone usually leave the nest to collect building materials and food for themselves, their slaves and larvæ. So that the masters in this country receive much less service from their slaves than they do in Switzerland.

By what steps the instinct of F. sanguinea originated I will not pretend to conjecture. But as ants, which are not slave-makers, will, as I have seen, carry off pupæ of other species, if scattered near their nests, it is possible that such pupæ originally stored as food might become developed; and the foreign ants thus unintentionally reared would then follow their proper instincts, and do what work they could. If their presence proved useful to the species which had seized them—if it were more advantageous to this species to capture workers than to procreate them—the habit of collecting pupæ originally for food might by natural selection be strengthened and rendered permanent for the very different purpose of raising slaves. When the instinct was once acquired, if carried out to a much less extent even than in our British F. sanguinea, which, as we have seen, is less aided by its slaves than the same species in Switzerland, I can see no difficulty in natural selection increasing and modifying the instinct—always supposing each modification to be of use to the species—until an ant was formed as abjectly dependent on its slaves as is the Formica rufescens.

Cell-making instinct of the Hive-Bee.—I will not here enter on minute details on this subject, but will merely give an outline of the conclusions at which I have arrived. He must be a dull man who can examine the exquisite structure of a comb, so beautifully adapted to its end, without enthusiastic admiration. We hear from mathematicians that bees have practically solved a recondite problem, and have made their cells of the proper shape to hold the greatest possible amount of honey, with the least possible consumption of precious wax in their construction. It has been remarked that a skilful workman, with fitting tools and measures, would find it very difficult to make cells of wax of the true form, though this is perfectly effected by a crowd of bees working in a dark hive. Grant whatever instincts you please, and it seems at first quite inconceivable how they can make all the necessary angles and planes, or even perceive when they are correctly made. But the difficulty is not nearly so great as it at first appears: all this beautiful work can be shown, I think, to follow from a few very simple instincts.

I was led to investigate this subject by Mr Waterhouse, who has shown that the form of the cell stands in close relation to the presence of adjoining cells; and the following view may, perhaps, be considered only as a modification of his theory. Let us look to the great principle of gradation, and see whether Nature does not reveal to us her method of work. At one end of a short series we have humble-bees, which use their old cocoons to hold honey, sometimes adding to them short tubes of wax, and likewise making separate and very irregular rounded cells of wax. At the other end of the series we have the cells of the hive-bee, placed in a double layer: each cell, as is well-known, is an hexagonal prism, with the basal edges of its six sides bevelled so as to fit on to a pyramid, formed of three rhombs. These rhombs have certain angles, and the three which form the pyramidal base of a single cell on one side of the comb, enter into the composition of the bases of three adjoining cells on the opposite side. In the series between the extreme perfection of the cells of the hive-bee and the simplicity of those of the humble-bee, we have the cells of the Mexican Melipona domestica, carefully described and figured by Pierre Huber. The Melipona itself is intermediate in structure between the hive and humble bee, but more nearly related to the latter: it forms a nearly regular waxen comb of cylindrical cells, in which the young are hatched, and, in addition, some large cells of wax for holding honey. These latter cells are nearly spherical and of nearly equal sizes, and are aggregated into an irregular mass. But the important point to notice, is that these cells are always made at that degree of nearness to each other, that they would have intersected or broken into each other, if the spheres had been completed; but this is never permitted, the bees building perfectly flat walls of wax between the spheres which thus tend to intersect. Hence each cell consists of an outer spherical portion and of two, three, or more perfectly flat surfaces, according as the cell adjoins two, three, or more other cells. When one cell comes into contact with three other cells, which, from the spheres being nearly of the same size, is very frequently and necessarily the case, the three flat surfaces are united into a pyramid; and this pyramid, as Huber has remarked, is manifestly a gross imitation of the three-sided pyramidal bases of the cell of the hive-bee. As in the cells of the hive-bee, so here, the three plane surfaces in any one cell necessarily enter into the construction of three adjoining cells. It is obvious that the Melipona saves wax by this manner of building; for the flat walls between the adjoining cells are not double, but are of the same thickness as the outer spherical portions, and yet each flat portion forms a part of two cells.

Reflecting on this case, it occurred to me that if the Melipona had made its spheres at some given distance from each other, and had made them of equal sizes and had arranged them symmetrically in a double layer, the resulting structure would probably have been as perfect as the comb of the hive-bee. Accordingly I wrote to Professor Miller, of Cambridge, and this geometer has kindly read over the following statement, drawn up from his information, and tells me that it is strictly correct:—

If a number of equal spheres be described with their centres placed in two parallel layers; with the centre of each sphere at the distance of radius $\times \sqrt{2}$, or radius $\times$ 1.41421 (or at some lesser distance), from the centres of the six surrounding spheres in the same layer; and at the same distance from the centres of the adjoining spheres in the other and parallel layer; then, if planes of intersection between the several spheres in both layers be formed, there will result a double layer of hexagonal prisms united together by pyramidal bases formed of three rhombs; and the rhombs and the sides of the hexagonal prisms will have every angle identically the same with the best measurements which have been made of the cells of the hive-bee.

Hence we may safely conclude that if we could slightly modify the instincts already possessed by the Melipona, and in themselves not very wonderful, this bee would make a structure as wonderfully perfect as that of the hive-bee. We must suppose the Melipona to make her cells truly spherical, and of equal sizes; and this would not be very surprising, seeing that she already does so to a certain extent, and seeing what perfectly cylindrical burrows in wood many insects can make, apparently by turning round on a fixed point. We must suppose the Melipona to arrange her cells in level layers, as she already does her cylindrical cells; and we must further suppose, and this is the greatest difficulty, that she can somehow judge accurately at what distance to stand from her fellow-labourers when several are making their spheres; but she is already so far enabled to judge of distance, that she always describes her spheres so as to intersect largely; and then she unites the points of intersection by perfectly flat surfaces. We have further to suppose, but this is no difficulty, that after hexagonal prisms have been formed by the intersection of adjoining spheres in the same layer, she can prolong the hexagon to any length requisite to hold the stock of honey; in the same way as the rude humble-bee adds cylinders of wax to the circular mouths of her old cocoons. By such modifications of instincts in themselves not very wonderful,—hardly more wonderful than those which guide a bird to make its nest,—I believe that the hive-bee has acquired, through natural selection, her inimitable architectural powers.

But this theory can be tested by experiment. Following the example of Mr Tegetmeier, I separated two combs, and put between them a long, thick, square strip of wax: the bees instantly began to excavate minute circular pits in it; and as they deepened these little pits, they made them wider and wider until they were converted into shallow basins, appearing to the eye perfectly true or parts of a sphere, and of about the diameter of a cell. It was most interesting to me to observe that wherever several bees had begun to excavate these basins near together, they had begun their work at such a distance from each other, that by the time the basins had acquired the above stated width (*i.e.* about the width of an ordinary cell), and were in depth about one sixth of the diameter of the sphere of which they formed a part, the rims of the basins intersected or broke into each other. As soon as this occurred, the bees ceased to excavate, and began to build up flat walls of wax on the lines of intersection between the basins, so that each hexagonal prism was built upon the scalloped edge of a smooth basin, instead of on the straight edges of a three-sided pyramid as in the case of ordinary cells.

I then put into the hive, instead of a thick, square piece of wax, a thin and narrow, knife-edged ridge, coloured with vermilion. The bees instantly began on both sides to excavate little basins near to each other, in the same way as before; but the ridge of wax was so thin, that the bottoms of the basins, if they had been excavated to the same depth as in the former experiment, would have broken into each other from the opposite sides. The bees, however, did not suffer this to happen, and they stopped their excavations in due time; so that the basins, as soon as they had been a little deepened, came to have flat bottoms; and these flat bottoms, formed by thin little plates of the vermilion wax having been left ungnawed, were situated, as far as the eye could judge, exactly along the planes of imaginary intersection between the basins on the opposite sides of the ridge of wax. In parts, only little bits, in other parts, large portions of a rhombic plate had been left between the opposed basins, but the work, from the unnatural state of things, had not been neatly performed. The bees must have worked at very nearly the same rate on the opposite sides of the ridge of vermilion wax, as they circularly gnawed away and deepened the basins on both sides, in order to have succeeded in thus leaving flat plates between the basins, by stopping work along the intermediate planes or planes of intersection.

Considering how flexible thin wax is, I do not see that there is any difficulty in the bees, whilst at work on the two sides of a strip of wax, perceiving when they have gnawed the wax away to the proper thinness, and then stopping their work. In ordinary combs it has appeared to me that the bees do not always succeed in working at exactly the same rate from the opposite sides; for I have noticed half-completed rhombs at the base of a just-commenced cell, which were slightly concave on one side, where I suppose that the bees had excavated too quickly, and convex on the opposed side, where the bees had worked less quickly. In one well-marked instance, I put the comb back into the hive, and allowed the bees to go on working for a short time, and again examined the cell, and I found that the rhombic plate had been completed, and had become *perfectly flat:* it was absolutely impossible, from the extreme thinness of the little rhombic plate, that they could have effected this by gnawing away the convex side; and I suspect that the bees in such cases stand in the opposed cells and push and bend the ductile and warm wax (which as I have tried is easily done) into its proper intermediate plane, and thus flatten it.

From the experiment of the ridge of vermilion wax, we can clearly see that if the bees were to build for themselves a thin wall of wax, they could make their cells of the proper shape, by standing at the proper distance from each other, by excavating at the same rate, and by endeavouring to make equal spherical hollows, but never allowing the spheres to break into each other. Now bees, as may be clearly seen by examining the edge of a growing comb, do make a rough, circumferential wall or rim all round the comb; and they gnaw into this from the opposite sides, always working circularly as they deepen each cell. They do not make the whole three-sided pyramidal base of any one cell at the same time, but only the one rhombic plate which stands on the extreme growing margin, or the two plates, as the case may be; and they never complete the upper edges of the rhombic plates, until the hexagonal walls are commenced. Some of these statements differ from those made by the justly celebrated elder Huber, but I am convinced of their accuracy; and if I had space, I could show that they are conformable with my theory.

Huber's statement that the very first cell is excavated out of a little parallel-sided wall of wax, is not, as far as I have seen, strictly correct; the first commencement having always been a little hood of wax; but I will not here enter on these details. We see how important a part excavation plays in the construction of the cells; but it would be a great error to suppose that the bees cannot build up a rough wall of wax in the proper position—that is, along the plane of intersection between two adjoining spheres. I have several specimens showing clearly that they can do this. Even in the rude circumferential rim or wall of wax round a growing comb, flexures may sometimes be observed, corresponding in position to the planes of the rhombic basal plates of future cells. But the rough wall of wax has in every case to be finished off, by being largely gnawed away on both sides. The manner in which the bees build is curious; they always make the first rough wall from ten to twenty times thicker than the excessively thin finished wall of the cell, which will ultimately be left. We shall understand how they work, by supposing masons first to pile up a broad ridge of cement, and then to begin cutting it away equally on both sides near the ground, till a smooth, very thin wall is left in the middle; the masons always piling up the cut-away cement, and adding fresh cement, on the summit of the ridge. We shall thus have a thin wall steadily growing upward; but always crowned by a gigantic coping. From all the cells, both those just commenced and those completed, being thus crowned by a strong coping of wax, the bees can cluster and crawl over the comb without injuring the delicate hexagonal walls, which are only about one four-hundredth of an inch in thickness; the plates of the pyramidal basis being about twice as thick. By this singular manner of building, strength is continually given to the comb, with the utmost ultimate economy of wax.

It seems at first to add to the difficulty of understanding how the cells are made, that a multitude of bees all work together; one bee after working a short time at one cell going to another, so that, as Huber has stated, a score of individuals work even at the commencement of the first cell. I was able practically to show this fact, by covering the edges of the hexagonal walls of a single cell, or the extreme margin of the circumferential rim of a growing comb, with an extremely thin layer of melted vermilion wax; and I invariably found that the colour was most delicately diffused by the bees—as delicately as a painter could have done with his brush—by atoms of the coloured wax having been taken from the spot on which it had been placed, and worked into the growing edges of the cells all round. The work of construction seems to be a sort of balance struck between many bees, all instinctively standing at the same relative distance from each other, all trying to sweep equal spheres, and then building up, or leaving ungnawed, the planes of intersection between these spheres. It was really curious to note in cases of difficulty, as when two pieces of comb met at an angle, how often the bees would pull down and rebuild in different ways the same cell, sometimes recurring to a shape which they had at first rejected.

When bees have a place on which they can stand in their proper positions for working,—for instance, on a slip of wood, placed directly under the middle of a comb growing downwards so that the comb has to be built over one face of the slip—in this case the bees can lay the foundations of one wall of a new hexagon, in its strictly proper place, projecting beyond the other completed cells. It suffices that the bees should be enabled to stand at their proper relative distances from each other and from the walls of the last completed cells, and then, by striking imaginary spheres, they can build up a wall intermediate between two adjoining spheres; but, as far as I have seen, they never gnaw away and finish off the angles of a cell till a large part both of that cell and of the adjoining cells has been built. This capacity in bees of laying down under certain circumstances a rough wall in its proper place between two just-commenced cells, is important, as it bears on a fact, which seems at first quite subversive of the foregoing theory; namely, that the cells on the extreme margin of wasp-combs are sometimes strictly hexagonal; but I have not space here to enter on this subject. Nor does there seem to me any great difficulty in a single insect (as in the case of a queen-wasp) making hexagonal cells, if she work alternately on the inside and outside of two or three cells commenced at the same time, always standing at the proper relative distance from the parts of the cells just begun, sweeping spheres or cylinders, and building up intermediate planes. It is even conceivable that an insect might, by fixing on a point at which to commence a cell, and then moving outside, first to one point, and then to five other points, at the proper relative distances from the central point and from each other, strike the planes of intersection, and so make an isolated hexagon: but I am not aware that any such case has been observed; nor would any good be derived from a single hexagon being built, as in its construction more materials would be required than for a cylinder.

As natural selection acts only by the accumulation of slight modifications of structure or instinct, each profitable to the individual under its conditions of life, it may reasonably be asked, how a long and graduated succession of modified architectural instincts, all tending towards the present perfect plan of construction, could have profited the progenitors of the hive-bee? I think the answer is not difficult: it is known that bees are often hard pressed to get sufficient nectar; and I am informed by Mr Tegetmeier that it has been experimentally found that no less than from twelve to fifteen pounds of dry sugar are consumed by a hive of bees for the secretion of each pound of wax; so that a prodigious quantity of fluid nectar must be collected and consumed by the bees in a hive for the secretion of the wax necessary for the construction of their combs. Moreover, many bees have to remain idle for many days during the process of secretion. A large store of honey is indispensable to support a large stock of bees during the winter; and the security of the hive is known mainly to depend on a large number of bees being supported. Hence the saving of wax by largely saving honey must be a most important element of success in any family of bees. Of course the success of any species of bee may be dependent on the number of its parasites or other enemies, or on quite distinct causes, and so be altogether independent of the quantity of honey which the bees could collect. But let us suppose that this latter circumstance determined, as it probably often does determine, the numbers of a humble-bee which could exist in a country; and let us further suppose that the community lived throughout the winter, and consequently required a store of honey: there can in this case be no doubt that it would be an advantage to our humble-bee, if a slight modification of her instinct led her to make her waxen cells near together, so as to intersect a little; for a wall in common even to two adjoining cells, would save some little wax. Hence it would continually be more and more advantageous to our humble-bee, if she were to make her cells more and more regular, nearer together, and aggregated into a mass, like the cells of the Melipona; for in this case a large part of the bounding surface of each cell would serve to bound other cells, and much wax would be saved. Again, from the same cause, it would be advantageous to the Melipona, if she were to make her cells closer together, and more regular in every way than at present; for then, as we have seen, the spherical surfaces would wholly disappear, and would all be replaced by plane surfaces; and the Melipona would make a comb as perfect as that of the hive-bee. Beyond this stage of perfection in architecture, natural selection could not lead; for the comb of the hive-bee, as far as we can see, is absolutely perfect in economising wax.

Thus, as I believe, the most wonderful of all known instincts, that of the hive-bee, can be explained by natural selection having taken advantage of numerous, successive, slight modifications of simpler instincts; natural selection having by slow degrees, more and more perfectly, led the bees to sweep equal spheres at a given distance from each other in a double layer, and to build up and excavate the wax along the planes of intersection. The bees, of course, no more knowing that they swept their spheres at one particular distance from each other, than they know what are the several angles of the hexagonal prisms and of the basal rhombic plates. The motive power of the process of natural selection having been economy of wax; that individual swarm which wasted least honey in the secretion of wax, having succeeded best, and having transmitted by inheritance its newly acquired economical instinct to new swarms, which in their turn will have had the best chance of succeeding in the struggle for existence.

No doubt many instincts of very difficult explanation could be opposed to the theory of natural selection,—cases, in which we cannot see how an instinct could possibly have originated; cases, in which no intermediate gradations are known to exist; cases of instinct of apparently such trifling importance, that they could hardly have been acted on by natural selection; cases of instincts almost identically the same in animals so remote in the scale of nature, that we cannot account for their similarity by inheritance from a common parent, and must therefore believe that they have been acquired by independent acts of natural selection. I will not here enter on these several cases, but will confine myself to one special difficulty, which at first appeared to me insuperable, and actually fatal to my whole theory. I allude to the neuters or sterile females in insect-communities: for these neuters often differ widely in instinct and in structure from both the males and fertile females, and yet, from being sterile, they cannot propagate their kind.

The subject well deserves to be discussed at great length, but I will here take only a single case, that of working or sterile ants. How the workers have been rendered sterile is a difficulty; but not much greater than that of any other striking modification of structure; for it can be shown that some insects and other articulate animals in a state of nature occasionally become sterile; and if such insects had been social, and it had been profitable to the community that a number should have been annually born capable of work, but incapable of procreation, I can see no very great difficulty in this being effected by natural selection. But I must pass over this preliminary difficulty. The great difficulty lies in the working ants differing widely from both the males and the fertile females in structure, as in the shape of the thorax and in being destitute of wings and sometimes of eyes, and in instinct. As far as instinct alone is concerned, the prodigious difference in this respect between the workers and the perfect females, would have been far better exemplified by the hive-bee. If a working ant or other neuter insect had been an animal in the ordinary state, I should have unhesitatingly assumed that all its characters had been slowly acquired through natural selection; namely, by an individual having been born with some slight profitable modification of structure, this being inherited by its offspring, which again varied and were again selected, and so onwards. But with the working ant we have an insect differing greatly from its parents, yet absolutely sterile; so that it could never have transmitted successively acquired modifications of structure or instinct to its progeny. It may well be asked how is it possible to reconcile this case with the theory of natural selection?

First, let it be remembered that we have innumerable instances, both in our domestic productions and in those in a state of nature, of all sorts of differences of structure which have become correlated to certain ages, and to either sex. We have differences correlated not only to one sex, but to that short period alone when the reproductive system is active, as in the nuptial plumage of many birds, and in the hooked jaws of the male salmon. We have even slight differences in the horns of different breeds of cattle in relation to an artificially imperfect state of the male sex; for oxen of certain breeds have longer horns than in other breeds, in comparison with the horns of the bulls or cows of these same breeds. Hence I can see no real difficulty in any character having become correlated with the sterile condition of certain members of insect-communities: the difficulty lies in understanding how such correlated modifications of structure could have been slowly accumulated by natural selection.

This difficulty, though appearing insuperable, is lessened, or, as I believe, disappears, when it is remembered that selection may be applied to the family, as well as to the individual, and may thus gain the desired end. Thus, a well-flavoured vegetable is cooked, and the individual is destroyed; but the horticulturist sows seeds of the same stock, and confidently expects to get nearly the same variety: breeders of cattle wish the flesh and fat to be well marbled together; the animal has been slaughtered, but the breeder goes with confidence to the same family. I have such faith in the powers of selection, that I do not doubt that a breed of cattle, always yielding oxen with extraordinarily long horns, could be slowly formed by carefully watching which individual bulls and cows, when matched, produced oxen with the longest horns; and yet no one ox could ever have propagated its kind. Thus I believe it has been with social insects: a slight modification of structure, or instinct, correlated with the sterile condition of certain members of the community, has been advantageous to the community: consequently the fertile males and females of the same community flourished, and transmitted to their fertile offspring a tendency to produce sterile members having the same modification. And I believe that this process has been repeated, until that prodigious amount of difference between the fertile and sterile females of the same species has been produced, which we see in many social insects.

But we have not as yet touched on the climax of the difficulty; namely, the fact that the neuters of several ants differ, not only from the fertile females and males, but from each other, sometimes to an almost incredible degree, and are thus divided into two or even three castes. The castes, moreover, do not generally graduate into each other, but are perfectly well defined; being as distinct from each other, as are any two species of the same genus, or rather as any two genera of the same family. Thus in Eciton, there are working and soldier neuters, with jaws and instincts extraordinarily different: in Cryptocerus, the workers of one caste alone carry a wonderful sort of shield on their heads, the use of which is quite unknown: in the Mexican Myrmecocystus, the workers of one caste never leave the nest; they are fed by the workers of another caste, and they have an enormously developed abdomen which secretes a sort of honey, supplying the place of that excreted by the aphides, or the domestic cattle as they may be called, which our European ants guard or imprison.

It will indeed be thought that I have an overweening confidence in the principle of natural selection, when I do not admit that such wonderful and well-established facts at once annihilate my theory. In the simpler case of neuter insects all of one caste or of the same kind, which have been rendered by natural selection, as I believe to be quite possible, different from the fertile males and females,—in this case, we may safely conclude from the analogy of ordinary variations, that each successive, slight, profitable modification did not probably at first appear in all the individual neuters in the same nest, but in a few alone; and that by the long-continued selection of the fertile parents which produced most neuters with the profitable modification, all the neuters ultimately came to have the desired character. On this view we ought occasionally to find neuter-insects of the same species, in the same nest, presenting gradations of structure; and this we do find, even often, considering how few neuter-insects out of Europe have been carefully examined. Mr F. Smith has shown how surprisingly the neuters of several British ants differ from each other in size and sometimes in colour; and that the extreme forms can sometimes be perfectly linked together by individuals taken out of the same nest: I have myself compared perfect gradations of this kind. It often happens that the larger or the smaller sized workers are the most numerous; or that both large and small are numerous, with those of an intermediate size scanty in numbers. Formica flava has larger and smaller workers, with some of intermediate size; and, in this species, as Mr F. Smith has observed, the larger workers have simple eyes (ocelli), which though small can be plainly distinguished, whereas the smaller workers have their ocelli rudimentary. Having carefully dissected several specimens of these workers, I can affirm that the eyes are far more rudimentary in the smaller workers than can be accounted for merely by their proportionally lesser size; and I fully believe, though I dare not assert so positively, that the workers of intermediate size have their ocelli in an exactly intermediate condition. So that we here have two bodies of sterile workers in the same nest, differing not only in size, but in their organs of vision, yet connected by some few members in an intermediate condition. I may digress by adding, that if the smaller workers had been the most useful to the community, and those males and females had been continually selected, which produced more and more of the smaller workers, until all the workers had come to be in this condition; we should then have had a species of ant with neuters very nearly in the same condition with those of Myrmica. For the workers of Myrmica have not even rudiments of ocelli, though the male and female ants of this genus have well-developed ocelli.

I may give one other case: so confidently did I expect to find gradations in important points of structure between the different castes of neuters in the same species, that I gladly availed myself of Mr F. Smith's offer of numerous specimens from the same nest of the driver ant (Anomma) of West Africa. The reader will perhaps best appreciate the amount of difference in these workers, by my giving not the actual measurements, but a strictly accurate illustration: the difference was the same as if we were to see a set of workmen building a house of whom many were five feet four inches high, and many sixteen feet high; but we must suppose that the larger workmen had heads four instead of three times as big as those of the smaller men, and jaws nearly five times as big. The jaws, moreover, of the working ants of the several sizes differed wonderfully in shape, and in the form and number of the teeth. But the important fact for us is, that though the workers can be grouped into castes of different sizes, yet they graduate insensibly into each other, as does the widely-different structure of their jaws. I speak confidently on this latter point, as Mr Lubbock made drawings for me with the camera lucida of the jaws which I had dissected from the workers of the several sizes.

With these facts before me, I believe that natural selection, by acting on the fertile parents, could form a species which should regularly produce neuters, either all of large size with one form of jaw, or all of small size with jaws having a widely different structure; or lastly, and this is our climax of difficulty, one set of workers of one size and structure, and simultaneously another set of workers of a different size and structure;—a graduated series having been first formed, as in the case of the driver ant, and then the extreme forms, from being the most useful to the community, having been produced in greater and greater numbers through the natural selection of the parents which generated them; until none with an intermediate structure were produced.

Thus, as I believe, the wonderful fact of two distinctly defined castes of sterile workers existing in the same nest, both widely different from each other and from their parents, has originated. We can see how useful their production may have been to a social community of insects, on the same principle that the division of labour is useful to civilised man. As ants work by inherited instincts and by inherited organs or tools, and not by acquired knowledge and manufactured instruments, a perfect division of labour could be effected with them only by the workers being sterile; for had they been fertile, they would have intercrossed, and their instincts and structure would have become blended. And nature has, as I believe, effected this admirable division of labour in the communities of ants, by the means of natural selection. But I am bound to confess, that, with all my faith in this principle, I should never have anticipated that natural selection could have been efficient in so high a degree, had not the case of these neuter insects convinced me of the fact. I have, therefore, discussed this case, at some little but wholly insufficient length, in order to show the power of natural selection, and likewise because this is by far the most serious special difficulty, which my theory has encountered. The case, also, is very interesting, as it proves that with animals, as with plants, any amount of modification in structure can be effected by the accumulation of numerous, slight, and as we must call them accidental, variations, which are in any manner profitable, without exercise or habit having come into play. For no amount of exercise, or habit, or volition, in the utterly sterile members of a community could possibly affect the structure or instincts of the fertile members, which alone leave descendants. I am surprised that no one has advanced this demonstrative case of neuter insects, against the well-known doctrine of Lamarck.

Summary.—I have endeavoured briefly in this chapter to show that the mental qualities of our domestic animals vary, and that the variations are inherited. Still more briefly I have attempted to show that instincts vary slightly in a state of nature. No one will dispute that instincts are of the highest importance to each animal. Therefore I can see no difficulty, under changing conditions of life, in natural selection accumulating slight modifications of instinct to any extent, in any useful direction. In some cases habit or use and disuse have probably come into play. I do not pretend that the facts given in this chapter strengthen in any great degree my theory; but none of the cases of difficulty, to the best of my judgment, annihilate it. On the other hand, the fact that instincts are not always absolutely perfect and are liable to mistakes;—that no instinct has been produced for the exclusive good of other animals, but that each animal takes advantage of the instincts of others;—that the canon in natural history, of 'Natura non facit saltum,' is applicable to instincts as well as to corporeal structure, and is plainly explicable on the foregoing views, but is otherwise inexplicable,—all tend to corroborate the theory of natural selection.

This theory is, also, strengthened by some few other facts in regard to instincts; as by that common case of closely allied, but certainly distinct, species, when inhabiting distant parts of the world and living under considerably different conditions of life, yet often retaining nearly the same instincts. For instance, we can understand on the principle of inheritance, how it is that the thrush of South America lines its nest with mud, in the same peculiar manner as does our British thrush: how it is that the male wrens (Troglodytes) of North America, build 'cock-nests,' to roost in, like the males of our distinct Kitty-wrens,—a habit wholly unlike that of any other known bird. Finally, it may not be a logical deduction, but to my imagination it is far more satisfactory to look at such instincts as the young cuckoo ejecting its foster-brothers,—ants making slaves,—the larvæ of ichneumonidæ feeding within the live bodies of caterpillars,—not as specially endowed or created instincts, but as small consequences of one general law, leading to the advancement of all organic beings, namely, multiply, vary, let the strongest live and the weakest die.

CHAPTER VIII

HYBRIDISM

Distinction between the sterility of first crosses and of hybrids—Sterility various in degree, not universal, affected by close interbreeding, removed by domestication—Laws governing the sterility of hybrids—Sterility not a special endowment, but incidental on other differences—Causes of the sterility of first crosses and of hybrids—Parallelism between the effects of changed conditions of life and crossing—Fertility of varieties when crossed and of their mongrel offspring not universal—Hybrids and mongrels compared independently of their fertility—Summary.

THE view generally entertained by naturalists is that species, when intercrossed, have been specially endowed with the quality of sterility, in order to prevent the confusion of all organic forms. This view certainly seems at first probable, for species within the same country could hardly have kept distinct had they been capable of crossing freely. The importance of the fact that hybrids are very generally sterile, has, I think, been much underrated by some late writers. *On the theory of natural selection the case is especially important, inasmuch as the sterility of hybrids could not possibly be of any advantage to them, and therefore could not have been acquired by the continued preservation of successive profitable degrees of sterility.* I hope, however, to be able to show that sterility is not a specially acquired or endowed quality, but is incidental on other acquired differences.

In treating this subject, two classes of facts, to a large extent fundamentally different, have generally been confounded together; namely, the sterility of two species when first crossed, and the sterility of the hybrids produced from them.

Pure species have of course their organs of reproduction in a perfect condition, yet when intercrossed they produce either few or no offspring. Hybrids, on the other hand, have their reproductive organs functionally impotent, as may be clearly seen in the state of the male element in both plants and animals; though the organs themselves are perfect in structure, as far as the microscope reveals. In the first case the two sexual elements which go to form the embryo are perfect; in the second case they are either not at all developed, or are imperfectly developed. This distinction is important, when the cause of the sterility, which is common to the two cases, has to be considered. The distinction has probably been slurred over, owing to the sterility in both cases being looked on as a special endowment, beyond the province of our reasoning powers.

The fertility of varieties, that is of the forms known or believed to have descended from common parents, when intercrossed, and likewise the fertility of their mongrel offspring, is, on my theory, of equal importance with the sterility of species; for it seems to make a broad and clear distinction between varieties and species.

First, for the sterility of species when crossed and of their hybrid offspring. It is impossible to study the several memoirs and works of those two conscientious and admirable observers, Kölreuter and Gärtner, who almost devoted their lives to this subject, without being deeply impressed with the high generality of some degree of sterility. Kölreuter makes the rule universal; but then he cuts the knot, for in ten cases in which he found two forms, considered by most authors as distinct species, quite fertile together, he unhesitatingly ranks them as varieties. Gärtner, also, makes the rule equally universal; and he disputes the entire fertility of Kölreuter's ten cases. But in these and in many other cases, Gärtner is obliged carefully to count the seeds, in order to show that there is any degree of sterility. He always compares the maximum number of seeds produced by two species when crossed and by their hybrid offspring, with the average number produced by both pure parent-species in a state of nature. But a serious cause of error seems to me to be here introduced: a plant to be hybridised must be castrated, and, what is often more important, must be secluded in order to prevent pollen being brought to it by insects from other plants. Nearly all the plants experimentised on by Gärtner were potted, and apparently were kept in a chamber in his house. That these processes are often injurious to the fertility of a plant cannot be doubted; for Gärtner gives in his table about a score of cases of plants which he castrated, and artificially fertilised with their own pollen, and (excluding all cases such as the Leguminosæ, in which there is an acknowledged difficulty in the manipulation) half of these twenty plants had their fertility in some degree impaired. Moreover, as Gärtner during several years repeatedly crossed the primrose and cowslip, which we have such good reason to believe to be varieties, and only once or twice succeeded in getting fertile seed; as he found the common red and blue pimpernels (Anagallis arvensis and cœrulea), which the best botanists rank as varieties, absolutely sterile together; and as he came to the same conclusion in several other analogous cases; it seems to me that we may well be permitted to doubt whether many other species are really so sterile, when intercrossed, as Gärtner believes.

It is certain, on the one hand, that the sterility of various species when crossed is so different in degree and graduates away so insensibly, and, on the other hand, that the fertility of pure species is so easily affected by various circumstances, that for all practical purposes it is most difficult to say where perfect fertility ends and sterility begins. I think no better evidence of this can be required than that the two most experienced observers who have ever lived, namely, Kölreuter and Gärtner, should have arrived at diametrically opposite conclusions in regard to the very same species. It is also most instructive to compare—but I have not space here to enter on details—the evidence advanced by our best botanists on the question whether certain doubtful forms should be ranked as species or varieties, with the evidence from fertility adduced by different hybridisers, or by the same author, from experiments made during different years. It can thus be shown that neither sterility nor fertility affords any clear distinction between species and varieties; but that the evidence from this source graduates away, and is doubtful in the same degree as is the evidence derived from other constitutional and structural differences.

In regard to the sterility of hybrids in successive generations; though Gärtner was enabled to rear some hybrids, carefully guarding them from a cross with either pure parent, for six or seven, and in one case for ten generations, yet he asserts positively that their fertility never increased, but generally greatly decreased. I do not doubt that this is usually the case, and that the fertility often suddenly decreases in the first few generations. Nevertheless I believe that in all these experiments the fertility has been diminished by an independent cause, namely, from close interbreeding. I have collected so large a body of facts, showing that close interbreeding lessens fertility, and, on the other hand, that an occasional cross with a distinct individual or variety increases fertility, that I cannot doubt the correctness of this almost universal belief amongst breeders. Hybrids are seldom raised by experimentalists in great numbers; and as the parent-species, or other allied hybrids, generally grow in the same garden, the visits of insects must be carefully prevented during the flowering season: hence hybrids will generally be fertilised during each generation by their own individual pollen; and I am convinced that this would be injurious to their fertility, already lessened by their hybrid origin. I am strengthened in this conviction by a remarkable statement repeatedly made by Gärtner, namely, that if even the less fertile hybrids be artificially fertilised with hybrid pollen of the same kind, their fertility, notwithstanding the frequent ill effects of manipulation, sometimes decidedly increases, and goes on increasing. Now, in artificial fertilisation pollen is as often taken by chance (as I know from my own experience) from the anthers of another flower, as from the anthers of the flower itself which is to be fertilised; so that a cross between two flowers, though probably on the same plant, would be thus effected. Moreover, whenever complicated experiments are in progress, so careful an observer as Gärtner would have castrated his hybrids, and this would have insured in each generation a cross with a pollen from a distinct flower, either from the same plant or from another plant of the same hybrid nature. And thus, the strange fact of the increase of fertility in the successive generations of *artificially fertilised* hybrids may, I believe, be accounted for by close interbreeding having been avoided.

Now let us turn to the results arrived at by the third most experienced hybridiser, namely, the Hon. and Revd W. Herbert. He is as emphatic in his conclusion that some hybrids are perfectly fertile—as fertile as the pure parent-species—as are Kölreuter and Gärtner that some degree of sterility between distinct species is a universal law of nature. He experimentised on some of the very same species as did Gärtner. The difference in their results may, I think, be in part accounted for by Herbert's great horticultural skill, and by his having hot-houses at his command. Of his many important statements I will here give only a single one as an example, namely, that 'every ovule in a pod of Crinum capense fertilised by C. revolutum produced a plant, which (he says) I never saw to occur in a case of its natural fecundation.' So that we here have perfect, or even more than commonly perfect, fertility in a first cross between two distinct species.

This case of the Crinum leads me to refer to a most singular fact, namely, that there are individual plants of certain species of Lobelia and of some other genera, which can be far more easily fertilised by the pollen of another and distinct species, than by their own pollen; and all the individuals of nearly all the species of Hippeastrum seem to be in this predicament. For these plants have been found to yield seed to the pollen of a distinct species, though quite sterile with their own pollen, notwithstanding that their own pollen was found to be perfectly good, for it fertilised distinct species. So that certain individual plants and all the individuals of certain species can actually be hybridised much more readily than they can be self-fertilised! For instance, a bulb of Hippeastrum aulicum produced four flowers; three were fertilised by Herbert with their own pollen, and the fourth was subsequently fertilised by the pollen of a compound hybrid descended from three other and distinct species: the result was that 'the ovaries of the three first flowers soon ceased to grow, and after a few days perished entirely, whereas the pod impregnated by the pollen of the hybrid made vigorous growth and rapid progress to maturity, and bore good seed, which vegetated freely.' In a letter to me, in 1839, Mr Herbert told me that he had then tried the experiment during five years, and he continued to try it during several subsequent years, and always with the same result. This result has, also, been confirmed by other observers in the case of Hippeastrum with its sub-genera, and in the case of some other genera, as

Lobelia, Passiflora, and Verbascum. Although the plants in these experiments appeared perfectly healthy, and although both the ovules and pollen of the same flower were perfectly good with respect to other species, yet as they were functionally imperfect in their mutual self-action, we must infer that the plants were in an unnatural state. Nevertheless these facts show on what slight and mysterious causes the lesser or greater fertility of species when crossed, in comparison with the same species when self-fertilised, sometimes depends.

The practical experiments of horticulturists, though not made with scientific precision, deserve some notice. It is notorious in how complicated a manner the species of Pelargonium, Fuchsia, Calceolaria, Petunia, Rhododendron, etc., have been crossed, yet many of these hybrids seed freely. For instance, Herbert asserts that a hybrid from Calceolaria integrifolia and plantaginea, species most widely dissimilar in general habit, 'reproduced itself as perfectly as if it had been a natural species from the mountains of Chile'. I have taken some pains to ascertain the degree of fertility of some of the complex crosses of Rhododendrons, and I am assured that many of them are perfectly fertile. Mr C. Noble, for instance, informs me that he raises stocks for grafting from a hybrid between Rhod. Ponticum and Catawbiense, and that this hybrid 'seeds as freely as it is possible to imagine'. Had hybrids, when fairly treated, gone on decreasing in fertility in each successive generation, as Gärtner believes to be the case, the fact would have been notorious to nurserymen. Horticulturists raise large beds of the same hybrids, and such alone are fairly treated, for by insect agency the several individuals of the same hybrid variety are allowed to freely cross with each other, and the injurious influence of close interbreeding is thus prevented. Any one may readily convince himself of the efficiency of insect-agency by examining the flowers of the more sterile kinds of hybrid rhododendrons, which produce no pollen, for he will find on their stigmas plenty of pollen brought from other flowers.

In regard to animals, much fewer experiments have been carefully tried than with plants. If our systematic arrangements can be trusted, that is if the genera of animals are as distinct from each other, as are the genera of plants, then we may infer that animals more widely separated in the scale of nature can be more easily crossed than in the case of plants; but the hybrids themselves are, I think, more sterile. I doubt whether any case of a perfectly fertile hybrid animal can be considered as thoroughly well authenticated. It should, however, be borne in mind that, owing to few animals breeding freely under confinement, few experiments have been fairly tried: for instance, the canary-bird has been crossed with nine other finches, but as not one of these nine species breeds freely in confinement, we have no right to expect that the first crosses between them and the canary, or that their hybrids, should be perfectly fertile. Again, with respect to the fertility in successive generations of the more fertile hybrid animals, I hardly know of an instance in which two families of the same hybrid have been raised at the same time from different parents, so as to avoid the ill effects of close interbreeding. On the contrary, brothers and sisters have usually been crossed in each successive generation, in opposition to the constantly repeated admonition of every breeder. And in this case, it is not at all surprising that the inherent sterility in the hybrids should have gone on increasing. If we were to act thus, and pair brothers and sisters in the case of any pure animal, which from any cause had the least tendency to sterility, the breed would assuredly be lost in a very few generations.

Although I do not know of any thoroughly well-authenticated cases of perfectly fertile hybrid animals, I have some reason to believe that the hybrids from Cervulus vaginalis and Reevesii, and from Phasianus colchicus with P. torquatus and with P. versicolor are perfectly fertile. There is no doubt that these three pheasants, namely, the common, the true ring-necked, and the Japan, intercross, and are becoming blended together in the woods of several parts of England. The hybrids from the common and Chinese geese (A. cygnoides), species which are so different that they are generally ranked in distinct genera, have often bred in this country with either pure parent, and in one single instance they have bred *inter se*. This was effected by Mr Eyton, who raised two hybrids from the same parents but from different hatches; and from these two birds he raised no less than eight hybrids (grandchildren of the pure geese) from one nest. In India, however, these cross-bred geese must be far more fertile; for I am assured by two eminently capable judges, namely Mr Blyth and Captain Hutton, that whole flocks of these crossed geese are kept in various parts of the country; and as they are kept for profit, where neither pure parent-species exists, they must certainly be highly fertile.

A doctrine which originated with Pallas, has been largely accepted by modern naturalists; namely, that most of our domestic animals have descended from two or more wild species, since commingled by intercrossing. On this view, the aboriginal species must either at first have produced quite fertile hybrids, or the hybrids must have become in subsequent generations quite fertile under domestication. This latter alternative seems to me the most probable, and I am inclined to believe in its truth, although it rests on no direct evidence. I believe, for instance, that our dogs have descended from several wild stocks; yet, with perhaps the exception of certain indigenous domestic dogs of South America, all are quite fertile together; and analogy makes me greatly doubt, whether the several aboriginal species would at first have freely bred together and have produced quite fertile hybrids. So again there is reason to believe that our European and the humped Indian cattle are quite fertile together; but from facts communicated to me by Mr Blyth, I think they must be considered as distinct species. On this view of the origin of many of our domestic animals, we must either give up the belief of the almost universal sterility of distinct species of animals when crossed; or we must look at sterility, not as an indelible characteristic, but as one capable of being removed by domestication.

Finally, looking to all the ascertained facts on the intercrossing of plants and animals, it may be concluded that some degree of sterility, both in first crosses and in hybrids, is an extremely general result; but that it cannot, under our present state of knowledge, be considered as absolutely universal.

Laws governing the Sterility of first Crosses and of Hybrids.—We will now consider a little more in detail the circumstances and rules governing the sterility of first crosses and of hybrids. Our chief object will be to see whether or not the rules indicate that species have specially been endowed with this quality, in order to prevent their crossing and blending together in utter confusion. The following rules and conclusions are chiefly drawn up from Gärtner's admirable work on the hybridisation of plants. I have taken much pains to ascertain how far the rules apply to animals, and considering how scanty our knowledge is in regard to hybrid animals, I have been surprised to find how generally the same rules apply to both kingdoms.

It has been already remarked, that the degree of fertility, both of first crosses and of hybrids, graduates from zero to perfect fertility. It is surprising in how many curious ways this gradation can be shown to exist; but only the barest outline of the facts can here be given. When pollen from a plant of one family is placed on the stigma of a plant of a distinct family, it exerts no more influence than so much inorganic dust. From this absolute zero of fertility, the pollen of different species of the same genus applied to the stigma of some one species, yields a perfect gradation in the number of seeds produced, up to nearly complete or even quite complete fertility; and, as we have seen, in certain abnormal cases, even to an excess of fertility, beyond that which the plant's own pollen will produce. So in hybrids themselves, there are some which never have produced, and probably never would produce, even with the pollen of either pure parent, a single fertile seed: but in some of these cases a first trace of fertility may be detected, by the pollen of one of the pure parent-species causing the flower of the hybrid to wither earlier than it otherwise would have done; and the early withering of the flower is well known to be a sign of incipient fertilisation. From this extreme degree of sterility we have self-fertilised hybrids producing a greater and greater number of seeds up to perfect fertility.

Hybrids from two species which are very difficult to cross, and which rarely produce any offspring, are generally very sterile; but the parallelism between the difficulty of making a first cross, and the sterility of the hybrids thus produced—two classes of facts which are generally confounded together—is by no means strict. There are many cases, in which two pure species can be united with unusual facility, and produce numerous hybrid-offspring, yet these hybrids are remarkably sterile. On the other hand, there are species which can be crossed very rarely, or with extreme difficulty, but the hybrids, when at last produced, are very fertile. Even within the limits of the same genus, for instance in Dianthus, these two opposite cases occur.

The fertility, both of first crosses and of hybrids, is more easily affected by unfavourable conditions, than is the fertility of pure species. But the degree of fertility is likewise innately variable; for it is not always the same when the same two species are crossed under the same circumstances, but depends in part upon the constitution of the individuals which happen to have been chosen for the experiment. So it is with hybrids, for their degree of fertility is often found to differ greatly in the several individuals raised from seed out of the same capsule and exposed to exactly the same conditions.

By the term systematic affinity is meant, the resemblance between species in structure and in constitution, more especially in the structure of parts which are of high physiological importance and which differ little in the allied species. Now the fertility of first crosses between species, and of the hybrids produced from them, is largely governed by their systematic affinity. This is clearly shown by hybrids never having been raised between species ranked by systematists in distinct families; and on the other hand, by very closely allied species generally uniting with facility. But the correspondence between systematic affinity and the facility of crossing is by no means strict. A multitude of cases could be given of very closely allied species which will not unite, or only with extreme difficulty; and on the other hand of very distinct species which unite with the utmost facility. In the same family there may be a genus, as Dianthus, in which very many species can most readily be crossed; and another genus, as Silene, in which the most persevering efforts have failed to produce between extremely close species a single hybrid. Even within the limits of the same genus, we meet with this same difference; for instance, the many species of Nicotiana have been more largely crossed than the species of almost any other genus; but Gärtner found that

N. acuminata, which is not a particularly distinct species, obstinately failed to fertilise, or to be fertilised by, no less than eight other species of Nicotiana. Very many analogous facts could be given.

No one has been able to point out what kind, or what amount, of difference in any recognisable character is sufficient to prevent two species crossing. It can be shown that plants most widely different in habit and general appearance, and having strongly marked differences in every part of the flower, even in the pollen, in the fruit, and in the cotyledons, can be crossed. Annual and perennial plants, deciduous and evergreen trees, plants inhabiting different stations and fitted for extremely different climates, can often be crossed with ease.

By a reciprocal cross between two species, I mean the case, for instance, of a stallion-horse being first crossed with a female-ass, and then a male-ass with a mare: these two species may then be said to have been reciprocally crossed. There is often the widest possible difference in the facility of making reciprocal crosses. Such cases are highly important, for they prove that the capacity in any two species to cross is often completely independent of their systematic affinity, or of any recognisable difference in their whole organisation. On the other hand, these cases clearly show that the capacity for crossing is connected with constitutional differences imperceptible by us, and confined to the reproductive system. This difference in the result of reciprocal crosses between the same two species was long ago observed by Kölreuter. To give an instance: Mirabilis jalapa can easily be fertilised by the pollen of M. longiflora, and the hybrids thus produced are sufficiently fertile; but Kölreuter tried more than two hundred times, during eight following years, to fertilise reciprocally M. longiflora with the pollen of M. jalapa, and utterly failed. Several other equally striking cases could be given. Thuret has observed the same fact with certain sea-weeds or Fuci. Gärtner, moreover, found that this difference of facility in making reciprocal crosses is extremely common in a lesser degree. He has observed it even between forms so closely related (as Matthiola annua and glabra) that many botanists rank them only as varieties. It is also a remarkable fact, that hybrids raised from reciprocal crosses, though of course compounded of the very same two species, the one species having first been used as the father and then as the mother, generally differ in fertility in a small, and occasionally in a high degree.

Several other singular rules could be given from Gärtner: for instance, some species have a remarkable power of crossing with other species; other species of the same genus have a remarkable power of impressing their likeness on their hybrid offspring; but these two powers do not at all necessarily go together. There are certain hybrids which instead of having, as is usual, an intermediate character between their two parents, always closely resemble one of them; and such hybrids, though externally so like one of their pure parent-species, are with rare exceptions extremely sterile. So again amongst hybrids which are usually intermediate in structure between their parents, exceptional and abnormal individuals sometimes are born, which closely resemble one of their pure parents; and these hybrids are almost always utterly sterile, even when the other hybrids raised from seed from the same capsule have a considerable degree of fertility. These facts show how completely fertility in the hybrid is independent of its external resemblance to either pure parent.

Considering the several rules now given, which govern the fertility of first crosses and of hybrids, we see that when forms, which must be considered as good and distinct species, are united, their fertility graduates from zero to perfect fertility, or even to fertility under certain conditions in excess. That their fertility, besides being eminently susceptible to favourable and unfavourable conditions, is innately variable. That it is by no means always the same in degree in the first cross and in the hybrids produced from this cross. That the fertility of hybrids is not related to the degree in which they resemble in external appearance either parent. And lastly, that the facility of making a first cross between any two species is not always governed by their systematic affinity or degree of resemblance to each other. This latter statement is clearly proved by reciprocal crosses between the same two species, for according as the one species or the other is used as the father or the mother, there is generally some difference, and occasionally the widest possible difference, in the facility of effecting a union. The hybrids, moreover, produced from reciprocal crosses often differ in fertility.

Now do these complex and singular rules indicate that species have been endowed with sterility simply to prevent their becoming confounded in nature? I think not. For why should the sterility be so extremely different in degree, when various species are crossed, all of which we must suppose it would be equally important to keep from blending together? Why should the degree of sterility be innately variable in the individuals of the same species? Why should some species cross with facility, and yet produce very sterile hybrids; and other species cross with extreme difficulty, and yet produce fairly fertile hybrids? Why should there often be so great a difference in the result of a reciprocal cross between the same two species? Why, it may even be asked, has the production of hybrids been permitted? to grant to species the special power of producing hybrids, and then to stop their further propagation by different degrees of sterility, not strictly related to the facility of the first union between their parents, seems to be a strange arrangement.

The foregoing rules and facts, on the other hand, appear to me clearly to indicate that the sterility both of first crosses and of hybrids is simply incidental or dependent on unknown differences, chiefly in the reproductive systems, of the species which are crossed. The differences being of so peculiar and limited a nature, that, in reciprocal crosses between two species the male sexual element of the one will often freely act on the female sexual element of the other, but not in a reversed direction. It will be advisable to explain a little more fully by an example what I mean by sterility being incidental on other differences, and not a specially endowed quality. As the capacity of one plant to be grafted or budded on another is so entirely unimportant for its welfare in a state of nature, I presume that no one will suppose that this capacity is a *specially* endowed quality, but will admit that it is incidental on differences in the laws of growth of the two plants. We can sometimes see the reason why one tree will not take on another, from differences in their rate of growth, in the hardness of their wood, in the period of the flow or nature of their sap, etc.; but in a multitude of cases we can assign no reason whatever. Great diversity in the size of two plants, one being woody and the other herbaceous, one being evergreen and the other deciduous, and adaptation to widely different climates, does not always prevent the two grafting together. As in hybridisation, so with grafting, the capacity is limited by systematic affinity, for no one has been able to graft trees together belonging to quite distinct families; and, on the other hand, closely allied species, and varieties of the same species, can usually, but not invariably, be grafted with ease. But this capacity, as in hybridisation, is by no means absolutely governed by systematic affinity. Although many distinct genera within the same family have been grafted together, in other cases species of the same genus will not take on each other. The pear can be grafted far more readily on the quince, which is ranked as a distinct genus, than on the apple, which is a member of the same genus. Even different varieties of the pear take with different degrees of facility on the quince; so do different varieties of the apricot and peach on certain varieties of the plum.

As Gärtner found that there was sometimes an innate difference in different *individuals* of the same two species in crossing; so Sagaret believes this to be the case with different individuals of the same two species in being grafted together. As in reciprocal crosses, the facility of effecting an union is often very far from equal, so it sometimes is in grafting; the common gooseberry, for instance, cannot be grafted on the currant, whereas the currant will take, though with difficulty, on the gooseberry.

We have seen that the sterility of hybrids, which have their reproductive organs in an imperfect condition, is a very different case from the difficulty of uniting two pure species, which have their reproductive organs perfect; yet these two distinct cases run to a certain extent parallel. Something analogous occurs in grafting; for Thouin found that three species of Robinia, which seeded freely on their own roots, and which could be grafted with no great difficulty on another species, when thus grafted were rendered barren. On the other hand, certain species of Sorbus, when grafted on other species, yielded twice as much fruit as when on their own roots. We are reminded by this latter fact of the extraordinary case of Hippeastrum, Lobelia, etc., which seeded much more freely when fertilised with the pollen of distinct species, than when self-fertilised with their own pollen.

We thus see, that although there is a clear and fundamental difference between the mere adhesion of grafted stocks, and the union of the male and female elements in the act of reproduction, yet that there is a rude degree of parallelism in the results of grafting and of crossing distinct species. And as we must look at the curious and complex laws governing the facility with which trees can be grafted on each other as incidental on unknown differences in their vegetative systems, so I believe that the still more complex laws governing the facility of first crosses, are incidental on unknown differences, chiefly in their reproductive systems. These differences, in both cases, follow to a certain extent, as might have been expected, systematic affinity, by which every kind of resemblance and dissimilarity between organic beings is attempted to be expressed. The facts by no means seem to me to indicate that the greater or lesser difficulty of either grafting or crossing together various species has been a special endowment; although in the case of crossing, the difficulty is as important for the endurance and stability of specific forms, as in the case of grafting it is unimportant for their welfare.

Causes of the Sterility of first Crosses and of Hybrids.—We may now look a little closer at the probable causes of the sterility of first crosses and of hybrids. These two cases are fundamentally different, for, as just remarked, in the union of two pure species the male and female sexual elements are perfect, whereas in hybrids they are imperfect. Even in first crosses, the greater or lesser difficulty in effecting a union apparently depends on several distinct causes. There must sometimes be a physical impossibility in the male element reaching the ovule, as would be the case with a plant having a pistil too long for the pollen-tubes to reach the ovarium. It has also been observed that when pollen of one species is placed on the stigma of a distinctly allied species, though the pollen tubes protrude, they do not penetrate the stigmatic surface. Again, the male element may reach the female element, but be incapable of causing an embryo to be developed, as seems to have been the case with some of Thuret's experiments on Fuci. No explanation can be given of these facts, any more than why certain trees cannot be grafted on others. Lastly, an embryo may be developed, and then perish at an early period. This latter alternative has not been sufficiently attended to; but I believe, from observations communicated to me by Mr Hewitt, who has had great experience in hybridising gallinaceous birds, that the early death of the embryo is a very frequent cause of sterility in first crosses. I was at first very unwilling to believe in this view; as hybrids, when once born, are generally healthy and long-lived, as we see in the case of the common mule. Hybrids, however, are differently circumstanced before and after birth: when born and living in a country where their two parents can live, they are generally placed under suitable conditions of life. But a hybrid partakes of only half of the nature and constitution of its mother, and therefore before birth, as long as it is nourished within its mother's womb or within the egg or seed produced by the mother, it may be exposed to conditions in some degree unsuitable, and consequently be liable to perish at an early period; more especially as all very young beings seem eminently sensitive to injurious or unnatural conditions of life.

In regard to the sterility of hybrids, in which the sexual elements are imperfectly developed, the case is very different. I have more than once alluded to a large body of facts, which I have collected, showing that when animals and plants are removed from their natural conditions, they are extremely liable to have their reproductive systems seriously affected. This, in fact, is the great bar to the domestication of animals. Between the sterility thus superinduced and that of hybrids, there are many points of similarity. In both cases the sterility is independent of general health, and is often accompanied by excess of size or great luxuriance. In both cases, the sterility occurs in various degrees; in both, the male element is the most liable to be affected; but sometimes the female more than the male. In both, the tendency goes to a certain extent with systematic affinity, for whole groups of animals and plants are rendered impotent by the same unnatural conditions; and whole groups of species tend to produce sterile hybrids. On the other hand, one species in a group will sometimes resist great changes of conditions with unimpaired fertility; and certain species in a group will produce unusually fertile hybrids. No one can tell, till he tries, whether any particular animal will breed under confinement or any exotic plant seed freely under culture; nor can he tell, till he tries, whether any two species of a genus will produce more or less sterile hybrids. Lastly, when organic beings are placed during several generations under conditions not natural to them, they are extremely liable to vary, which is due, as I believe, to their reproductive systems having been specially affected, though in a lesser degree than when sterility ensues. So it is with hybrids, for hybrids in successive generations are eminently liable to vary, as every experimentalist has observed.

Thus we see that when organic beings are placed under new and unnatural conditions, and when hybrids are produced by the unnatural crossing of two species, the reproductive system, independently of the general state of health, is affected by sterility in a very similar manner. In the one case, the conditions of life have been disturbed, though often in so slight a degree as to be inappreciable by us; in the other case, or that of hybrids, the external conditions have remained the same, but the organisation has been disturbed by two different structures and constitutions having been blended into one. For it is scarcely possible that two organisations should be compounded into one, without some disturbance occurring in the development, or periodical action, or mutual relation of the different parts and organs one to another, or to the conditions of life. When hybrids are able to breed *inter se*, they transmit to their offspring from generation to generation the same compounded organisation, and hence we need not be surprised that their sterility, though in some degree variable, rarely diminishes.

It must, however, be confessed that we cannot understand, excepting on vague hypotheses, several facts with respect to the sterility of hybrids; for instance, the unequal fertility of hybrids produced from reciprocal crosses; or the increased sterility in those hybrids which occasionally and exceptionally resemble closely either pure parent. Nor do I pretend that the foregoing remarks go to the root of the matter: no explanation is offered why an organism, when placed under unnatural conditions, is rendered sterile. All that I have attempted to show, is that in two cases, in some respects allied, sterility is the common result,—in the one case from the conditions of life having been disturbed, in the other case from the organisation having been disturbed by two organisations having been compounded into one.

It may seem fanciful, but I suspect that a similar parallelism extends to an allied yet very different class of facts. It is an old and almost universal belief, founded, I think, on a considerable body of evidence, that slight changes in the conditions of life are beneficial to all living things. We see this acted on by farmers and gardeners in their frequent exchanges of seed, tubers, etc., from one soil or climate to another, and back again. During the convalescence of animals, we plainly see that great benefit is derived from almost any change in the habits of life. Again, both with plants and animals, there is abundant evidence, that a cross between very distinct individuals of the same species, that is between members of different strains or sub-breeds, gives vigour and fertility to the offspring. I believe, indeed, from the facts alluded to in our fourth chapter, that a certain amount of crossing is indispensable even with hermaphrodites; and that close interbreeding continued during several generations between the nearest relations, especially if these be kept under the same conditions of life, always induces weakness and sterility in the progeny.

Hence it seems that, on the one hand, slight changes in the conditions of life benefit all organic beings, and on the other hand, that slight crosses, that is crosses between the males and females of the same species which have varied and become slightly different, give vigour and fertility to the offspring. But we have seen that greater changes, or changes of a particular nature, often render organic beings in some degree sterile; and that greater crosses, that is crosses between males and females which have become widely or specifically different, produce hybrids which are generally sterile in some degree. I cannot persuade myself that this parallelism is an accident or an illusion. Both series of facts seem to be connected together by some common but unknown bond, which is essentially related to the principle of life.

Fertility of Varieties when crossed, and of their Mongrel offspring.—It may be urged, as a most forcible argument, that there must be some essential distinction between species and varieties, and that there must be some error in all the foregoing remarks, inasmuch as varieties, however much they may differ from each other in external appearance, cross with perfect facility, and yield perfectly fertile offspring. I fully admit that this is almost invariably the case. But if we look to varieties produced under nature, we are immediately involved in hopeless difficulties; for if two hitherto reputed varieties be found in any degree sterile together, they are at once ranked by most naturalists as species. For instance, the blue and red pimpernel, the primrose and cowslip, which are considered by many of our best botanists as varieties, are said by Gärtner not to be quite fertile when crossed, and he consequently ranks them as undoubted species. If we thus argue in a circle, the fertility of all varieties produced under nature will assuredly have to be granted.

If we turn to varieties, produced, or supposed to have been produced, under domestication, we are still involved in doubt. For when it is stated, for instance, that the German Spitz dog unites more easily than other dogs with foxes, or that certain South American indigenous domestic dogs do not readily cross with European dogs, the explanation which will occur to every one, and probably the true one, is that these dogs have descended from several aboriginally distinct species. Nevertheless the perfect fertility of so many domestic varieties, differing widely from each other in appearance, for instance of the pigeon or of the cabbage, is a remarkable fact; more especially when we reflect how many species there are, which, though resembling each other most closely, are utterly sterile when intercrossed. Several considerations, however, render the fertility of domestic varieties less remarkable than at first appears. It can, in the first place, be clearly shown that mere external dissimilarity between two species does not determine their greater or lesser degree of sterility when crossed; and we may apply the same rule to domestic varieties. In the second place, some eminent naturalists believe that a long course of domestication tends to eliminate sterility in the successive generations of hybrids which were at first only slightly sterile; and if this be so, we surely ought not to expect to find sterility both appearing and disappearing under nearly the same conditions of life. Lastly, and this seems to me by far the most important consideration, new races of animals and plants are produced under domestication by man's methodical and unconscious power of selection, for his own use and pleasure: he neither wishes to select, nor could select, slight differences in the reproductive system, or other constitutional differences correlated with the reproductive system. He supplies his several varieties with the same food; treats them in nearly the same manner, and does not wish to alter their general habits of life. Nature acts uniformly and slowly during vast periods of time on the whole organisation, in any way which may be for each creature's own good; and thus she may, either directly, or more probably indirectly, through correlation, modify the reproductive system in the several descendants from any one species. Seeing this difference in the process of selection, as carried on by man and nature, we need not be surprised at some difference in the result.

I have as yet spoken as if the varieties of the same species were invariably fertile when intercrossed. But it seems to me impossible to resist the evidence of the existence of a certain amount of sterility in the few following cases, which I will briefly abstract. The evidence is at least as good as that from which we believe in the sterility of a multitude of species. The evidence is, also, derived from hostile witnesses, who in all other cases consider fertility and sterility as safe criterions of specific distinction. Gärtner kept during several years a dwarf kind of maize with yellow seeds, and a tall variety with red seeds, growing near each other in his garden; and although these plants have separated sexes, they never naturally crossed. He then fertilised thirteen flowers of the one with the pollen of the other; but only a single head produced any seed, and this one head produced only five grains. Manipulation in this case could not have been injurious, as the plants have separated sexes. No one, I believe, has suspected that these varieties of maize are distinct species; and it is important to notice that the hybrid plants thus raised were themselves *perfectly* fertile; so that even Gärtner did not venture to consider the two varieties as specifically distinct.

Girou de Buzareingues crossed three varieties of gourd, which like the maize has separated sexes, and he asserts that their mutual fertilisation is by so much the less easy as their differences are greater. How far these experiments may be trusted, I know not; but the forms experimentised on, are ranked by Sagaret, who mainly founds his classification by the test of infertility, as varieties.

The following case is far more remarkable, and seems at first quite incredible; but it is the result of an astonishing number of experiments made during many years on nine species of Verbascum, by so good an observer and so hostile a witness as Gärtner: namely, that yellow and white varieties of the same species of Verbascum when intercrossed produce less seed, than do either coloured varieties when fertilised with pollen from their own coloured flowers. Moreover, he asserts that when yellow and white varieties of one species are crossed with yellow and white varieties of a *distinct* species, more seed is produced by the crosses between the similarly coloured flowers, than between those which are differently coloured. Yet these varieties of Verbascum present no other difference besides the mere colour of the flower; and one variety can sometimes be raised from the seed of the other.

From observations which I have made on certain varieties of hollyhock, I am inclined to suspect that they present analogous facts.

Kölreuter, whose accuracy has been confirmed by every subsequent observer, has proved the remarkable fact, that one variety of the common tobacco is more fertile, when crossed with a widely distinct species, than are the other varieties. He experimentised on five forms, which are commonly reputed to be varieties, and which he tested by the severest trial, namely, by reciprocal crosses, and he found their mongrel offspring perfectly fertile. But one of these five varieties, when used either as father or mother, and crossed with the Nicotiana glutinosa, always yielded hybrids not so sterile as those which were produced from the four other varieties when crossed with N. glutinosa. Hence the reproductive system of this one variety must have been in some manner and in some degree modified.

From these facts; from the great difficulty of ascertaining the infertility of varieties in a state of nature, for a supposed variety if infertile in any degree would generally be ranked as species; from man selecting only external characters in the production of the most distinct domestic varieties, and from not wishing or being able to produce recondite and functional differences in the reproductive system; from these several considerations and facts, I do not think that the very general fertility of varieties can be proved to be of universal occurrence, or to form a fundamental distinction between varieties and species. The general fertility of varieties does not seem to me sufficient to overthrow the view which I have taken with respect to the very general, but not invariable, sterility of first crosses and of hybrids, namely, that it is not a special endowment, but is incidental on slowly acquired modifications, more especially in the reproductive systems of the forms which are crossed.

Hybrids and Mongrels compared, independently of their fertility.—Independently of the question of fertility, the offspring of species when crossed and of varieties when crossed may be compared in several other respects. Gärtner, whose strong wish was to draw a marked line of distinction between species and varieties, could find very few and, as it seems to me, quite unimportant differences between the so-called hybrid offspring of species, and the so-called mongrel offspring of varieties. And, on the other hand, they agree most closely in very many important respects.

I shall here discuss this subject with extreme brevity. The most important distinction is, that in the first generation mongrels are more variable than hybrids; but Gärtner admits that hybrids from species which have long been cultivated are often variable in the first generation; and I have myself seen striking instances of this fact. Gärtner further admits that hybrids between very closely allied species are more variable than those from very distinct species; and this shows that the difference in the degree of variability graduates away. When mongrels and the more fertile hybrids are propagated for several generations an extreme amount of variability in their offspring is notorious; but some few cases both of hybrids and mongrels long retaining uniformity of character could be given. The variability, however, in the successive generations of mongrels is, perhaps, greater than in hybrids.

This greater variability of mongrels than of hybrids does not seem to me at all surprising. For the parents of mongrels are varieties, and mostly domestic varieties (very few experiments having been tried on natural varieties), and this implies in most cases that there has been recent variability; and therefore we might expect that such variability would often continue and be super-added to that arising from the mere act of crossing. The slight degree of variability in hybrids from the first cross or in the first generation, in contrast with their extreme variability in the succeeding generations, is a curious fact and deserves attention. For it bears on and corroborates the view which I have taken on the cause of ordinary variability; namely, that it is due to the reproductive system being eminently sensitive to any change in the conditions of life, being thus often rendered either impotent or at least incapable of its proper function of producing offspring identical with the parent-form. Now hybrids in the first generation are descended from species (excluding those long cultivated) which have not had their reproductive systems in any way affected, and they are not variable; but hybrids themselves have their reproductive systems seriously affected, and their descendants are highly variable.

But to return to our comparison of mongrels and hybrids: Gärtner states that mongrels are more liable than hybrids to revert to either parent-form; but this, if it be true, is certainly only a difference in degree. Gärtner further insists that when any two species, although most closely allied to each other, are crossed with a third species, the hybrids are widely different from each other; whereas if two very distinct varieties of one species are crossed with another species, the hybrids do not differ much. But this conclusion, as far as I can make out, is founded on a single experiment; and seems directly opposed to the results of several experiments made by Kölreuter.

These alone are the unimportant differences, which Gärtner is able to point out, between hybrid and mongrel plants. On the other hand, the resemblance in mongrels and in hybrids to their respective parents, more especially in hybrids produced from nearly related species, follows according to Gärtner the same laws. When two species are crossed, one has sometimes a prepotent power of impressing its likeness on the hybrid; and so I believe it to be with varieties of plants. With animals one variety certainly often has this prepotent power over another variety. Hybrid plants produced from a reciprocal cross, generally resemble each other closely; and so it is with mongrels from a reciprocal cross. Both hybrids and mongrels can be reduced to either pure parent-form, by repeated crosses in successive generations with either parent.

These several remarks are apparently applicable to animals; but the subject is here excessively complicated, partly owing to the existence of secondary sexual characters; but more especially owing to prepotency in transmitting likeness running more strongly in one sex than in the other, both when one species is crossed with another, and when one variety is crossed with another variety. For instance, I think those authors are right, who maintain that the ass has a prepotent power over the horse, so that both the mule and the hinny more resemble the ass than the horse; but that the prepotency runs more strongly in the male-ass than in the female, so that the mule, which is the offspring of the male-ass and mare, is more like an ass, than is the hinny, which is the offspring of the female-ass and stallion.

Much stress has been laid by some authors on the supposed fact, that mongrel animals alone are born closely like one of their parents; but it can be shown that this does sometimes occur with hybrids; yet I grant much less frequently with hybrids than with mongrels. Looking to the cases which I have collected of cross-bred animals closely resembling one parent, the resemblances seem chiefly confined to characters almost monstrous in their nature, and which have suddenly appeared—such as albinism, melanism, deficiency of tail or horns, or additional fingers and toes; and do not relate to characters which have been slowly acquired by selection. Consequently, sudden reversions to the perfect character of either parent would be more likely to occur with mongrels, which are descended from varieties often suddenly produced and semi-monstrous in character, than with hybrids, which are descended from species slowly and naturally produced. On the whole I entirely agree with Dr Prosper Lucas, who, after arranging an enormous body of facts with respect to animals, comes to the conclusion, that the laws of resemblance of the child to its parents are the same, whether the two parents differ much or little from each other, namely in the union of individuals of the same variety, or of different varieties, or of distinct species.

Laying aside the question of fertility and sterility, in all other respects there seems to be a general and close similarity in the offspring of crossed species, and of crossed varieties. If we look at species as having been specially created, and at varieties as having been produced by secondary laws, this similarity would be an astonishing fact. But it harmonises perfectly with the view that there is no essential distinction between species and varieties.

Summary of Chapter.—First crosses between forms sufficiently distinct to be ranked as species, and their hybrids, are very generally, but not universally, sterile. The sterility is of all degrees, and is often so slight that the two most careful experimentalists who have ever lived, have come to diametrically opposite conclusions in ranking forms by this test. The sterility is innately variable in individuals of the same species, and is eminently susceptible of favourable and unfavourable conditions. The degree of sterility does not strictly follow systematic affinity, but is governed by several curious and complex laws. It is generally different, and sometimes widely different, in reciprocal crosses between the same two species. It is not always equal in degree in a first cross and in the hybrid produced from this cross.

In the same manner as in grafting trees, the capacity of one species or variety to take on another, is incidental on generally unknown differences in their vegetative systems, so in crossing, the greater or less facility of one species to unite with another, is incidental on unknown differences in their reproductive systems. There is no more reason to think that species have been specially endowed with various degrees of sterility to prevent them crossing and blending in nature, than to think that trees have been specially endowed with various and somewhat analogous degrees of difficulty in being grafted together in order to prevent them becoming inarched in our forests.

The sterility of first crosses between pure species, which have their reproductive systems perfect, seems to depend on several circumstances; in some cases largely on the early death of the embryo. The sterility of hybrids, which have their reproductive systems imperfect, and which have had this system and their whole organisation disturbed by being compounded of two distinct species, seems closely allied to that sterility which so frequently affects pure species, when their natural conditions of life have been disturbed. This view is supported by a parallelism of another kind;—namely, that the crossing of forms only slightly different is favourable to the vigour and fertility of their offspring; and that slight changes in the conditions of life are apparently favourable to the vigour and fertility of all organic beings. It is not surprising that the degree of difficulty in uniting two species, and the degree of sterility of their hybrid-offspring should generally correspond, though due to distinct causes; for both depend on the amount of difference of some kind between the species which are crossed. Nor is it surprising that the facility of effecting a first cross, the fertility of the hybrids produced from it, and the capacity of being grafted together—though this latter capacity evidently depends on widely different circumstances—should all run, to a certain extent, parallel with the systematic affinity of the forms which are subjected to experiment; for systematic affinity attempts to express all kinds of resemblance between all species.

First crosses between forms known to be varieties, or sufficiently alike to be considered as varieties, and their mongrel offspring, are very generally, but not quite universally, fertile. Nor is this nearly general and perfect fertility surprising, when we remember how liable we are to argue in a circle with respect to varieties in a state of nature; and when we remember that the greater number of varieties have been produced under domestication by the selection of mere external differences, and not of differences in the reproductive system. In all other respects, excluding fertility, there is a close general resemblance between hybrids and mongrels. Finally, then, the facts briefly given in this chapter do not seem to me opposed to, but even rather to support the view, that there is no fundamental distinction between species and varieties.

CHAPTER IX

ON THE IMPERFECTION OF THE GEOLOGICAL RECORD

On the absence of intermediate varieties at the present day—On the nature of extinct intermediate varieties; on their number—On the vast lapse of time, as inferred from the rate of deposition and of denudation—On the poorness of our palæontological collections—On the intermittence of geological formations—On the absence of intermediate varieties in any one formation—On the sudden appearance of groups of species—On their sudden appearance in the lowest known fossiliferous strata.

In the sixth chapter I enumerated the chief objections which might be justly urged against the views maintained in this volume. Most of them have now been discussed. One, namely the distinctness of specific forms, and their not being blended together by innumerable transitional links, is a very obvious difficulty. I assigned reasons why such links do not commonly occur at the present day, under the circumstances apparently most favourable for their presence, namely on an extensive and continuous area with graduated physical conditions. I endeavoured to show, that the life of each species depends in a more important manner on the presence of other already defined organic forms, than on climate; and, therefore, that the really governing conditions of life do not graduate away quite insensibly like heat or moisture. I endeavoured, also, to show that intermediate varieties, from existing in lesser numbers than the forms which they connect, will generally be beaten out and exterminated during the course of further modification and improvement. The main cause, however, of innumerable intermediate links not now occurring everywhere throughout nature depends on the very process of natural selection, through which new varieties continually take the places of and exterminate their parent-forms. But just in proportion as this process of extermination has acted on an enormous scale, so must the number of intermediate varieties, which have formerly existed on the earth, be truly enormous. Why then is not every geological formation and every stratum full of such intermediate links? Geology assuredly does not reveal any such finely graduated organic chain; and this, perhaps, is the most obvious and gravest objection which can be urged against my theory. The explanation lies, as I believe, in the extreme imperfection of the geological record.

In the first place it should always be borne in mind what sort of intermediate forms must, on my theory, have formerly existed. I have found it difficult, when looking at any two species, to avoid picturing to myself forms *directly* intermediate between them. But this is a wholly false view; we should always look for forms intermediate between each species and a common but unknown progenitor; and the progenitor will generally have differed in some respects from all its modified descendants. To give a simple illustration: the fantail and pouter pigeons have both descended from the rock-pigeon; if we possessed all the intermediate varieties which have ever existed, we should have an extremely close series between both and the rock-pigeon; but we should have no varieties directly intermediate between the fantail and pouter; none, for instance, combining a tail somewhat expanded with a crop somewhat enlarged, the characteristic features of these two breeds. These two breeds, moreover, have become so much modified, that if we had no historical or indirect evidence regarding their origin, it would not have been possible to have determined from a mere comparison of their structure with that of the rock-pigeon, whether they had descended from this species or from some other allied species, such as C. oenas.

So with natural species, if we look to forms very distinct, for instance to the horse and tapir, we have no reason to suppose that links ever existed directly intermediate between them, but between each and an unknown common parent. The common parent will have had in its whole organisation much general resemblance to the tapir and to the horse; but in some points of structure may have differed considerably from both, even perhaps more than they differ from each other. Hence in all such cases, we should be unable to recognise the parent-form of any two or more species, even if we closely compared the structure of the parent with that of its modified descendants, unless at the same time we had a nearly perfect chain of the intermediate links.

It is just possible by my theory, that one of two living forms might have descended from the other; for instance, a horse from a tapir; and in this case *direct* intermediate links will have existed between them. But such a case would imply that one form had remained for a very long period unaltered, whilst its descendants had undergone a vast amount of change; and the principle of competition between organism and organism, between child and parent, will render this a very rare event; for in all cases the new and improved forms of life tend to supplant the old and unimproved forms.

By the theory of natural selection all living species have been connected with the parent-species of each genus, by differences not greater than we see between the varieties of the same species at the present day; and these parent-species, now generally extinct, have in their turn been similarly connected with more ancient species; and so on backwards, always converging to the common ancestor of each great class. So that the number of intermediate and transitional links, between all living and extinct species, must have been inconceivably great. But assuredly, if this theory be true, such have lived upon this earth.

On the lapse of Time.—Independently of our not finding fossil remains of such infinitely numerous connecting links, it may be objected, that time will not have sufficed for so great an amount of organic change, all changes having been effected very slowly through natural selection. It is hardly possible for me even to recall to the reader, who may not be a practical geologist, the facts leading the mind feebly to comprehend the lapse of time. He who can read Sir Charles Lyell's grand work on the *Principles of Geology*, which the future historian will recognise as having produced a revolution in natural science, yet does not admit how incomprehensively vast have been the past periods of time, may at once close this volume. Not that it suffices to study the *Principles of Geology*, or to read special treatises by different observers on separate formations, and to mark how each author attempts to give an inadequate idea of the duration of each formation or even each stratum. A man must for years examine for himself great piles of superimposed strata, and watch the sea at work grinding down old rocks and making fresh sediment, before he can hope to comprehend anything of the lapse of time, the monuments of which we see around us.

It is good to wander along lines of sea-coast, when formed of moderately hard rocks, and mark the process of degradation. The tides in most cases reach the cliffs only for a short time twice a day, and the waves eat into them only when they are charged with sand or pebbles; for there is good evidence that pure water can effect little or nothing in wearing away rock. At last the base of the cliff is undermined, huge fragments fall down, and these remaining fixed, have to be worn away, atom by atom, until reduced in size they can be rolled about by the waves, and then are more quickly ground into pebbles, sand, or mud. But how often do we see along the bases of retreating cliffs rounded boulders, all thickly clothed by marine productions, showing how little they are abraded and how seldom they are rolled about! Moreover, if we follow for a few miles any line of rocky cliff, which is undergoing degradation, we find that it is only here and there, along a short length or round a promontory, that the cliffs are at the present time suffering. The appearance of the surface and the vegetation show that elsewhere years have elapsed since the waters washed their base.

He who most closely studies the action of the sea on our shores, will, I believe, be most deeply impressed with the slowness with which rocky coasts are worn away. The observations on this head by Hugh Miller, and by that excellent observer Mr Smith of Jordan Hill, are most impressive. With the mind thus impressed, let any one examine beds of conglomerate many thousand feet in thickness, which, though probably formed at a quicker rate than many other deposits, yet, from being formed of worn and rounded pebbles, each of which bears the stamp of time, are good to show how slowly the mass has been accumulated. In the Cordillera I estimated one pile of conglomerate at ten thousand feet in thickness. Let the observer remember Lyell's profound remark that the thickness and extent of sedimentary formations are the result and measure of the degradation which the earth's crust has elsewhere suffered. And what an amount of degradation is implied by the sedimentary deposits of many countries! Professor Ramsay has given me the maximum thickness, in most cases from actual measurement, in a few cases from estimate, of each formation in different parts of Great Britain; and this is the result:—

	Feet
Palæozoic strata (not including igneous beds)	57,154
Secondary strata	13,190
Tertiary strata	2,240

—making altogether 72,584 feet; that is, very nearly thirteen and three-quarters British miles. Some of the formations, which are represented in England by thin beds, are thousands of feet in thickness on the Continent. Moreover, between each successive formation, we have, in the opinion of most geologists, enormously long blank periods. So that the lofty pile of sedimentary rocks in Britain, gives but an inadequate idea of the time which has elapsed during their accumulation; yet what time this must have consumed! Good observers have estimated that sediment is deposited by the great Mississippi river at the rate of only 600 feet in a hundred thousand years. This estimate has no pretension to strict exactness; yet, considering over what wide spaces very fine sediment is transported by the currents of the sea, the process of accumulation in any one area must be extremely slow.

But the amount of denudation which the strata have in many places suffered, independently of the rate of accumulation of the degraded matter, probably offers the best evidence of the lapse of time. I remember having been much struck with the evidence of denudation, when viewing volcanic islands, which have been worn by the waves and pared all round into perpendicular cliffs of one or two thousand feet in height; for the gentle slope of the lava-streams, due to their formerly liquid state, showed at a glance how far the hard, rocky beds had once extended into the open ocean. The same story is still more plainly told by faults,—those great cracks along which the strata have been upheaved on one side, or thrown down on the other, to the height or depth of thousands of feet; for since the crust cracked, the surface of the land has been so completely planed down by the action of the sea, that no trace of these vast dislocations is externally visible.

The Craven fault, for instance, extends for upwards of 30 miles, and along this line the vertical displacement of the strata has varied from 600 to 3000 feet. Prof. Ramsay has published an account of a downthrow in Anglesea of 2300 feet; and he informs me that he fully believes there is one in Merionethshire of 12,000 feet; yet in these cases there is nothing on the surface to show such prodigious movements; the pile of rocks on the one or other side having been smoothly swept away. The consideration of these facts impresses my mind almost in the same manner as does the vain endeavour to grapple with the idea of eternity.

I am tempted to give one other case, the well-known one of the denudation of the Weald. Though it must be admitted that the denudation of the Weald has been a mere trifle, in comparison with that which has removed masses of our palæozoic strata, in parts ten thousand feet in thickness, as shown in Prof. Ramsay's masterly memoir on this subject: yet it is an admirable lesson to stand on the intermediate hilly country and look on the one hand at the North Downs, and on the other hand at the South Downs; for, remembering that at no great distance to the west the northern and southern escarpments meet and close, one can safely picture to oneself the great dome of rocks which must have covered up the Weald within so limited a period as since the latter part of the Chalk formation. The distance from the northern to the southern Downs is about 22 miles, and the thickness of the several formations is on an average about 1100 feet, as I am informed by Prof. Ramsay. But if, as some geologists suppose, a range of older rocks underlies the Weald, on the flanks of which the overlying sedimentary deposits might have accumulated in thinner masses than elsewhere, the above estimate would be erroneous; but this source of doubt probably would not greatly affect the estimate as applied to the western extremity of the district. If, then, we knew the rate at which the sea commonly wears away a line of cliff of any given height, we could measure the time requisite to have denuded the Weald. This, of course, cannot be done; but we may, in order to form some crude notion on the subject, assume that the sea would eat into cliffs 500 feet in height at the rate of one inch in a century. This will at first appear much too small an allowance; but it is the same as if we were to assume a cliff one yard in height to be eaten back along a whole line of coast at the rate of one yard in nearly every twenty-two years. I doubt whether any rock, even as soft as chalk, would yield at this rate excepting on the most exposed coasts; though no doubt the degradation of a lofty cliff would be more rapid from the breakage of the fallen fragments. On the other hand, I do not believe that any line of coast, ten or twenty miles in length, ever suffers degradation at the same time along its whole indented length; and we must remember that almost all strata contain harder layers or nodules, which from long resisting attrition form a breakwater at the base. We may at least confidently believe that no rocky coast 500 feet in height commonly yields at the rate of a foot per century; for this would be the same in amount as a cliff one yard in height retreating twelve yards in twenty-two years; and no one, I think, who has carefully observed the shape of old fallen fragments at the base of cliffs, will admit any near approach to such rapid wearing away. Hence, under ordinary circumstances, I should infer that for a cliff 500 feet in height, a denudation of one inch per century for the whole length would be a sufficient allowance. At this rate, on the above data, the denudation of the Weald must have required 306,662,400 years; or say three hundred million years. But perhaps it would be safer to allow two or three inches per century, and this would reduce the number of years to one hundred and fifty or one hundred million years.

The action of fresh water on the gently inclined Wealden district, when upraised, could hardly have been great, but it would somewhat reduce the above estimate. On the other hand, during oscillations of level, which we know this area has undergone, the surface may have existed for millions of years as land, and thus have escaped the action of the sea: when deeply submerged for perhaps equally long periods, it would, likewise, have escaped the action of the coast-waves. So that it is not improbable that a longer period than 300 million years has elapsed since the latter part of the Secondary period.

I have made these few remarks because it is highly important for us to gain some notion, however imperfect, of the lapse of years. During each of these years, over the whole world, the land and the water has been peopled by hosts of living forms. What an infinite number of generations, which the mind cannot grasp, must have succeeded each other in the long roll of years! Now turn to our richest geological museums, and what a paltry display we behold!

On the poorness of our Palæontological collections.—That our palæontological collections are very imperfect, is admitted by every one. The remark of that admirable palæontologist, the late Edward Forbes, should not be forgotten, namely, that numbers of our fossil species are known and named from single and often broken specimens, or from a few specimens collected on some one spot. Only a small portion of the surface of the earth has been geologically explored, and no part with sufficient care, as the important discoveries made every year in Europe prove. No organism wholly soft can be preserved. Shells and bones will decay and disappear when left on the bottom of the sea, where sediment is not accumulating. I believe we are continually taking a most erroneous view, when we tacitly admit to ourselves that sediment is being deposited over nearly the whole bed of the sea, at a rate sufficiently quick to embed and preserve fossil remains. Throughout an enormously large proportion of the ocean, the bright blue tint of the water bespeaks its purity. The many cases on record of a formation conformably covered, after an enormous interval of time, by another and later formation, without the underlying bed having suffered in the interval any wear and tear, seem explicable only on the view of the bottom of the sea not rarely lying for ages in an unaltered condition. The remains which do become embedded, if in sand or gravel, will when the beds are upraised generally be dissolved by the percolation of rain-water. I suspect that but few of the very many animals which live on the beach between high and low watermark are preserved. For instance, the several species of the Chthamalinæ (a sub-family of sessile cirripedes) coat the rocks all over the world in infinite numbers: they are all strictly littoral, with the exception of a single Mediterranean species, which inhabits deep water and has been found fossil in Sicily, whereas not one other species has hitherto been found in any tertiary formation: yet it is now known that the genus Chthamalus existed during the chalk period. The molluscan genus Chiton offers a partially analogous case.

With respect to the terrestrial productions which lived during the Secondary and Palæozoic periods, it is superfluous to state that our evidence from fossil remains is fragmentary in an extreme degree. For instance, not a land shell is known belonging to either of these vast periods, with the exception of one species discovered by Sir C. Lyell and Dr Dawson in the carboniferous strata of North America, of which shell several specimens have now been collected. In regard to mammiferous remains, a single glance at the historical table published in the Supplement to Lyell's Manual, will bring home the truth, how accidental and rare is their preservation, far better than pages of detail. Nor is their rarity surprising, when we remember how large a proportion of the bones of tertiary mammals have been discovered either in caves or in lacustrine deposits; and that not a cave or true lacustrine bed is known belonging to the age of our secondary or palæozoic formations.

But the imperfection in the geological record mainly results from another and more important cause than any of the foregoing; namely, from the several formations being separated from each other by wide intervals of time. When we see the formations tabulated in written works, or when we follow them in nature, it is difficult to avoid believing that they are closely consecutive. But we know, for instance, from Sir R. Murchison's great work on Russia, what wide gaps there are in that country between the superimposed formations; so it is in North America, and in many other parts of the world. The most skilful geologist, if his attention had been exclusively confined to these large territories, would never have suspected that during the periods which were blank and barren in his own country, great piles of sediment, charged with new and peculiar forms of life, had elsewhere been accumulated. And if in each separate territory, hardly any idea can be formed of the length of time which has elapsed between the consecutive formations, we may infer that this could nowhere be ascertained. The frequent and great changes in the mineralogical composition of consecutive formations, generally implying great changes in the geography of the surrounding lands, whence the sediment has been derived, accords with the belief of vast intervals of time having elapsed between each formation.

But we can, I think, see why the geological formations of each region are almost invariably intermittent; that is, have not followed each other in close sequence. Scarcely any fact struck me more when examining many hundred miles of the South American coasts, which have been upraised several hundred feet within the recent period, than the absence of any recent deposits sufficiently extensive to last for even a short geological period. Along the whole west coast, which is inhabited by a peculiar marine fauna, tertiary beds are so poorly developed, that no record of several successive and peculiar marine faunas will probably be preserved to a distant age. A little reflection will explain why along the rising coast of the western side of South America, no extensive formations with recent or tertiary remains can anywhere be found, though the supply of sediment must for ages have been great, from the enormous degradation of the coast-rocks and from muddy streams entering the sea. The explanation, no doubt, is, that the littoral and sub-littoral deposits are continually worn away, as soon as they are brought up by the slow and gradual rising of the land within the grinding action of the coast-waves.

We may, I think, safely conclude that sediment must be accumulated in extremely thick, solid, or extensive masses, in order to withstand the incessant action of the waves, when first upraised and during subsequent oscillations of level. Such thick and extensive accumulations of sediment may be formed in two ways; either, in profound depths of the sea, in which case, judging from the researches of E. Forbes, we may conclude that the bottom will be inhabited by extremely few animals, and the mass when upraised will give a most imperfect record of the forms of life which then existed; or, sediment may be accumulated to any thickness and extent over a shallow bottom, if it continue slowly to subside. In this latter case, as long as the rate of subsidence and supply of sediment nearly balance each other, the sea will remain shallow and favourable for life, and thus a fossiliferous formation thick enough, when upraised, to resist any amount of degradation, may be formed.

I am convinced that all our ancient formations which are rich in fossils have thus been formed during subsidence. Since publishing my views on this subject in 1845, I have watched the progress of Geology, and have been surprised to note how author after author, in treating of this or that great formation, has come to the conclusion that it was accumulated during subsidence. I may add, that the only ancient tertiary formation on the west coast of South America, which has been bulky enough to resist such degradation as it has as yet suffered, but which will hardly last to a distant geological age, was certainly deposited during a downward oscillation of level, and thus gained considerable thickness.

All geological facts tell us plainly that each area has undergone numerous slow oscillations of level, and apparently these oscillations have affected wide spaces. Consequently formations rich in fossils and sufficiently thick and extensive to resist subsequent degradation, may have been formed over wide spaces during periods of subsidence, but only where the supply of sediment was sufficient to keep the sea shallow and to embed and preserve the remains before they had time to decay. On the other hand, as long as the bed of the sea remained stationary, *thick* deposits could not have been accumulated in the shallow parts, which are the most favourable to life. Still less could this have happened during the alternate periods of elevation; or, to speak more accurately, the beds which were then accumulated will have been destroyed by being upraised and brought within the limits of the coast-action.

Thus the geological record will almost necessarily be rendered intermittent. I feel much confidence in the truth of these views, for they are in strict accordance with the general principles inculcated by Sir C. Lyell; and E. Forbes subsequently but independently arrived at a similar conclusion.

One remark is here worth a passing notice. During periods of elevation the area of the land and of the adjoining shoal parts of the sea will be increased, and new stations will often be formed;—all circumstances most favourable, as previously explained, for the formation of new varieties and species; but during such periods there will generally be a blank in the geological record. On the other hand, during subsidence, the inhabited area and number of inhabitants will decrease (excepting the productions on the shores of a continent when first broken up into an archipelago), and consequently during subsidence, though there will be much extinction, fewer new varieties or species will be formed; and it is during these very periods of subsidence, that our great deposits rich in fossils have been accumulated. Nature may almost be said to have guarded against the frequent discovery of her transitional or linking forms.

From the foregoing considerations it cannot be doubted that the geological record, viewed as a whole, is extremely imperfect; but if we confine our attention to any one formation, it becomes more difficult to understand, why we do not therein find closely graduated varieties between the allied species which lived at its commencement and at its close. Some cases are on record of the same species presenting distinct varieties in the upper and lower parts of the same formation, but, as they are rare, they may be here passed over. Although each formation has indisputably required a vast number of years for its deposition, I can see several reasons why each should not include a graduated series of links between the species which then lived; but I can by no means pretend to assign due proportional weight to the following considerations.

Although each formation may mark a very long lapse of years, each perhaps is short compared with the period requisite to change one species into another. I am aware that two palæontologists, whose opinions are worthy of much deference, namely Bronn and Woodward, have concluded that the average duration of each formation is twice or thrice as long as the average duration of specific forms. But insuperable difficulties, as it seems to me, prevent us coming to any just conclusion on this head. When we see a species first appearing in the middle of any formation, it would be rash in the extreme to infer that it had not elsewhere previously existed. So again when we find a species disappearing before the uppermost layers have been deposited, it would be equally rash to suppose that it then became wholly extinct. We forget how small the area of Europe is compared with the rest of the world; nor have the several stages of the same formation throughout Europe been correlated with perfect accuracy.

With marine animals of all kinds, we may safely infer a large amount of migration during climatal and other changes; and when we see a species first appearing in any formation, the probability is that it only then first immigrated into that area. It is well known, for instance, that several species appeared somewhat earlier in the palæozoic beds of North America than in those of Europe; time having apparently been required for their migration from the American to the European seas. In examining the latest deposits of various quarters of the world, it has everywhere been noted, that some few still existing species are common in the deposit, but have become extinct in the immediately surrounding sea; or, conversely, that some are now abundant in the neighbouring sea, but are rare or absent in this particular deposit. It is an excellent lesson to reflect on the ascertained amount of migration of the inhabitants of Europe during the Glacial period, which forms only a part of one whole geological period; and likewise to reflect on the great changes of level, on the inordinately great change of climate, on the prodigious lapse of time, all included within this same glacial period. Yet it may be doubted whether in any quarter of the world, sedimentary deposits, *including fossil remains*, have gone on accumulating within the same area during the whole of this period. It is not, for instance, probable that sediment was deposited during the whole of the glacial period near the mouth of the Mississippi, within that limit of depth at which marine animals can flourish; for we know what vast geographical changes occurred in other parts of America during this space of time. When such beds as were deposited in shallow water near the mouth of the Mississippi during some part of the glacial period shall have been upraised, organic remains will probably first appear and disappear at different levels, owing to the migration of species and to geographical changes. And in the distant future, a geologist examining these beds, might be tempted to conclude that the average duration of life of the embedded fossils had been less than that of the glacial period, instead of having been really far greater, that is extending from before the glacial epoch to the present day.

In order to get a perfect gradation between two forms in the upper and lower parts of the same formation, the deposit must have gone on accumulating for a very long period, in order to have given sufficient time for the slow process of variation; hence the deposit will generally have to be a very thick one; and the species undergoing modification will have had to live on the same area throughout this whole time. But we have seen that a thick fossiliferous formation can only be accumulated during a period of subsidence; and to keep the depth approximately the same, which is necessary in order to enable the same species to live on the same space, the supply of sediment must nearly have counterbalanced the amount of subsidence. But this same movement of subsidence will often tend to sink the area whence the sediment is derived, and thus diminish the supply whilst the downward movement continues. In fact, this nearly exact balancing between the supply of sediment and the amount of subsidence is probably a rare contingency; for it has been observed by more than one palæontologist, that very thick deposits are usually barren of organic remains, except near their upper or lower limits.

It would seem that each separate formation, like the whole pile of formations in any country, has generally been intermittent in its accumulation. When we see, as is so often the case, a formation composed of beds of different mineralogical composition, we may reasonably suspect that the process of deposition has been much interrupted, as a change in the currents of the sea and a supply of sediment of a different nature will generally have been due to geographical changes requiring much time. Nor will the closest inspection of a formation give any idea of the time which its deposition has consumed. Many instances could be given of beds only a few feet in thickness, representing formations, elsewhere thousands of feet in thickness, and which must have required an enormous period for their accumulation; yet no one ignorant of this fact would have suspected the vast lapse of time represented by the thinner formation. Many cases could be given of the lower beds of a formation having been upraised, denuded, submerged, and then re-covered by the upper beds of the same formation,—facts, showing what wide, yet easily overlooked, intervals have occurred in its accumulation. In other cases we have the plainest evidence in great fossilised trees, still standing upright as they grew, of many long intervals of time and changes of level during the process of deposition, which would never even have been suspected, had not the trees chanced to have been preserved: thus Messrs Lyell and Dawson found carboniferous beds 1400 feet thick in Nova Scotia, with ancient root-bearing strata, one above the other, at no less than sixty-eight different levels. Hence, when the same species occur at the bottom, middle, and top of a formation, the probability is that they have not lived on the same spot during the whole period of deposition, but have disappeared and reappeared, perhaps many times, during the same geological period. So that if such species were to undergo a considerable amount of modification during any one geological period, a section would not probably include all the fine intermediate gradations which must on my theory have existed between them, but abrupt, though perhaps very slight, changes of form.

It is all-important to remember that naturalists have no golden rule by which to distinguish species and varieties; they grant some little variability to each species, but when they meet with a somewhat greater amount of difference between any two forms, they rank both as species, unless they are enabled to connect them together by close intermediate gradations. And this from the reasons just assigned we can seldom hope to effect in any one geological section. Supposing B and C to be two species, and a third, A, to be found in an underlying bed; even if A were strictly intermediate between B and C, it would simply be ranked as a third and distinct species, unless at the same time it could be most closely connected with either one or both forms by intermediate varieties. Nor should it be forgotten, as before explained, that A might be the actual progenitor of B and C, and yet might not at all necessarily be strictly intermediate between them in all points of structure. So that we might obtain the parent-species and its several modified descendants from the lower and upper beds of a formation, and unless we obtained numerous transitional gradations, we should not recognise their relationship, and should consequently be compelled to rank them all as distinct species.

It is notorious on what excessively slight differences many palæontologists have founded their species; and they do this the more readily if the specimens come from different sub-stages of the same formation. Some experienced conchologists are now sinking many of the very fine species of D'Orbigny and others into the rank of varieties; and on this view we do find the kind of evidence of change which on my theory we ought to find. Moreover, if we look to rather wider intervals, namely, to distinct but consecutive stages of the same great formation, we find that the embedded fossils, though almost universally ranked as specifically different, yet are far more closely allied to each other than are the species found in more widely separated formations; but to this subject I shall have to return in the following chapter.

One other consideration is worth notice: with animals and plants that can propagate rapidly and are not highly locomotive, there is reason to suspect, as we have formerly seen, that their varieties are generally at first local; and that such local varieties do not spread widely and supplant their parent-forms until they have been modified and perfected in some considerable degree. According to this view, the chance of discovering in a formation in any one country all the early stages of transition between any two forms, is small, for the successive changes are supposed to have been local or confined to some one spot. Most marine animals have a wide range; and we have seen that with plants it is those which have the widest range, that oftenest present varieties; so that with shells and other marine animals, it is probably those which have had the widest range, far exceeding the limits of the known geological formations of Europe, which have oftenest given rise, first to local varieties and ultimately to new species; and this again would greatly lessen the chance of our being able to trace the stages of transition in any one geological formation.

It should not be forgotten, that at the present day, with perfect specimens for examination, two forms can seldom be connected by intermediate varieties and thus proved to be the same species, until many specimens have been collected from many places; and in the case of fossil species this could rarely be effected by palæontologists. We shall, perhaps, best perceive the improbability of our being enabled to connect species by numerous, fine, intermediate, fossil links, by asking ourselves whether, for instance, geologists at some future period will be able to prove, that our different breeds of cattle, sheep, horses, and dogs have descended from a single stock or from several aboriginal stocks; or, again, whether certain sea-shells inhabiting the shores of North America, which are ranked by some conchologists as distinct species from their European representatives, and by other conchologists as only varieties, are really varieties or are, as it is called, specifically distinct. This could be effected only by the future geologist discovering in a fossil state numerous intermediate gradations; and such success seems to me improbable in the highest degree.

Geological research, though it has added numerous species to existing and extinct genera, and has made the intervals between some few groups less wide than they otherwise would have been, yet has done scarcely anything in breaking down the distinction between species, by connecting them together by numerous, fine, intermediate varieties; and this not having been effected, is probably the gravest and most obvious of all the many objections which may be urged against my views. Hence it will be worth while to sum up the foregoing remarks, under an imaginary illustration. The Malay Archipelago is of about the size of Europe from the North Cape to the Mediterranean, and from Britain to Russia; and therefore equals all the geological formations which have been examined with any accuracy, excepting those of the United States of America. I fully agree with Mr Godwin-Austen, that the present condition of the Malay Archipelago, with its numerous large islands separated by wide and shallow seas, probably represents the former state of Europe, whilst most of our formations were accumulating. The Malay Archipelago is one of the richest regions of the whole world in organic beings; yet if all the species were to be collected which have ever lived there, how imperfectly would they represent the natural history of the world!

But we have every reason to believe that the terrestrial productions of the archipelago would be preserved in an excessively imperfect manner in the formations which we suppose to be there accumulating. I suspect that not many of the strictly littoral animals, or of those which lived on naked submarine rocks, would be embedded; and those embedded in gravel or sand, would not endure to a distant epoch. Wherever sediment did not accumulate on the bed of the sea, or where it did not accumulate at a sufficient rate to protect organic bodies from decay, no remains could be preserved.

I believe that fossiliferous formations could be formed in the archipelago, of thickness sufficient to last to an age as distant in futurity, as the secondary formations lie in the past, only during periods of subsidence. These periods of subsidence would be separated from each other by enormous intervals, during which the area would be either stationary or rising; whilst rising, each fossiliferous formation would be destroyed, almost as soon as accumulated, by the incessant coast-action, as we now see on the shores of South America. During the periods of subsidence there would probably be much extinction of life; during the periods of elevation, there would be much variation, but the geological record would then be least perfect.

It may be doubted whether the duration of any one great period of subsidence over the whole or part of the archipelago, together with a contemporaneous accumulation of sediment, would *exceed* the average duration of the same specific forms; and these contingencies are indispensable for the preservation of all the transitional gradations between any two or more species. If such gradations were not fully preserved, transitional varieties would merely appear as so many distinct species. It is, also, probable that each great period of subsidence would be interrupted by oscillations of level, and that slight climatal changes would intervene during such lengthy periods; and in these cases the inhabitants of the archipelago would have to migrate, and no closely consecutive record of their modifications could be preserved in any one formation.

Very many of the marine inhabitants of the archipelago now range thousands of miles beyond its confines; and analogy leads me to believe that it would be chiefly these far-ranging species which would oftenest produce new varieties; and the varieties would at first generally be local or confined to one place, but if possessed of any decided advantage, or when further modified and improved, they would slowly spread and supplant their parent-forms. When such varieties returned to their ancient homes, as they would differ from their former state, in a nearly uniform, though perhaps extremely slight degree, they would, according to the principles followed by many palæontologists, be ranked as new and distinct species.

If then, there be some degree of truth in these remarks, we have no right to expect to find in our geological formations, an infinite number of those fine transitional forms, which on my theory assuredly have connected all the past and present species of the same group into one long and branching chain of life. We ought only to look for a few links, some more closely, some more distantly related to each other; and these links, let them be ever so close, if found in different stages of the same formation, would, by most palæontologists, be ranked as distinct species. But I do not pretend that I should ever have suspected how poor a record of the mutations of life, the best preserved geological section presented, had not the difficulty of our not discovering innumerable transitional links between the species which appeared at the commencement and close of each formation, pressed so hardly on my theory.

On the sudden appearance of whole groups of Allied Species.—The abrupt manner in which whole groups of species suddenly appear in certain formations, has been urged by several palæontologists—for instance, by Agassiz, Pictet, and by none more forcibly than by Professor Sedgwick—as a fatal objection to the belief in the transmutation of species. If numerous species, belonging to the same genera or families, have really started into life all at once, the fact would be fatal to the theory of descent with slow modification through natural selection.

For the development of a group of forms, all of which have descended from some one progenitor, must have been an extremely slow process; and the progenitors must have lived long ages before their modified descendants. But we continually overrate the perfection of the geological record, and falsely infer, because certain genera or families have not been found beneath a certain stage, that they did not exist before that stage. We continually forget how large the world is, compared with the area over which our geological formations have been carefully examined; we forget that groups of species may elsewhere have long existed and have slowly multiplied before they invaded the ancient archipelagoes of Europe and of the United States. We do not make due allowance for the enormous intervals of time, which have probably elapsed between our consecutive formations,—longer perhaps in most cases than the time required for the accumulation of each formation. These intervals will have given time for the multiplication of species from some one or some few parent-forms; and in the succeeding formation such species will appear as if suddenly created.

I may here recall a remark formerly made, namely that it might require a long succession of ages to adapt an organism to some new and peculiar line of life, for instance to fly through the air; but that when this had been effected, and a few species had thus acquired a great advantage over other organisms, a comparatively short time would be necessary to produce many divergent forms, which would be able to spread rapidly and widely throughout the world.

I will now give a few examples to illustrate these remarks, and to show how liable we are to error in supposing that whole groups of species have suddenly been produced. I may recall the well-known fact that in geological treatises, published not many years ago, the great class of mammals was always spoken of as having abruptly come in at the commencement of the tertiary series. And now one of the richest known accumulations of fossil mammals, for its thickness, belongs to the middle of the secondary series; and one true mammal has been discovered in the new red sandstone at nearly the commencement of this great series. Cuvier used to urge that no monkey occurred in any tertiary stratum; but now extinct species have been discovered in India, South America, and in Europe even as far back as the eocene stage. Had it not been for the rare accident of the preservation of footsteps in the new red sandstone of the United States, who would have ventured to suppose that, besides reptiles, no less than at least thirty kinds of birds, some of gigantic size, existed during that period? Not a fragment of bone has been discovered in these beds. Notwithstanding that the number of joints shown in the fossil impressions correspond with the number in the several toes of living birds' feet, some authors doubt whether the animals which left the impressions were really birds. Until quite recently these authors might have maintained, and some have maintained, that the whole class of birds came suddenly into existence during an early tertiary period; but now we know, on the authority of Professor Owen (as may be seen in Lyell's *Manual*), that a bird certainly lived during the deposition of the upper greensand.

I may give another instance, which from having passed under my own eyes has much struck me. In a memoir on Fossil Sessile Cirripedes, I have stated that, from the number of existing and extinct tertiary species; from the extraordinary abundance of the individuals of many species all over the world, from the Arctic regions to the equator, inhabiting various zones of depths from the upper tidal limits to 50 fathoms; from the perfect manner in which specimens are preserved in the oldest tertiary beds; from the ease with which even a fragment of a valve can be recognised; from all these circumstances, I inferred that had sessile cirripedes existed during the secondary periods, they would certainly have been preserved and discovered; and as not one species had then been discovered in beds of this age, I concluded that this great group had been suddenly developed at the commencement of the tertiary series. This was a sore trouble to me, adding as I thought one more instance of the abrupt appearance of a great group of species. But my work had hardly been published, when a skilful palæontologist, M. Bosquet, sent me a drawing of a perfect specimen of an unmistakable sessile cirripede, which he had himself extracted from the chalk of Belgium. And, as if to make the case as striking as possible, this sessile cirripede was a Chthamalus, a very common, large, and ubiquitous genus, of which not one specimen has as yet been found even in any tertiary stratum. Hence we now positively know that sessile cirripedes existed during the secondary period; and these cirripedes might have been the progenitors of our many tertiary and existing species.

The case most frequently insisted on by palæontologists of the apparently sudden appearance of a whole group of species, is that of the teleostean fishes, low down in the Chalk period. This group includes the large majority of existing species. Lately, Professor Pictet has carried their existence one sub-stage further back; and some palæontologists believe that certain much older fishes, of which the affinities are as yet imperfectly known, are really teleostean. Assuming, however, that the whole of them did appear, as Agassiz believes, at the commencement of the chalk formation, the fact would certainly be highly remarkable; but I cannot see that it would be an insuperable difficulty on my theory, unless it could likewise be shown that the species of this group appeared suddenly and simultaneously throughout the world at this same period. It is almost superfluous to remark that hardly any fossil-fish are known from south of the equator; and by running through Pictet's Palæontology it will be seen that very few species are known from several formations in Europe. Some few families of fish now have a confined range; the teleostean fish might formerly have had a similarly confined range, and after having been largely developed in some one sea, might have spread widely. Nor have we any right to suppose that the seas of the world have always been so freely open from south to north as they are at present. Even at this day, if the Malay Archipelago were converted into land, the tropical parts of the Indian Ocean would form a large and perfectly enclosed basin, in which any great group of marine animals might be multiplied; and here they would remain confined, until some of the species became adapted to a cooler climate, and were enabled to double the southern capes of Africa or Australia, and thus reach other and distant seas.

From these and similar considerations, but chiefly from our ignorance of the geology of other countries beyond the confines of Europe and the United States; and from the revolution in our palæontological ideas on many points, which the discoveries of even the last dozen years have effected, it seems to me to be about as rash in us to dogmatise on the succession of organic beings throughout the world, as it would be for a naturalist to land for five minutes on some one barren point in Australia, and then to discuss the number and range of its productions.

On the sudden appearance of groups of Allied Species in the lowest known fossiliferous strata.—There is another and allied difficulty, which is much graver. I allude to the manner in which numbers of species of the same group, suddenly appear in the lowest known fossiliferous rocks. Most of the arguments which have convinced me that all the existing species of the same group have descended from one progenitor, apply with nearly equal force to the earliest known species. For instance, I cannot doubt that all the Silurian trilobites have descended from some one crustacean, which must have lived long before the Silurian age, and which probably differed greatly from any known animal. Some of the most ancient Silurian animals, as the Nautilus, Lingula, etc., do not differ much from living species; and it cannot on my theory be supposed, that these old species were the progenitors of all the species of the orders to which they belong, for they do not present characters in any degree intermediate between them. If, moreover, they had been the progenitors of these orders, they would almost certainly have been long ago supplanted and exterminated by their numerous and improved descendants.

Consequently, if my theory be true, it is indisputable that before the lowest Silurian stratum was deposited, long periods elapsed, as long as, or probably far longer than, the whole interval from the Silurian age to the present day; and that during these vast, yet quite unknown, periods of time, the world swarmed with living creatures.

To the question why we do not find records of these vast primordial periods, I can give no satisfactory answer. Several of the most eminent geologists, with Sir R. Murchison at their head, are convinced that we see in the organic remains of the lowest Silurian stratum the dawn of life on this planet. Other highly competent judges, as Lyell and the late E. Forbes, dispute this conclusion. We should not forget that only a small portion of the world is known with accuracy. M. Barrande has lately added another and lower stage to the Silurian system, abounding with new and peculiar species. Traces of life have been detected in the Longmynd beds, beneath Barrande's so-called primordial zone. The presence of phosphatic nodules and bituminous matter in some of the lowest azoic rocks, probably indicates the former existence of life at these periods. But the difficulty of understanding the absence of vast piles of fossiliferous strata, which on my theory no doubt were somewhere accumulated before the Silurian epoch, is very great. If these most ancient beds had been wholly worn away by denudation, or obliterated by metamorphic action, we ought to find only small remnants of the formations next succeeding them in age, and these ought to be very generally in a metamorphosed condition. But the descriptions which we now possess of the Silurian deposits over immense territories in Russia and in North America, do not support the view, that the older a formation is, the more it has always suffered the extremity of denudation and metamorphism.

The case at present must remain inexplicable; and may be truly urged as a valid argument against the views here entertained. To show that it may hereafter receive some explanation, I will give the following hypothesis. From the nature of the organic remains which do not appear to have inhabited profound depths, in the several formations of Europe and of the United States; and from the amount of sediment, miles in thickness, of which the formations are composed, we may infer that from first to last large islands or tracts of land, whence the sediment was derived, occurred in the neighbourhood of the existing continents of Europe and North America. But we do not know what was the state of things in the intervals between the successive formations; whether Europe and the United States during these intervals existed as dry land, or as a submarine surface near land, on which sediment was not deposited, or as the bed of an open and unfathomable sea.

Looking to the existing oceans, which are thrice as extensive as the land, we see them studded with many islands; but not one oceanic island is as yet known to afford even a remnant of any palæozoic or secondary formation. Hence we may perhaps infer, that during the palæozoic and secondary periods, neither continents nor continental islands existed where our oceans now extend; for had they existed there, palæozoic and secondary formations would in all probability have been accumulated from sediment derived from their wear and tear; and would have been at least partially upheaved by the oscillations of level, which we may fairly conclude must have intervened during these enormously long periods. If then we may infer anything from these facts, we may infer that where our oceans now extend, oceans have extended from the remotest period of which we have any record; and on the other hand, that where continents now exist, large tracts of land have existed, subjected no doubt to great oscillations of level, since the earliest silurian period. The coloured map appended to my volume on Coral Reefs, led me to conclude that the great oceans are still mainly areas of subsidence, the great archipelagoes still areas of oscillations of level, and the continents areas of elevation. But have we any right to assume that things have thus remained from the beginning of this world? Our continents seem to have been formed by a preponderance, during many oscillations of level, of the force of elevation; but may not the areas of preponderant movement have changed in the lapse of ages? At a period immeasurably antecedent to the silurian epoch, continents may have existed where oceans are now spread out; and clear and open oceans may have existed where our continents now stand. Nor should we be justified in assuming that if, for instance, the bed of the Pacific Ocean were now converted into a continent, we should there find formations older than the silurian strata, supposing such to have been formerly deposited; for it might well happen that strata which had subsided some miles nearer to the centre of the earth, and which had been pressed on by an enormous weight of superincumbent water, might have undergone far more metamorphic action than strata which have always remained nearer to the surface. The immense areas in some parts of the world, for instance in South America, of bare metamorphic rocks, which must have been heated under great pressure, have always seemed to me to require some special explanation; and we may perhaps believe that we see in these large areas, the many formations long anterior to the silurian epoch in a completely metamorphosed condition.

The several difficulties here discussed, namely our not finding in the successive formations infinitely numerous transitional links between the many species which now exist or have existed; the sudden manner in which whole groups of species appear in our European formations; the almost entire absence, as at present known, of fossiliferous formations beneath the Silurian strata, are all undoubtedly of the gravest nature. We see this in the plainest manner by the fact that all the most eminent palæontologists, namely Cuvier, Agassiz, Barrande, Falconer, E. Forbes, etc., and all our greatest geologists, as Lyell, Murchison, Sedgwick, etc., have unanimously, often vehemently, maintained the immutability of species. But I have reason to believe that one great authority, Sir Charles Lyell, from further reflection entertains grave doubts on this subject. I feel how rash it is to differ from these authorities, to whom, with others, we owe all our knowledge. Those who think the natural geological record in any degree perfect, and who do not attach much weight to the facts and arguments of other kinds given in this volume, will undoubtedly at once reject my theory. For my part, following out Lyell's metaphor, I look at the natural geological record, as a history of the world imperfectly kept, and written in a changing dialect; of this history we possess the last volume alone, relating only to two or three countries. Of this volume, only here and there a short chapter has been preserved; and of each page, only here and there a few lines. Each word of the slowly-changing language, in which the history is supposed to be written, being more or less different in the interrupted succession of chapters, may represent the apparently abruptly changed forms of life, entombed in our consecutive, but widely separated, formations. On this view, the difficulties above discussed are greatly diminished, or even disappear.

CHAPTER X

ON THE GEOLOGICAL SUCCESSION OF ORGANIC BEINGS

On the slow and successive appearance of new species—On their different rates of change—Species once lost do not reappear—Groups of species follow the same general rules in their appearance and disappearance as do single species—On Extinction—On simultaneous changes in the forms of life throughout the world—On the affinities of extinct species to each other and to living species—On the state of development of ancient forms—On the succession of the same types within the same areas—Summary of preceding and present chapters.

LET us now see whether the several facts and rules relating to the geological succession of organic beings, better accord with the common view of the immutability of species, or with that of their slow and gradual modification, through descent and natural selection.

New species have appeared very slowly, one after another, both on the land and in the waters. Lyell has shown that it is hardly possible to resist the evidence on this head in the case of the several tertiary stages; and every year tends to fill up the blanks between them, and to make the percentage system of lost and new forms more gradual. In some of the most recent beds, though undoubtedly of high antiquity if measured by years, only one or two species are lost forms, and only one or two are new forms, having here appeared for the first time, either locally, or, as far as we know, on the face of the earth. If we may trust the observations of Philippi in Sicily, the successive changes in the marine inhabitants of that island have been many and most gradual. The secondary formations are more broken; but, as Bronn has remarked, neither the appearance nor disappearance of their many now extinct species has been simultaneous in each separate formation.

Species of different genera and classes have not changed at the same rate, or in the same degree. In the oldest tertiary beds a few living shells may still be found in the midst of a multitude of extinct forms. Falconer has given a striking instance of a similar fact, in an existing crocodile associated with many strange and lost mammals and reptiles in the sub-Himalayan deposits. The Silurian Lingula differs but little from the living species of this genus; whereas most of the other Silurian Molluscs and all the Crustaceans have changed greatly. The productions of the land seem to change at a quicker rate than those of the sea, of which a striking instance has lately been observed in Switzerland. There is some reason to believe that organisms, considered high in the scale of nature, change more quickly than those that are low: though there are exceptions to this rule. The amount of organic change, as Pictet has remarked, does not strictly correspond with the succession of our geological formations; so that between each two consecutive formations, the forms of life have seldom changed in exactly the same degree. Yet if we compare any but the most closely related formations, all the species will be found to have undergone some change. When a species has once disappeared from the face of the earth, we have reason to believe that the same identical form never reappears. The strongest apparent exception to this latter rule, is that of the so-called 'colonies' of M. Barrande, which intrude for a period in the midst of an older formation, and then allow the pre-existing fauna to reappear; but Lyell's explanation, namely, that it is a case of temporary migration from a distinct geographical province, seems to me satisfactory.

These several facts accord well with my theory. I believe in no fixed law of development, causing all the inhabitants of a country to change abruptly, or simultaneously, or to an equal degree. The process of modification must be extremely slow. The variability of each species is quite independent of that of all others. Whether such variability be taken advantage of by natural selection, and whether the variations be accumulated to a greater or lesser amount, thus causing a greater or lesser amount of modification in the varying species, depends on many complex contingencies,—on the variability being of a beneficial nature, on the power of intercrossing, on the rate of breeding, on the slowly changing physical conditions of the country, and more especially on the nature of the other inhabitants with which the varying species comes into competition. Hence it is by no means surprising that one species should retain the same identical form much longer than others; or, if changing, that it should change less. We see the same fact in geographical distribution; for instance, in the land-shells and coleopterous insects of Madeira having come to differ considerably from their nearest allies on the continent of Europe, whereas the marine shells and birds have remained unaltered. We can perhaps understand the apparently quicker rate of change in terrestrial and in more highly organised productions compared with marine and lower productions, by the more complex relations of the higher beings to their organic and inorganic conditions of life, as explained in a former chapter. When many of the inhabitants of a country have become modified and improved, we can understand, on the principle of competition, and on that of the many all-important relations of organism to organism, that any form which does not become in some degree modified and improved, will be liable to be exterminated. Hence we can see why all the species in the same region do at last, if we look to wide enough intervals of time, become modified; for those which do not change will become extinct.

In members of the same class the average amount of change, during long and equal periods of time, may, perhaps, be nearly the same; but as the accumulation of long-enduring fossiliferous formations depends on great masses of sediment having been deposited on areas whilst subsiding, our formations have been almost necessarily accumulated at wide and irregularly intermittent intervals; consequently the amount of organic change exhibited by the fossils embedded in consecutive formations is not equal. Each formation, on this view, does not mark a new and complete act of creation, but only an occasional scene, taken almost at hazard, in a slowly changing drama.

We can clearly understand why a species when once lost should never reappear, even if the very same conditions of life, organic and inorganic, should recur. For though the offspring of one species might be adapted (and no doubt this has occurred in innumerable instances) to fill the exact place of another species in the economy of nature, and thus supplant it; yet the two forms—the old and the new—would not be identically the same; for both would almost certainly inherit different characters from their distinct progenitors. For instance, it is just possible, if our fantail-pigeons were all destroyed, that fanciers, by striving during long ages for the same object, might make a new breed hardly distinguishable from our present fantail; but if the parent rock-pigeon were also destroyed, and in nature we have every reason to believe that the parent-form will generally be supplanted and exterminated by its improved offspring, it is quite incredible that a fantail, identical with the existing breed, could be raised from any other species of pigeon, or even from the other well-established races of the domestic pigeon, for the newly-formed fantail would be almost sure to inherit from its new progenitor some slight characteristic differences.

Groups of species, that is, genera and families, follow the same general rules in their appearance and disappearance as do single species, changing more or less quickly, and in a greater or lesser degree. A group does not reappear after it has once disappeared; or its existence, as long as it lasts, is continuous. I am aware that there are some apparent exceptions to this rule, but the exceptions are surprisingly few, so few that E. Forbes, Pictet, and Woodward (though all strongly opposed to such views as I maintain) admit its truth; and the rule strictly accords with my theory. For as all the species of the same group have descended from some one species, it is clear that as long as any species of the group have appeared in the long succession of ages, so long must its members have continuously existed, in order to have generated either new and modified or the same old and unmodified forms. Species of the genus Lingula, for instance, must have continuously existed by an unbroken succession of generations, from the lowest Silurian stratum to the present day.

We have seen in the last chapter that the species of a group sometimes falsely appear to have come in abruptly; and I have attempted to give an explanation of this fact, which if true would have been fatal to my views. But such cases are certainly exceptional; the general rule being a gradual increase in number, till the group reaches its maximum, and then, sooner or later, it gradually decreases. If the number of the species of a genus, or the number of the genera of a family, be represented by a vertical line of varying thickness, crossing the successive geological formations in which the species are found, the line will sometimes falsely appear to begin at its lower end, not in a sharp point, but abruptly; it then gradually thickens upwards, sometimes keeping for a space of equal thickness, and ultimately thins out in the upper beds, marking the decrease and final extinction of the species. This gradual increase in number of the species of a group is strictly conformable with my theory; as the species of the same genus, and the genera of the same family, can increase only slowly and progressively; for the process of modification and the production of a number of allied forms must be slow and gradual,—one species giving rise first to two or three varieties, these being slowly converted into species, which in their turn produce by equally slow steps other species, and so on, like the branching of a great tree from a single stem, till the group becomes large.

On Extinction.—We have as yet spoken only incidentally of the disappearance of species and of groups of species. On the theory of natural selection the extinction of old forms and the production of new and improved forms are intimately connected together. The old notion of all the inhabitants of the earth having been swept away at successive periods by catastrophes, is very generally given up, even by those geologists, as Elie de Beaumont, Murchison, Barrande, etc., whose general views would naturally lead them to this conclusion. On the contrary, we have every reason to believe, from the study of the tertiary formations, that species and groups of species gradually disappear, one after another, first from one spot, then from another, and finally from the world. Both single species and whole groups of species last for very unequal periods; some groups, as we have seen, having endured from the earliest known dawn of life to the present day; some having disappeared before the close of the palæozoic period. No fixed law seems to determine the length of time during which any single species or any single genus endures. There is reason to believe that the complete extinction of the species of a group is generally a slower process than their production: if the appearance and disappearance of a group of species be represented, as before, by a vertical line of varying thickness, the line is found to taper more gradually at its upper end, which marks the progress of extermination, than at its lower end, which marks the first appearance and increase in numbers of the species. In some cases, however, the extermination of whole groups of beings, as of ammonites towards the close of the secondary period, has been wonderfully sudden.

The whole subject of the extinction of species has been involved in the most gratuitous mystery. Some authors have even supposed that as the individual has a definite length of life, so have species a definite duration. No one I think can have marvelled more at the extinction of species, than I have done. When I found in La Plata the tooth of a horse embedded with the remains of Mastodon, Megatherium, Toxodon, and other extinct monsters, which all co-existed with still living shells at a very late geological period, I was filled with astonishment; for seeing that the horse, since its introduction by the Spaniards into South America, has run wild over the whole country and has increased in numbers at an unparalleled rate, I asked myself what could so recently have exterminated the former horse under conditions of life apparently so favourable. But how utterly groundless was my astonishment! Professor Owen soon perceived that the tooth, though so like that of the existing horse, belonged to an extinct species. Had this horse been still living, but in some degree rare, no naturalist would have felt the least surprise at its rarity; for rarity is the attribute of a vast number of species of all classes, in all countries. If we ask ourselves why this or that species is rare, we answer that something is unfavourable in its conditions of life; but what that something is, we can hardly ever tell. On the supposition of the fossil horse still existing as a rare species, we might have felt certain from the analogy of all other mammals, even of the slow-breeding elephant, and from the history of the naturalisation of the domestic horse in South America, that under more favourable conditions it would in a very few years have stocked the whole continent. But we could not have told what the unfavourable conditions were which checked its increase, whether some one or several contingencies, and at what period of the horse's life, and in what degree, they severally acted. If the conditions had gone on, however slowly, becoming less and less favourable, we assuredly should not have perceived the fact, yet the fossil horse would certainly have become rarer and rarer, and finally extinct;—its place being seized on by some more successful competitor.

It is most difficult always to remember that the increase of every living being is constantly being checked by unperceived injurious agencies; and that these same unperceived agencies are amply sufficient to cause rarity, and finally extinction. We see in many cases in the more recent tertiary formations, that rarity precedes extinction; and we know that this has been the progress of events with those animals which have been exterminated, either locally or wholly, through man's agency. I may repeat what I published in 1845, namely, that to admit that species generally become rare before they become extinct—to feel no surprise at the rarity of a species, and yet to marvel greatly when it ceases to exist, is much the same as to admit that sickness in the individual is the forerunner of death—to feel no surprise at sickness, but when the sick man dies, to wonder and to suspect that he died by some unknown deed of violence.

The theory of natural selection is grounded on the belief that each new variety, and ultimately each new species, is produced and maintained by having some advantage over those with which it comes into competition; and the consequent extinction of less-favoured forms almost inevitably follows. It is the same with our domestic productions: when a new and slightly improved variety has been raised, it at first supplants the less improved varieties in the same neighbourhood; when much improved it is transported far and near, like our short-horn cattle, and takes the place of other breeds in other countries. Thus the appearance of new forms and the disappearance of old forms, both natural and artificial, are bound together. In certain flourishing groups, the number of new specific forms which have been produced within a given time is probably greater than that of the old specific forms which have been exterminated; but we know that the number of species has not gone on indefinitely increasing, at least during the later geological periods, so that looking to later times we may believe that the production of new forms has caused the extinction of about the same number of old forms.

The competition will generally be most severe, as formerly explained and illustrated by examples, between the forms which are most like each other in all respects. Hence the improved and modified descendants of a species will generally cause the extermination of the parent-species; and if many new forms have been developed from any one species, the nearest allies of that species, *i.e.* the species of the same genus, will be the most liable to extermination. Thus, as I believe, a number of new species descended from one species, that is a new genus, comes to supplant an old genus, belonging to the same family. But it must often have happened that a new species belonging to some one group will have seized on the place occupied by a species belonging to a distinct group, and thus caused its extermination; and if many allied forms be developed from the successful intruder, many will have to yield their places; and it will generally be allied forms, which will suffer from some inherited inferiority in common. But whether it be species belonging to the same or to a distinct class, which yield their places to other species which have been modified and improved, a few of the sufferers may often long be preserved, from being fitted to some peculiar line of life, or from inhabiting some distant and isolated station, where they have escaped severe competition. For instance, a single species of Trigonia, a great genus of shells in the secondary formations, survives in the Australian seas; and a few members of the great and almost extinct group of Ganoid fishes still inhabit our fresh waters. Therefore the utter extinction of a group is generally, as we have seen, a slower process than its production.

With respect to the apparently sudden extermination of whole families or orders, as of Trilobites at the close of the palæozoic period and of Ammonites at the close of the secondary period, we must remember what has been already said on the probable wide intervals of time between our consecutive formations; and in these intervals there may have been much slow extermination. Moreover, when by sudden immigration or by unusually rapid development, many species of a new group have taken possession of a new area, they will have exterminated in a correspondingly rapid manner many of the old inhabitants; and the forms which thus yield their places will commonly be allied, for they will partake of some inferiority in common.

Thus, as it seems to me, the manner in which single species and whole groups of species become extinct, accords well with the theory of natural selection. We need not marvel at extinction; if we must marvel, let it be at our presumption in imagining for a moment that we understand the many complex contingencies, on which the existence of each species depends. If we forget for an instant, that each species tends to increase inordinately, and that some check is always in action, yet seldom perceived by us, the whole economy of nature will be utterly obscured. Whenever we can precisely say why this species is more abundant in individuals than that; why this species and not another can be naturalised in a given country; then, and not till then, we may justly feel surprised why we cannot account for the extinction of this particular species or group of species.

On the Forms of Life changing almost simultaneously throughout the World.—Scarcely any palæontological discovery is more striking than the fact, that the forms of life change almost simultaneously throughout the world. Thus our European Chalk formation can be recognised in many distant parts of the world, under the most different climates, where not a fragment of the mineral chalk itself can be found; namely, in North America, in equatorial South America, in Tierra del Fuego, at the Cape of Good Hope, and in the peninsula of India. For at these distant points, the organic remains in certain beds present an unmistakable degree of resemblance to those of the Chalk. It is not that the same species are met with; for in some cases not one species is identically the same, but they belong to the same families, genera, and sections of genera, and sometimes are similarly characterised in such trifling points as mere superficial sculpture. Moreover other forms, which are not found in the Chalk of Europe, but which occur in the formations either above or below, are similarly absent at these distant points of the world. In the several successive palæozoic formations of Russia, Western Europe and North America, a similar parallelism in the forms of life has been observed by several authors: so it is, according to Lyell, with the several European and North American tertiary deposits. Even if the few fossil species which are common to the Old and New Worlds be kept wholly out of view, the general parallelism in the successive forms of life, in the stages of the widely separated palæozoic and tertiary periods, would still be manifest, and the several formations could be easily correlated.

These observations, however, relate to the marine inhabitants of distant parts of the world: we have not sufficient data to judge whether the productions of the land and of fresh water change at distant points in the same parallel manner. We may doubt whether they have thus changed: if the Megatherium, Mylodon, Macrauchenia, and Toxodon had been brought to Europe from La Plata, without any information in regard to their geological position, no one would have suspected that they had co-existed with still living sea-shells; but as these anomalous monsters co-existed with the Mastodon and Horse, it might at least have been inferred that they had lived during one of the later tertiary stages.

When the marine forms of life are spoken of as having changed simultaneously throughout the world, it must not be supposed that this expression relates to the same thousandth or hundred-thousandth year, or even that it has a very strict geological sense; for if all the marine animals which live at the present day in Europe, and all those that lived in Europe during the pleistocene period (an enormously remote period as measured by years, including the whole glacial epoch), were to be compared with those now living in South America or in Australia, the most skilful naturalist would hardly be able to say whether the existing or the pleistocene inhabitants of Europe resembled most closely those of the southern hemisphere. So, again, several highly competent observers believe that the existing productions of the United States are more closely related to those which lived in Europe during certain later tertiary stages, than to those which now live here; and if this be so, it is evident that fossiliferous beds deposited at the present day on the shores of North America would hereafter be liable to be classed with somewhat older European beds. Nevertheless, looking to a remotely future epoch, there can, I think, be little doubt that all the more modern *marine* formations, namely, the upper pliocene, the pleistocene and strictly modern beds, of Europe, North and South America, and Australia, from containing fossil remains in some degree allied, and from not including those forms which are only found in the older underlying deposits, would be correctly ranked as simultaneous in a geological sense.

The fact of the forms of life changing simultaneously, in the above large sense, at distant parts of the world, has greatly struck those admirable observers, MM. de Verneuil and d'Archiac. After referring to the parallelism of the palæozoic forms of life in various parts of Europe, they add, 'If struck by this strange sequence, we turn our attention to North America, and there discover a series of analogous phenomena, it will appear certain that all these modifications of species, their extinction, and the introduction of new ones, cannot be owing to mere changes in marine currents or other causes more or less local and temporary, but depend on general laws which govern the whole animal kingdom.' M. Barrande has made forcible remarks to precisely the same effect. It is, indeed, quite futile to look to changes of currents, climate, or other physical conditions, as the cause of these great mutations in the forms of life throughout the world, under the most different climates. We must, as Barrande has remarked, look to some special law. We shall see this more clearly when we treat of the present distribution of organic beings, and find how slight is the relation between the physical conditions of various countries, and the nature of their inhabitants.

This great fact of the parallel succession of the forms of life throughout the world, is explicable on the theory of natural selection. New species are formed by new varieties arising, which have some advantage over older forms; and those forms, which are already dominant, or have some advantage over the other forms in their own country, would naturally oftenest give rise to new varieties or incipient species; for these latter must be victorious in a still higher degree in order to be preserved and to survive. We have distinct evidence on this head, in the plants which are dominant, that is, which are commonest in their own homes, and are most widely diffused, having produced the greatest number of new varieties. It is also natural that the dominant, varying, and far-spreading species, which already have invaded to a certain extent the territories of other species, should be those which would have the best chance of spreading still further, and of giving rise in new countries to new varieties and species. The process of diffusion may often be very slow, being dependent on climatal and geographical changes, or on strange accidents, but in the long run the dominant forms will generally succeed in spreading. The diffusion would, it is probable, be slower with the terrestrial inhabitants of distinct continents than with the marine inhabitants of the continuous sea. We might therefore expect to find, as we apparently do find, a less strict degree of parallel succession in the productions of the land than of the sea.

Dominant species spreading from any region might encounter still more dominant species, and then their triumphant course, or even their existence, would cease. We know not at all precisely what are all the conditions most favourable for the multiplication of new and dominant species; but we can, I think, clearly see that a number of individuals, from giving a better chance of the appearance of favourable variations, and that severe competition with many already existing forms, would be highly favourable, as would be the power of spreading into new territories. A certain amount of isolation, recurring at long intervals of time, would probably be also favourable, as before explained. One quarter of the world may have been most favourable for the production of new and dominant species on the land, and another for those in the waters of the sea. If two great regions had been for a long period favourably circumstanced in an equal degree, whenever their inhabitants met, the battle would be prolonged and severe; and some from one birthplace and some from the other might be victorious. But in the course of time, the forms dominant in the highest degree, wherever produced, would tend everywhere to prevail. As they prevailed, they would cause the extinction of other and inferior forms; and as these inferior forms would be allied in groups by inheritance, whole groups would tend slowly to disappear; though here and there a single member might long be enabled to survive.

Thus, as it seems to me, the parallel, and, taken in a large sense, simultaneous, succession of the same forms of life throughout the world, accords well with the principle of new species having been formed by dominant species spreading widely and varying; the new species thus produced being themselves dominant owing to inheritance, and to having already had some advantage over their parents or over other species; these again spreading, varying, and producing new species. The forms which are beaten and which yield their places to the new and victorious forms, will generally be allied in groups, from inheriting some inferiority in common; and therefore as new and improved groups spread throughout the world, old groups will disappear from the world; and the succession of forms in both ways will everywhere tend to correspond.

There is one other remark connected with this subject worth making. I have given my reasons for believing that all our greater fossiliferous formations were deposited during periods of subsidence; and that blank intervals of vast duration occurred during the periods when the bed of the sea was either stationary or rising, and likewise when sediment was not thrown down quickly enough to embed and preserve organic remains. During these long and blank intervals I suppose that the inhabitants of each region underwent a considerable amount of modification and extinction, and that there was much migration from other parts of the world. As we have reason to believe that large areas are affected by the same movement, it is probable that strictly contemporaneous formations have often been accumulated over very wide spaces in the same quarter of the world; but we are far from having any right to conclude that this has invariably been the case, and that large areas have invariably been affected by the same movements. When two formations have been deposited in two regions during nearly, but not exactly the same period, we should find in both, from the causes explained in the foregoing paragraphs, the same general succession in the forms of life; but the species would not exactly correspond; for there will have been a little more time in the one region than in the other for modification, extinction, and immigration.

I suspect that cases of this nature occur in Europe. Mr Prestwich, in his admirable Memoirs on the eocene deposits of England and France, is able to draw a close general parallelism between the successive stages in the two countries; but when he compares certain stages in England with those in France, although he finds in both a curious accordance in the numbers of the species belonging to the same genera, yet the species themselves differ in a manner very difficult to account for, considering the proximity of the two areas,—unless, indeed, it be assumed that an isthmus separated two seas inhabited by distinct, but contemporaneous, faunas. Lyell has made similar observations on some of the later tertiary formations. Barrande, also, shows that there is a striking general parallelism in the successive Silurian deposits of Bohemia and Scandinavia; nevertheless he finds a surprising amount of difference in the species. If the several formations in these regions have not been deposited during the same exact periods,—a formation in one region often corresponding with a blank interval in the other,—and if in both regions the species have gone on slowly changing during the accumulation of the several formations and during the long intervals of time between them; in this case, the several formations in the two regions could be arranged in the same order, in accordance with the general succession of the form of life, and the order would falsely appear to be strictly parallel; nevertheless the species would not all be the same in the apparently corresponding stages in the two regions.

On the Affinities of extinct Species to each other, and to living forms.—Let us now look to the mutual affinities of extinct and living species. They all fall into one grand natural system; and this fact is at once explained on the principle of descent. The more ancient any form is, the more, as a general rule, it differs from living forms. But, as Buckland long ago remarked, all fossils can be classed either in still existing groups, or between them. That the extinct forms of life help to fill up the wide intervals between existing genera, families, and orders, cannot be disputed. For if we confine our attention either to the living or to the extinct alone, the series is far less perfect than if we combine both into one general system. With respect to the Vertebrata, whole pages could be filled with striking illustrations from our great palæontologist, Owen, showing how extinct animals fall in between existing groups. Cuvier ranked the Ruminants and Pachyderms, as the two most distinct orders of mammals; but Owen has discovered so many fossil links, that he has had to alter the whole classification of these two orders; and has placed certain pachyderms in the same sub-order with ruminants: for example, he dissolves by fine gradations the apparently wide difference between the pig and the camel. In regard to the Invertebrata, Barrande, and a higher authority could not be named, asserts that he is every day taught that Palæozoic animals, though belonging to the same orders, families, or genera with those living at the present day, were not at this early epoch limited in such distinct groups as they now are.

Some writers have objected to any extinct species or group of species being considered as intermediate between living species or groups. If by this term it is meant that an extinct form is directly intermediate in all its characters between two living forms, the objection is probably valid. But I apprehend that in a perfectly natural classification many fossil species would have to stand between living species, and some extinct genera between living general, even between genera belonging to distinct families. The most common case, especially with respect to very distinct groups, such as fish and reptiles, seems to be, that supposing them to be distinguished at the present day from each other by a dozen characters, the ancient members of the same two groups would be distinguished by a somewhat lesser number of characters, so that the two groups, though formerly quite distinct, at that period made some small approach to each other.

It is a common belief that the more ancient a form is, by so much the more it tends to connect by some of its characters groups now widely separated from each other. This remark no doubt must be restricted to those groups which have undergone much change in the course of geological ages; and it would be difficult to prove the truth of the proposition, for every now and then even a living animal, as the Lepidosiren, is discovered having affinities directed towards very distinct groups. Yet if we compare the older Reptiles and Batrachians, the older Fish, the older Cephalopods, and the eocene Mammals, with the more recent members of the same classes, we must admit that there is some truth in the remark.

Let us see how far these several facts and inferences accord with the theory of descent with modification. As the subject is somewhat complex, I must request the reader to turn to the diagram in the preliminary. We may suppose that the numbered letters represent genera, and the dotted lines diverging from them the species in each genus. The diagram is much too simple, too few genera and too few species being given, but this is unimportant for us. The horizontal lines may represent successive geological formations, and all the forms beneath the uppermost line may be considered as extinct. The three existing genera, a^{14}, q^{14}, p^{14}, will form a small family; b^{14} and f^{14} a closely allied family or sub-family; and o^{14}, e^{14}, m^{14}, a third family. These three families, together with the many extinct genera on the several lines of descent diverging from the parent-form (A), will form an order; for all will have inherited something in common from their ancient and common progenitor. On the principle of the continued tendency to divergence of character, which was formerly illustrated by this diagram, the more recent any form is, the more it will generally differ from its ancient progenitor. Hence we can understand the rule that the most ancient fossils differ most from existing forms. We must not, however, assume that divergence of character is a necessary contingency; it depends solely on the descendants from a species being thus enabled to seize on many and different places in the economy of nature. Therefore it is quite possible, as we have seen in the case of some Silurian forms, that a species might go on being slightly modified in relation to its slightly altered conditions of life, and yet retain throughout a vast period the same general characteristics. This is represented in the diagram by the letter f^{14}.

All the many forms, extinct and recent, descended from (A), make, as before remarked, one order; and this order, from the continued effects of extinction and divergence of character, has become divided into several sub-families and families, some of which are supposed to have perished at different periods, and some to have endured to the present day.

By looking at the diagram we can see that if many of the extinct forms, supposed to be embedded in the successive formations, were discovered at several points low down in the series, the three existing families on the uppermost line would be rendered less distinct from each other. If, for instance, the genera a^1, a^5, a^{10}, f^8, m^3, m^6, m^9, were disinterred, these three families would be so closely linked together that they probably would have to be united into one great family, in nearly the same manner as has occurred with ruminants and pachyderms. Yet he who objected to call the extinct genera, which thus linked the living genera of three families together, intermediate in character, would be justified, as they are intermediate, not directly, but only by a long and circuitous course through many widely different forms. If many extinct forms were to be discovered above one of the middle horizontal lines or geological formations—for instance, above No. VI.—but none from beneath this line, then only the two families on the left hand (namely, a^{14}, etc., and b^{14}, etc.) would have to be united into one family; and the two other families (namely, a^{14} to f^{14} now including five genera, and o^{14} to m^{14}) would yet remain distinct. These two families, however, would be less distinct from each other than they were before the discovery of the fossils. If, for instance, we suppose the existing genera of the two families to differ from each other by a dozen characters, in this case the genera, at the early period marked VI., would differ by a lesser number of characters; for at this early stage of descent they have not diverged in character from the common progenitor of the order, nearly so much as they subsequently diverged. Thus it comes that ancient and extinct genera are often in some slight degree intermediate in character between their modified descendants, or between their collateral relations.

In nature the case will be far more complicated than is represented in the diagram; for the groups will have been more numerous, they will have endured for extremely unequal lengths of time, and will have been modified in various degrees. As we possess only the last volume of the geological record, and that in a very broken condition, we have no right to expect, except in very rare cases, to fill up wide intervals in the natural system, and thus unite distinct families or orders. All that we have a right to expect, is that those groups, which have within known geological periods undergone much modification, should in the older formations make some slight approach to each other; so that the older members should differ less from each other in some of their characters than do the existing members of the same groups; and this by the concurrent evidence of our best palæontologists seems frequently to be the case.

Thus, on the theory of descent with modification, the main facts with respect to the mutual affinities of the extinct forms of life to each other and to living forms, seem to me explained in a satisfactory manner. And they are wholly inexplicable on any other view.

On this same theory, it is evident that the fauna of any great period in the earth's history will be intermediate in general character between that which preceded and that which succeeded it. Thus, the species which lived at the sixth great stage of descent in the diagram are the modified offspring of those which lived at the fifth stage, and are the parents of those which became still more modified at the seventh stage; hence they could hardly fail to be nearly intermediate in character between the forms of life above and below. We must, however, allow for the entire extinction of some preceding forms, and in any one region for the immigration of new forms from other regions, and for a large amount of modification, during the long and blank intervals between the successive formations. Subject to these allowances, the fauna of each geological period undoubtedly is intermediate in character, between the preceding and succeeding faunas. I need give only one instance, namely, the manner in which the fossils of the Devonian system, when this system was first discovered, were at once recognised by palæontologists as intermediate in character between those of the overlying carboniferous, and underlying

Silurian system. But each fauna is not necessarily exactly intermediate, as unequal intervals of time have elapsed between consecutive formations.

It is no real objection to the truth of the statement, that the fauna of each period as a whole is nearly intermediate in character between the preceding and succeeding faunas, that certain genera offer exceptions to the rule. For instance, mastodons and elephants, when arranged by Dr Falconer in two series, first according to their mutual affinities and then according to their periods of existence, do not accord in arrangement. The species extreme in character are not the oldest, or the most recent; nor are those which are intermediate in character, intermediate in age. But supposing for an instant, in this and other such cases, that the record of the first appearance and disappearance of the species was perfect, we have no reason to believe that forms successively produced necessarily endure for corresponding lengths of time: a very ancient form might occasionally last much longer than a form elsewhere subsequently produced, especially in the case of terrestrial productions inhabiting separated districts. To compare small things with great: if the principal living and extinct races of the domestic pigeon were arranged as well as they could be in serial affinity, this arrangement would not closely accord with the order in time of their production, and still less with the order of their disappearance; for the parent rock-pigeon now lives; and many varieties between the rock-pigeon and the carrier have become extinct; and carriers which are extreme in the important character of length of beak originated earlier than short-beaked tumblers, which are at the opposite end of the series in this same respect.

Closely connected with the statement, that the organic remains from an intermediate formation are in some degree intermediate in character, is the fact, insisted on by all palæontologists, that fossils from two consecutive formations are far more closely related to each other, than are the fossils from two remote formations. Pictet gives as a well-known instance, the general resemblance of the organic remains from the several stages of the Chalk formation, though the species are distinct in each stage. This fact alone, from its generality, seems to have shaken Professor Pictet in his firm belief in the immutability of species. He who is acquainted with the distribution of existing species over the globe, will not attempt to account for the close resemblance of the distinct species in closely consecutive formations, by the physical conditions of the ancient areas having remained nearly the same. Let it be remembered that the forms of life, at least those inhabiting the sea, have changed almost simultaneously throughout the world, and therefore under the most different climates and conditions. Consider the prodigious vicissitudes of climate during the pleistocene period, which includes the whole glacial period, and note how little the specific forms of the inhabitants of the sea have been affected.

On the theory of descent, the full meaning of the fact of fossil remains from closely consecutive formations, though ranked as distinct species, being closely related, is obvious. As the accumulation of each formation has often been interrupted, and as long blank intervals have intervened between successive formations, we ought not to expect to find, as I attempted to show in the last chapter, in any one or two formations all the intermediate varieties between the species which appeared at the commencement and close of these periods; but we ought to find after intervals, very long as measured by years, but only moderately long as measured geologically, closely allied forms, or, as they have been called by some authors, representative species; and these we assuredly do find. We find, in short, such evidence of the slow and scarcely sensible mutation of specific forms, as we have a just right to expect to find.

On the state of Development of Ancient Forms.—There has been much discussion whether recent forms are more highly developed than ancient. I will not here enter on this subject, for naturalists have not as yet defined to each other's satisfaction what is meant by high and low forms. The best definition probably is, that the higher forms have their organs more distinctly specialised for different functions; and as such division of physiological labour seems to be an advantage to each being, natural selection will constantly tend in so far to make the later and more modified forms higher than their early progenitors, or than the slightly modified descendants of such progenitors. In a more general sense the more recent forms must, on my theory, be higher than the more ancient; for each new species is formed by having had some advantage in the struggle for life over other and preceding forms. If under a nearly similar climate, the eocene inhabitants of one quarter of the world were put into competition with the existing inhabitants of the same or some other quarter, the eocene fauna or flora would certainly be beaten and exterminated; as would a secondary fauna by an eocene, and a palæozoic fauna by a secondary fauna. I do not doubt that this process of improvement has affected in a marked and sensible manner the organisation of the more recent and victorious forms of life, in comparison with the ancient and beaten forms; but I can see no way of testing this sort of progress. Crustaceans, for instance, not the highest in their own class, may have beaten the highest molluscs. From the extraordinary manner in which European productions have recently spread over New Zealand, and have seized on places which must have been previously occupied, we may believe, if all the animals and plants of Great Britain were set free in New Zealand, that in the course of time a multitude of British forms would become thoroughly naturalised there, and would exterminate many of the natives. On the other hand, from what we see now occurring in New Zealand, and from hardly a single inhabitant of the southern hemisphere having become wild in any part of Europe, we may doubt, if all the productions of New Zealand were set free in Great Britain, whether any considerable number would be enabled to seize on places now occupied by our native plants and animals. Under this point of view, the productions of Great Britain may be said to be higher than those of New Zealand. Yet the most skilful naturalist from an examination of the species of the two countries could not have foreseen this result.

Agassiz insists that ancient animals resemble to a certain extent the embryos of recent animals of the same classes; or that the geological succession of extinct forms is in some degree parallel to the embryological development of recent forms. I must follow Pictet and Huxley in thinking that the truth of this doctrine is very far from proved. Yet I fully expect to see it hereafter confirmed, at least in regard to subordinate groups, which have branched off from each other within comparatively recent times. For this doctrine of Agassiz accords well with the theory of natural selection. In a future chapter I shall attempt to show that the adult differs from its embryo, owing to variations supervening at a not early age, and being inherited at a corresponding age. This process, whilst it leaves the embryo almost unaltered, continually adds, in the course of successive generations, more and more difference to the adult.

Thus the embryo comes to be left as a sort of picture, preserved by nature, of the ancient and less modified condition of each animal. This view may be true, and yet it may never be capable of full proof. Seeing, for instance, that the oldest known mammals, reptiles, and fish strictly belong to their own proper classes, though some of these old forms are in a slight degree less distinct from each other than are the typical members of the same groups at the present day, it would be vain to look for animals having the common embryological character of the Vertebrata, until beds far beneath the lowest Silurian strata are discovered—a discovery of which the chance is very small.

On the Succession of the same Types within the same areas, during the later tertiary periods.—Mr Clift many years ago showed that the fossil mammals from the Australian caves were closely allied to the living marsupials of that continent. In South America, a similar relationship is manifest, even to an uneducated eye, in the gigantic pieces of armour like those of the armadillo, found in several parts of La Plata; and Professor Owen has shown in the most striking manner that most of the fossil mammals, buried there in such numbers, are related to South American types. This relationship is even more clearly seen in the wonderful collection of fossil bones made by MM. Lund and Clausen in the caves of Brazil. I was so much impressed with these facts that I strongly insisted, in 1839 and 1845, on this 'law of the succession of types',—on 'this wonderful relationship in the same continent between the dead and the living'.

Professor Owen has subsequently extended the same generalisation to the mammals of the Old World. We see the same law in this author's restorations of the extinct and gigantic birds of New Zealand. We see it also in the birds of the caves of Brazil. Mr Woodward has shown that the same law holds good with sea-shells, but from the wide distribution of most genera of milluscs, it is not well displayed by them. Other cases could be added, as the relation between the extinct and living land-shells of Madeira; and between the extinct and living brackish-water shells of the Aralo-Caspian Sea.

Now what does this remarkable law of the succession of the same types within the same areas mean? He would be a bold man, who after comparing the present climate of Australia and of parts of South America under the same latitude, would attempt to account, on the one hand, by dissimilar physical conditions for the dissimilarity of the inhabitants of these two continents, and, on the other hand, by similarity of conditions, for the uniformity of the same types in each during the later tertiary periods. Nor can it be pretended that it is an immutable law that marsupials should have been chiefly or solely produced in Australia; or that Edentata and other American types should have been solely produced in South America. For we know that Europe in ancient times was peopled by numerous marsupials; and I have shown in the publications above alluded to, that in America the law of distribution of terrestrial mammals was formerly different from what it now is. North America formerly partook strongly of the present character of the southern half of the continent; and the southern half was formerly more closely allied, than it is at present, to the northern half. In a similar manner we know from Falconer and Cautley's discoveries, that northern India was formerly more closely related in its mammals to Africa than it is at the present time. Analogous facts could be given in relation to the distribution of marine animals.

On the theory of descent with modification, the great law of the long enduring, but not immutable, succession of the same types within the same areas, is at once explained; for the inhabitants of each quarter of the world will obviously tend to leave in that quarter, during the next succeeding period of time, closely allied though in some degree modified descendants. If the inhabitants of one continent formerly differed greatly from those of another continent, so will their modified descendants still differ in nearly the same manner and degree. But after very long intervals of time and after great geographical changes, permitting much inter-migration, the feebler will yield to the more dominant forms, and there will be nothing immutable in the laws of past and present distribution.

It may be asked in ridicule, whether I suppose that the megatherium and other allied huge monsters have left behind them in South America, the sloth, armadillo, and anteater, as their degenerate descendants. This cannot for an instant be admitted. These huge animals have become wholly extinct, and have left no progeny. But in the caves of Brazil, there are many extinct species which are closely allied in size and in other characters to the species still living in South America; and some of these fossils may be the actual progenitors of living species. It must not be forgotten that, on my theory, all the species of the same genus have descended from some one species; so that if six genera, each having eight species, be found in one geological formation, and in the next succeeding formation there be six other allied or representative genera with the same number of species, then we may conclude that only one species of each of the six older genera has left modified descendants, constituting the six new genera. The other seven species of the old genera have all died out and have left no progeny. Or, which would probably be a far commoner case, two or three species of two or three alone of the six older genera will have been the parents of the six new genera; the other old species and the other whole old genera having become utterly extinct. In failing orders, with the genera and species decreasing in numbers, as apparently is the case of the Edentata of South America, still fewer genera and species will have left modified blood-descendants.

Summary of the preceding and present Chapters.—I have attempted to show that the geological record is extremely imperfect; that only a small portion of the globe has been geologically explored with care; that only certain classes of organic beings have been largely preserved in a fossil state; that the number both of specimens and of species, preserved in our museums, is absolutely as nothing compared with the incalculable number of generations which must have passed away even during a single formation; that, owing to subsidence being necessary for the accumulation of fossiliferous deposits thick enough to resist future degradation, enormous intervals of time have elapsed between the successive formations; that there has probably been more extinction during the periods of subsidence, and more variation during the periods of elevation, and during the latter the record will have been least perfectly kept; that each single formation has not been continuously deposited; that the duration of each formation is, perhaps, short compared with the average duration of specific forms; that migration has played an important part in the first appearance of new forms in any one area and formation; that widely ranging species are those which have varied most, and have oftenest given rise to new species; and that varieties have at first often been local. All these causes taken conjointly, must have tended to make the geological record extremely imperfect, and will to a large extent explain why we do not find interminable varieties, connecting together all the extinct and existing forms of life by the finest graduated steps.

He who rejects these views on the nature of the geological record, will rightly reject my whole theory. For he may ask in vain where are the numberless transitional links which must formerly have connected the closely allied or representative species, found in the several stages of the same great formation. He may disbelieve in the enormous intervals of time which have elapsed between our consecutive formations; he may overlook how important a part migration must have played, when the formations of any one great region alone, as that of Europe, are considered; he may urge the apparent, but often falsely apparent, sudden coming in of whole groups of species. He may ask where are the remains of those infinitely numerous organisms which must have existed long before the first bed of the Silurian system was deposited: I can answer this latter question only hypothetically, by saying that as far as we can see, where our oceans now extend they have for an enormous period extended, and where our oscillating continents now stand they have stood ever since the Silurian epoch; but that long before that period, the world may have presented a wholly different aspect; and that the older continents, formed of formations older than any known to us, may now all be in a metamorphosed condition, or may lie buried under the ocean.

Passing from these difficulties, all the other great leading facts in palæontology seem to me simply to follow on the theory of descent with modification through natural selection. We can thus understand how it is that new species come in slowly and successively; how species of different classes do not necessarily change together, or at the same rate, or in the same degree; yet in the long run that all undergo modification to some extent. The extinction of old forms is the almost inevitable consequence of the production of new forms. We can understand why when a species has once disappeared it never reappears. Groups of species increase in numbers slowly, and endure for unequal periods of time; for the process of modification is necessarily slow, and depends on many complex contingencies. The dominant species of the larger dominant groups tend to leave many modified descendants, and thus new sub-groups and groups are formed. As these are formed, the species of the less vigorous groups, from their inferiority inherited from a common progenitor, tend to become extinct together, and to leave no modified offspring on the face of the earth. But the utter extinction of a whole group of species may often be a very slow process, from the survival of a few descendants, lingering in protected and isolated situations. When a group has once wholly disappeared, it does not reappear; for the link of generation has been broken.

We can understand how the spreading of the dominant forms of life, which are those that oftenest vary, will in the long run tend to people the world with allied, but modified, descendants; and these will generally succeed in taking the places of those groups of species which are their inferiors in the struggle for existence. Hence, after long intervals of time, the productions of the world will appear to have changed simultaneously.

We can understand how it is that all the forms of life, ancient and recent, make together one grand system; for all are connected by generation. We can understand, from the continued tendency to divergence of character, why the more ancient a form is, the more it generally differs from those now living. Why ancient and extinct forms often tend to fill up gaps between existing forms, sometimes blending two groups previously classed as distinct into one; but more commonly only bringing them a little closer together. The more ancient a form is, the more often, apparently, it displays characters in some degree intermediate between groups now distinct; for the more ancient a form is, the more nearly it will be related to, and consequently resemble, the common progenitor of groups, since become widely divergent. Extinct forms are seldom directly intermediate between existing forms; but are intermediate only by a long and circuitous course through many extinct and very different forms. We can clearly see why the organic remains of closely consecutive formations are more closely allied to each other, than are those of remote formations; for the forms are more closely linked together by generation: we can clearly see why the remains of an intermediate formation are intermediate in character.

The inhabitants of each successive period in the world's history have beaten their predecessors in the race for life, and are, in so far, higher in the scale of nature; and this may account for that vague yet ill-defined sentiment, felt by many palæontologists, that organisation on the whole has progressed. If it should hereafter be proved that ancient animals resemble to a certain extent the embryos of more recent animals of the same class, the fact will be intelligible. The succession of the same types of structure within the same areas during the later geological periods ceases to be mysterious, and is simply explained by inheritance.

If then the geological record be as imperfect as I believe it to be, and it may at least be asserted that the record cannot be proved to be much more perfect, the main objections to the theory of natural selection are greatly diminished or disappear. On the other hand, all the chief laws of palæontology plainly proclaim, as it seems to me, that species have been produced by ordinary generation: old forms having been supplanted by new and improved forms of life, produced by the laws of variation still acting around us, and preserved by Natural Selection.

CHAPTER XI

GEOGRAPHICAL DISTRIBUTION

Present distribution cannot be accounted for by differences in physical conditions—Importance of barriers—Affinity of the productions of the same continent—Centres of creation—Means of dispersal, by changes of climate and of the level of the land, and by occasional means—Dispersal during the Glacial period co-extensive with the world.

In considering the distribution of organic beings over the face of the globe, the first great fact which strikes us is, that neither the similarity nor the dissimilarity of the inhabitants of various regions can be accounted for by their climatal and other physical conditions. Of late, almost every author who has studied the subject has come to this conclusion. The case of America alone would almost suffice to prove its truth: for if we exclude the northern parts where the circumpolar land is almost continuous, all authors agree that one of the most fundamental divisions in geographical distribution is that between the New and Old Worlds; yet if we travel over the vast American continent, from the central parts of the United States to its extreme southern point, we meet with the most diversified conditions; the most humid districts, arid deserts, lofty mountains, grassy plains, forests, marshes, lakes, and great rivers, under almost every temperature. There is hardly a climate or condition in the Old World which cannot be paralleled in the New—at least as closely as the same species generally require; for it is a most rare case to find a group of organisms confined to any small spot, having conditions peculiar in only a slight degree; for instance, small areas in the Old World could be pointed out hotter than any in the New World, yet these are not inhabited by a peculiar fauna or flora. Notwithstanding this parallelism in the conditions of the Old and New Worlds, how widely different are their living productions!

In the southern hemisphere, if we compare large tracts of land in Australia, South Africa, and western South America, between latitudes 25° and 35°, we shall find parts extremely similar in all their conditions, yet it would not be possible to point out three faunas and floras more utterly dissimilar. Or again we may compare the productions of South America south of lat. 35° with those north of 25°, which consequently inhabit a considerably different climate, and they will be found incomparably more closely related to each other, than they are to the productions of Australia or Africa under nearly the same climate. Analogous facts could be given with respect to the inhabitants of the sea.

A second great fact which strikes us in our general review is, that barriers of any kind, or obstacles to free migration, are related in a close and important manner to the differences between the productions of various regions. We see this in the great difference of nearly all the terrestrial productions of the New and Old Worlds, excepting in the northern parts, where the land almost joins, and where, under a slightly different climate, there might have been free migration for the northern temperate forms, as there now is for the strictly arctic productions. We see the same fact in the great difference between the inhabitants of Australia, Africa, and South America under the same latitude for these countries are almost as much isolated from each other as is possible. On each continent, also, we see the same fact; for on the opposite sides of lofty and continuous mountain-ranges, and of great deserts, and sometimes even of large rivers, we find different productions; though as mountain-chains, deserts, etc., are not as impassable, or likely to have endured so long as the oceans separating continents, the differences are very inferior in degree to those characteristic of distinct continents.

Turning to the sea, we find the same law. No two marine faunas are more distinct, with hardly a fish, shell, or crab in common, than those of the eastern and western shores of South and Central America; yet these great faunas are separated only by the narrow, but impassable, isthmus of Panama. Westward of the shores of America, a wide space of open ocean extends, with not an island as a halting-place for emigrants; here we have a barrier of another kind, and as soon as this is passed we meet in the eastern islands of the Pacific, with another and totally distinct fauna. So that here three marine faunas range far northward and southward, in parallel lines not far from each other, under corresponding climates; but from being separated from each other by impassable barriers, either of land or open sea, they are wholly distinct. On the other hand, proceeding still further westward from the eastern islands of the tropical parts of the Pacific, we encounter no impassable barriers, and we have innumerable islands as halting-places, or continuous coasts, until after travelling over a hemisphere we come to the shores of Africa; and over this vast space we meet with no well-defined and distinct marine faunas. Although hardly one shell, crab or fish is common to the above-named three approximate faunas of Eastern and Western America and the eastern Pacific islands, yet many fish range from the Pacific into the Indian Ocean, and many shells are common to the eastern islands of the Pacific and the eastern shores of Africa, on almost exactly opposite meridians of longitude.

A third great fact, partly included in the foregoing statements, is the affinity of the productions of the same continent or sea, though the species themselves are distinct at different points and stations. It is a law of the widest generality, and every continent offers innumerable instances. Nevertheless the naturalist in travelling, for instance, from north to south never fails to be struck by the manner in which successive groups of beings, specifically distinct, yet clearly related, replace each other. He hears from closely allied, yet distinct kinds of birds, notes nearly similar, and sees their nests similarly constructed, but not quite alike, with eggs coloured in nearly the same manner. The plains near the Straits of Magellan are inhabited by one species of Rhea (American ostrich), and northward the plains of La Plata by another species of the same genus; and not by a true ostrich or emu, like those found in Africa and Australia under the same latitude. On these same plains of La Plata, we see the agouti and bizcacha, animals having nearly the same habits as our hares and rabbits and belonging to the same order of Rodents, but they plainly display an American type of structure. We ascend the lofty peaks of the Cordillera and we find an alpine species of bizcacha; we look to the waters, and we do not find the beaver or musk-rat, but the coypu and capybara, rodents of the American type. Innumerable other instances could be given. If we look to the islands off the American shore, however much they may differ in geological structure, the inhabitants, though they may be all peculiar species, are essentially American. We may look back to past ages, as shown in the last chapter, and we find American types then prevalent on the American continent and in the American seas. We see in these facts some deep organic bond, prevailing throughout space and time, over the same areas of land and water, and independent of their physical conditions. The naturalist must feel little curiosity, who is not led to inquire what this bond is.

This bond, on my theory, is simply inheritance, that cause which alone, as far as we positively know, produces organisms quite like, or, as we see in the case of varieties, nearly like each other. The dissimilarity of the inhabitants of different regions may be attributed to modification through natural selection, and in a quite subordinate degree to the direct influence of different physical conditions. The degree of dissimilarity will depend on the migration of the more dominant forms of life from one region into another having been effected with more or less ease, at periods more or less remote;—on the nature and number of the former immigrants;—and on their action and reaction, in their mutual struggles for life;—the relation of organism to organism being, as I have already often remarked, the most important of all relations. Thus the high importance of barriers comes into play by checking migration; as does time for the slow process of modification through natural selection. Widely-ranging species, abounding in individuals, which have already triumphed over many competitors in their own widely-extended homes will have the best chance of seizing on new places, when they spread into new countries. In their new homes they will be exposed to new conditions, and will frequently undergo further modification and improvement; and thus they will become still further victorious, and will produce groups of modified descendants. On this principle of inheritance with modification, we can understand how it is that sections of genera, whole genera, and even families are confined to the same areas, as is so commonly and notoriously the case.

I believe, as was remarked in the last chapter, in no law of necessary development. As the variability of each species is an independent property, and will be taken advantage of by natural selection, only so far as it profits the individual in its complex struggle for life, so the degree of modification in different species will be no uniform quantity. If, for instance, a number of species, which stand in direct competition with each other, migrate in a body into a new and afterwards isolated country, they will be little liable to modification; for neither migration nor isolation in themselves can do anything. These principles come into play only by bringing organisms into new relations with each other, and in a lesser degree with the surrounding physical conditions. As we have seen in the last chapter that some forms have retained nearly the same character from an enormously remote geological period, so certain species have migrated over vast spaces, and have not become greatly modified.

On these views, it is obvious, that the several species of the same genus, though inhabiting the most distant quarters of the world, must originally have proceeded from the same source, as they have descended from the same progenitor. In the case of those species, which have undergone during whole geological periods but little modification, there is not much difficulty in believing that they may have migrated from the same region; for during the vast geographical and climatal changes which will have supervened since ancient times, almost any amount of migration is possible. But in many other cases, in which we have reason to believe that the species of a genus have been produced within comparatively recent times, there is great difficulty on this head. It is also obvious that the individuals of the same species, though now inhabiting distant and isolated regions, must have proceeded from one spot, where their parents were first produced: for, as explained in the last chapter, it is incredible that individuals identically the same should ever have been produced through natural selection from parents specifically distinct.

We are thus brought to the question which has been largely discussed by naturalists, namely, whether species have been created at one or more points of the earth's surface. Undoubtedly there are very many cases of extreme difficulty, in understanding how the same species could possibly have migrated from some one point to the several distant and isolated points, where now found. Nevertheless the simplicity of the view that each species was first produced within a single region captivates the mind. He who rejects it, rejects the *vera causa* of ordinary generation with subsequent migration, and calls in the agency of a miracle. It is universally admitted, that in most cases the area inhabited by a species is continuous; and when a plant or animal inhabits two points so distant from each other, or with an interval of such a nature, that the space could not be easily passed over by migration, the fact is given as something remarkable and exceptional. The capacity of migrating across the sea is more distinctly limited in terrestrial mammals, than perhaps in any other organic beings; and, accordingly, we find no inexplicable cases of the same mammal inhabiting distant points of the world. No geologist will feel any difficulty in such cases as Great Britain having been formerly united to Europe, and consequently possessing the same quadrupeds. But if the same species can be produced at two separate points, why do we not find a single mammal common to Europe and Australia or South America? The conditions of life are nearly the same, so that a multitude of European animals and plants have become naturalised in America and Australia; and some of the aboriginal plants are identically the same at these distant points of the northern and southern hemispheres? The answer, as I believe, is, that mammals have not been able to migrate, whereas some plants, from their varied means of dispersal, have migrated across the vast and broken interspace. The great and striking influence which barriers of every kind have had on distribution, is intelligible only on the view that the great majority of species have been produced on one side alone, and have not been able to migrate to the other side. Some few families, many sub-families, very many genera, and a still greater number of sections of genera are confined to a single region; and it has been observed by several naturalists, that the most natural genera, or those genera in which the species are most closely related to each other, are generally local, or confined to one area. What a strange anomaly it would be, if, when coming one step lower in the series, to the individuals of the same species, a directly opposite rule prevailed; and species were not local, but had been produced in two or more distinct areas!

Hence it seems to me, as it has to many other naturalists, that the view of each species having been produced in one area alone, and having subsequently migrated from that area as far as its powers of migration and subsistence under past and present conditions permitted, is the most probable. Undoubtedly many cases occur, in which we cannot explain how the same species could have passed from one point to the other. But the geographical and climatal changes, which have certainly occurred within recent geological times, must have interrupted or rendered discontinuous the formerly continuous range of many species. So that we are reduced to consider whether the exceptions to continuity of range are so numerous and of so grave a nature, that we ought to give up the belief, rendered probable by general considerations, that each species has been produced within one area, and has migrated thence as far as it could. It would be hopelessly tedious to discuss all the exceptional cases of the same species, now living at distant and separated points; nor do I for a moment pretend that any explanation could be offered of many such cases. But after some preliminary remarks, I will discuss a few of the most striking classes of facts; namely, the existence of the same species on the summits of distant mountain-ranges, and at distant points in the arctic and antarctic regions; and secondly (in the following chapter), the wide distribution of fresh-water productions; and thirdly, the occurrence of the same terrestrial species on islands and on the mainland, though separated by hundreds of miles of open sea. If the existence of the same species at distant and isolated points of the earth's surface, can in many instances be explained on the view of each species having migrated from a single birthplace; then, considering our ignorance with respect to former climatal and geographical changes and various occasional means of transport, the belief that this has been the universal law, seems to me incomparably the safest.

In discussing this subject, we shall be enabled at the same time to consider a point equally important for us, namely, whether the several distinct species of a genus, which on my theory have all descended from a common progenitor, can have migrated (undergoing modification during some part of their migration) from the area inhabited by their progenitor. If it can be shown to be almost invariably the case, that a region, of which most of its inhabitants are closely related to, or belong to the same genera with the species of a second region, has probably recieved at some former period immigrants from this other region, my theory will be strengthened; for we can clearly understand, on the principle of modification, why the inhabitants of a region should be related to those of another region, whence it has been stocked. A volcanic island, for instance, upheaved and formed at the distance of a few hundreds of miles from a continent, would probably receive from it in the course of time a few colonists, and their descendants, though modified, would still be plainly related by inheritance to the inhabitants of the continent. Cases of this nature are common, and are, as we shall hereafter more fully see, inexplicable on the theory of independent creation. This view of the relation of species in one region to those in another, does not differ much (by substituting the word variety for species) from that lately advanced in an ingenious paper by Mr Wallace, in which he concludes, that 'every species has come into existence coincident both in space and time with a pre-existing closely allied species.' And I now know from correspondence, that this coincidence he attributes to generation with modification.

The previous remarks on 'single and multiple centres of creation' do not directly bear on another allied question,—namely whether all the individuals of the same species have descended from a single pair, or single hermaphrodite, or whether, as some authors suppose, from many individuals simultaneously created. With those organic beings which never intercross (if such exist), the species, on my theory, must have descended from a succession of improved varieties, which will never have blended with other individuals or varieties, but will have supplanted each other; so that, at each successive stage of modification and improvement, all the individuals of each variety will have descended from a single parent. But in the majority of cases, namely, with all organisms which habitually unite for each birth, or which often intercross, I believe that during the slow process of modification the individuals of the species will have been kept nearly uniform by intercrossing; so that many individuals will have gone on simultaneously changing, and the whole amount of modification will not have been due, at each stage, to descent from a single parent. To illustrate what I mean: our English race-horses differ slightly from the horses of every other breed; but they do not owe their difference and superiority to descent from any single pair, but to continued care in selecting and training many individuals during many generations.

Before discussing the three classes of facts, which I have selected as presenting the greatest amount of difficulty on the theory of 'single centres of creation', I must say a few words on the means of dispersal.

Means of Disperal.—Sir C. Lyell and other authors have ably treated this subject. I can give here only the briefest abstract of the more important facts. Change of climate must have had a powerful influence on migration: a region when its climate was different may have been a high road for migration, but now be impassable; I shall, however, presently have to discuss this branch of the subject in some detail. Changes of level in the land must also have been highly influential: a narrow isthmus now separates two marine faunas; submerge it, or let it formerly have been submerged, and the two faunas will now blend or may formerly have blended: where the sea now extends, land may at a former period have connected islands or possibly even continents together, and thus have allowed terrestrial productions to pass from one to the other. No geologist will dispute that great mutations of level have occurred within the period of existing organisms. Edward Forbes insisted that all the islands in the Atlantic must recently have been connected with Europe or Africa, and Europe likewise with America. Other authors have thus hypothetically bridged over every ocean, and have united almost every island to some mainland. If indeed the arguments used by Forbes are to be trusted, it must be admitted that scarcely a single island exists which has not recently been united to some continent. This view cuts the Gordian knot of the dispersal of the same species to the most distant points, and removes many a difficulty: but to the best of my judgment we are not authorised in admitting such enormous geographical changes within the period of existing species. It seems to me that we have abundant evidence of great oscillations of level in our continents; but not of such vast changes in their position and extension, as to have united them within the recent period to each other and to the several intervening oceanic islands. I freely admit the former existence of many islands, now buried beneath the sea, which may have served as halting-places for plants and for many animals during their migration. In the coral-producing oceans such sunken islands are now marked, as I believe, by rings of coral or atolls standing over them. Whenever it is fully admitted, as I believe it will some day be, that each species has proceeded from a single birthplace, and when in the course of time we know something definite about the means of distribution, we shall be enabled to speculate with security on the former extension of the land. But I do not believe that it will ever be proved that within the recent period continents which are now quite separate, have been continuously, or almost continuously, united with each other, and with the many existing oceanic islands. Several facts in distribution,—such as the great difference in the marine faunas on the opposite sides of almost every continent,—the close relation of the tertiary inhabitants of several lands and even seas to their present inhabitants,—a certain degree of relation (as we shall hereafter see) between the distribution of mammals and the depth of the sea,—these and other such facts seem to me opposed to the admission of such prodigious geographical revolutions within the recent period, as are necessitated on the view advanced by Forbes and admitted by his many followers. The nature and relative proportions of the inhabitants of oceanic islands likewise seem to me opposed to the belief of their former continuity with continents. Nor does their almost universally volcanic composition favour the admission that they are the wrecks of sunken continents;—if they had originally existed as mountain-ranges on the land, some at least of the islands would have been formed, like other mountain-summits, of granite, metamorphic schists, old fossiliferous or other such rocks, instead of consisting of mere piles of volcanic matter.

I must now say a few words on what are called accidental means, but which more properly might be called occasional means of distribution. I shall here confine myself to plants. In botanical works, this or that plant is stated to be ill adapted for wide dissemination; but for transport across the sea, the greater or less facilities may be said to be almost wholly unknown. Until I tried, with Mr Berkeley's aid, a few experiments, it was not even known how far seeds could resist the injurious action of sea-water. To my surprise I found that out of 87 kinds, 64 germinated after an immersion of 28 days, and a few survived an immersion of 137 days. For convenience' sake I chiefly tried small seeds, without the capsule or fruit; and as all of these sank in a few days, they could not be floated across wide spaces of the sea, whether or not they were injured by the salt-water. Afterwards I tried some larger fruits, capsules, etc., and some of these floated for a long time. It is well known what a difference there is in the buoyancy of green and seasoned timber; and it occurred to me that floods might wash down plants or branches, and that these might be dried on the banks, and then by a fresh rise in the stream be washed into the sea. Hence I was led to dry stems and branches of 94 plants with ripe fruit, and to place them on sea-water. The majority sank quickly, but some which whilst green floated for a very short time, when dried floated much longer; for instance, ripe hazel-nuts sank immediately, but when dried they floated for 90 days, and afterwards when planted they germinated; an asparagus plant with ripe berries floated for 23 days, when dried it floated for 85 days, and the seeds afterwards germinated; the ripe seeds of Helosciadium sank in two days, when dried they floated for above 90 days, and afterwards germinated. Altogether out of the 94 dried plants, 18 floated for above 28 days, and some of the 18 floated for a very much longer period. So that as 64/87 seeds germinated after an immersion of 28 days; and as 18/94 plants with ripe fruit (but not all the same species as in the foregoing experiment) floated, after being dried, for above 28 days, as far as we may infer anything from these scanty facts, we may conclude that the seeds of 14/100 plants of any country might be floated by sea-currents during 28 days, and would retain their power of germination. In Johnston's Physical Atlas, the average rate of the several Atlantic currents is 33 miles per diem (some currents running at the rate of 60 miles per diem); on this average, the seeds of 14/100 plants belonging to one country might be floated across 924 miles of sea to another country; and when stranded, if blown to a favourable spot by an inland gale, they would germinate.

Subsequently to my experiments, M. Martens tried similar ones, but in a much better manner, for he placed the seeds in a box in the actual sea, so that they were alternately wet and exposed to the air like really floating plants. He tried 98 seeds, mostly different from mine; but he chose many large fruits and likewise seeds from plants which live near the sea; and this would have favoured the average length of their flotation and of their resistance to the injurious action of the salt-water. On the other hand he did not previously dry the plants or branches with the fruit; and this, as we have seen, would have caused some of them to have floated much longer. The result was that 18/98 of his seeds floated for 42 days, and were then capable of germination. But I do not doubt that plants exposed to the waves would float for a less time than those protected from violent movement as in our experiments. Therefore it would perhaps be safer to assume that the seeds of about 10/100 plants of a flora, after having been dried, could be floated across a space of sea 900 miles in width, and would then germinate. The fact of the larger fruits often floating longer than the small, is interesting; as plants with large seeds or fruit could hardly be transported by any other means; and Alph. de Candolle has shown that such plants generally have restricted ranges.

But seeds may be occasionally transported in another manner. Drift timber is thrown up on most islands, even on those in the midst of the widest oceans; and the natives of the coral-islands in the Pacific, procure stones for their tools, solely from the roots of drifted trees, these stones being a valuable royal tax. I find on examination, that when irregularly shaped stones are embedded in the roots of trees, small parcels of earth are very frequently enclosed in their interstices and behind them,—so perfectly that not a particle could be washed away in the longest transport: out of one small portion of earth thus *completely* enclosed by wood in an oak about 50 years old, three dicotyledonous plants germinated: I am certain of the accuracy of this observation. Again, I can show that the carcasses of birds, when floating on the sea, sometimes escape being immediately devoured; and seeds of many kinds in the crops of floating birds long retain their vitality: peas and vetches, for instance, are killed by even a few days' immersion in sea-water; but some taken out of the crop of a pigeon, which had floated on artificial salt-water for 30 days, to my surprise nearly all germinated.

Living birds can hardly fail to be highly effective agents in the transportation of seeds. I could give many facts showing how frequently birds of many kinds are blown by gales to vast distances across the ocean. We may I think safely assume that under such circumstances their rate of flight would often be 35 miles an hour; and some authors have given a far higher estimate. I have never seen an instance of nutritious seeds passing through the intestines of a bird; but hard seeds of fruit pass uninjured through even the digestive organs of a turkey. In the course of two months, I picked up in my garden 12 kinds of seeds, out of the excrement of small birds, and these seemed perfect, and some of them, which I tried, germinated. But the following fact is more important: the crops of birds do not secrete gastric juice, and do not in the least injure, as I know by trial, the germination of seeds; now after a bird has found and devoured a large supply of food, it is positively asserted that all the grains do not pass into the gizzard for 12 or even 18 hours. A bird in this interval might easily be blown to the distance of 500 miles, and hawks are known to look out for tired birds, and the contents of their torn crops might thus readily get scattered. Mr Brent informs me that a friend of his had to give up flying carrier-pigeons from France to England, as the hawks on the English coast destroyed so many on their arrival. Some hawks and owls bolt their prey whole, and after an interval of from twelve to twenty hours, disgorge pellets, which, as I know from experiments made in the Zoological Gardens, include seeds capable of germination. Some seeds of the oat, wheat, millet, canary, hemp, clover, and beet germinated after having been from twelve to twenty-one hours in the stomachs of different birds of prey; and two seeds of beet grew after having been thus retained for two days and fourteen hours. Fresh-water fish, I find, eat seeds of many land and water plants: fish are frequently devoured by birds, and thus the seeds might be transported from place to place. I forced many kinds of seeds into the stomachs of dead fish, and then gave their bodies to fishing-eagles, storks, and pelicans; these birds after an interval of many hours, either rejected the seeds in pellets or passed them in their excrement; and several of these seeds retained their power of germination. Certain seeds, however, were always killed by this process.

Although the beaks and feet of birds are generally quite clean, I can show that earth sometimes adheres to them: in one instance I removed twenty-two grains of dry argillaceous earth from one foot of a partridge, and in this earth there was a pebble quite as large as the seed of a vetch. Thus seeds might occasionally be transported to great distances; for many facts could be given showing that soil almost everywhere is charged with seeds. Reflect for a moment on the millions of quails which annually cross the Mediterranean; and can we doubt that the earth adhering to their feet would sometimes include a few minutes seeds? But I shall presently have to recur to this subject.

As icebergs are known to be sometimes loaded with earth and stones, and have even carried brushwood, bones, and the nest of a land-bird, I can hardly doubt that they must occasionally have transported seeds from one part to another of the arctic and antarctic regions, as suggested by Lyell; and during the Glacial period from one part of the now temperate regions to another. In the Azores, from the large number of the species of plants common to Europe, in comparison with the plants of other oceanic islands nearer to the mainland, and (as remarked by Mr H. C. Watson) from the somewhat northern character of the flora in comparison with the latitude,

I suspected that these islands had been partly stocked by ice-borne seeds, during the Glacial epoch. At my request Sir C. Lyell wrote to M. Hartung to inquire whether he had observed erratic boulders on these islands, and he answered that he had found large fragments of granite and other rocks, which do not occur in the archipelago. Hence we may safely infer that icebergs formerly landed their rocky burthens on the shores of these mid-ocean islands, and it is at least possible that they may have brought thither the seeds of northern plants.

Considering that the several above means of transport, and that several other means, which without doubt remain to be discovered, have been in action year after year, for centuries and tens of thousands of years, it would I think be a marvellous fact if many plants had not thus become widely transported. These means of transport are sometimes called accidental, but this is not strictly correct: the currents of the sea are not accidental, nor is the direction of prevalent gales of wind. It should be observed that scarcely any means of transport would carry seeds for very great distances; for seeds do not retain their vitality when exposed for a great length of time to the action of sea-water; nor could they be long carried in the crops or intestines of birds. These means, however, would suffice for occasional transport across tracts of sea some hundred miles in breadth, or from island to island, or from a continent to a neighbouring island, but not from one distant continent to another. The floras of distant continents would not by such means become mingled in any great degree; but would remain as distinct as we now see them to be. The currents, from their course would never bring seeds from North America to Britain, though they might and do bring seeds from the West Indies to our western shores, where, if not killed by so long an immersion in salt-water, they could not endure our climate. Almost every year, one or two land-birds are blown across the whole Atlantic Ocean, from North America to the western shores of Ireland and England; but seeds could be transported by these wanderers only by one means, namely, in dirt sticking to their feet, which is in itself a rare accident. Even in this case, how small would the chance be of a seed falling on favourable soil, and coming to maturity! But it would be a great error to argue that because a well-stocked island, like Great Britain, has not, as far as is known (and it would be very difficult to prove this), received within the last few centuries, through occasional means of transport, immigrants from Europe or any other continent, that a poorly-stocked island, though standing more remote from the mainland, would not receive colonists by similar means. I do not doubt that out of twenty seeds or animals transported to an island, even if far less well-stocked than Britain, scarcely more than one would be so well fitted to its new home, as to become naturalised. But this, as it seems to me, is no valid argument against what would be effected by occasional means of transport, during the long lapse of geological time, whilst an island was being upheaved and formed, and before it had become fully stocked with inhabitants. On almost bare land, with few or no destructive insects or birds living there, nearly every seed, which chanced to arrive, if fitted for the climate, would be sure to germinate and survive.

Dispersal during the Glacial period.—The identity of many plants and animals, on mountain-summits, separated from each other by hundreds of miles of lowlands, where the Alpine species could not possibly exist, is one of the most striking cases known of the same species living at distant points, without the apparent possibility of their having migrated from one to the other. It is indeed a remarkable fact to see so many of the same plants living on the snowy regions of the Alps or Pyrenees, and in the extreme northern parts of Europe; but it is far more remarkable, that the plants on the White Mountains, in the United States of America, are all the same with those of Labrador, and nearly all the same, as we hear from Asa Gray, with those on the loftiest mountains of Europe. Even as long ago as 1747, such facts led Gmelin to conclude that the same species must have been independently created at several distinct points; and we might have remained in this same belief, had not Agassiz and others called vivid attention to the Glacial period, which, as we shall immediately see, affords a simple explanation of these facts. We have evidence of almost every conceivable kind, organic and inorganic, that within a very recent geological period, central Europe and North America suffered under an Arctic climate. The ruins of a house burnt by fire do not tell their tale more plainly, than do the mountains of Scotland and Wales, with their scored flanks, polished surfaces, and perched boulders, of the icy streams with which their valleys were lately filled. So greatly has the climate of Europe changed, that in Northern Italy, gigantic moraines, left by old glaciers, are now clothed by the vine and maize. Throughout a large part of the United States, erratic boulders, and rocks scored by drifted icebergs and coast-ice, plainly reveal a former cold period.

The former influence of the glacial climate on the distribution of the inhabitants of Europe, as explained with remarkable clearness by Edward Forbes, is substantially as follows. But we shall follow the changes more readily, by supposing a new glacial period to come slowly on, and then pass away, as formerly occurred. As the cold came on, and as each more southern zone became fitted for arctic beings and ill-fitted for their former more temperate inhabitants, the latter would be supplanted and arctic productions would take their places. The inhabitants of the more temperate regions would at the same time travel southward, unless they were stopped by barriers, in which case they would perish. The mountains would become covered with snow and ice, and their former Alpine inhabitants would descend to the plains. By the time that the cold had reached its maximum, we should have a uniform arctic fauna and flora, covering the central parts of Europe, as far south as the Alps and Pyrenees, and even stretching into Spain. The now temperate regions of the United States would likewise be covered by arctic plants and animals, and these would be nearly the same with those of Europe; for the present circumpolar inhabitants, which we suppose to have everywhere travelled southward, are remarkably uniform round the world. We may suppose that the Glacial period came on a little earlier or later in North America than in Europe, so will the southern migration there have been a little earlier or later; but this will make no difference in the final result.

As the warmth returned, the arctic forms would retreat northward, closely followed up in their retreat by the productions of the more temperate regions. And as the snow melted from the bases of the mountains, the arctic forms would seize on the cleared and thawed ground, always ascending higher and higher, as the warmth increased, whilst their brethren were pursuing their northern journey. Hence, when the warmth had fully returned, the same arctic species, which had lately lived in a body together on the lowlands of the Old and New Worlds, would be left isolated on distant mountain-summits (having been exterminated on all lesser heights) and in the arctic regions of both hemispheres.

Thus we can understand the identity of many plants at points so immensely remote as on the mountains of the United States and of Europe. We can thus also understand the fact that the Alpine plants of each mountain-range are more especially related to the arctic forms living due north or nearly due north of them: for the migration as the cold came on, and the re-migration on the returning warmth, will generally have been due south and north. The Alpine plants, for example, of Scotland, as remarked by Mr H. C. Watson, and those of the Pyrenees, as remarked by Ramond, are more especially allied to the plants of northern Scandinavia; those of the United States to Labrador; those of the mountains of Siberia to the arctic regions of that country. These views, grounded as they are on the perfectly well-ascertained occurrence of a former Glacial period, seem to me to explain in so satisfactory a manner the present distribution of the Alpine and Arctic productions of Europe and America, that when in other regions we find the same species on distant mountain-summits, we may almost conclude without other evidence, that a colder climate permitted their former migration across the low intervening tracts, since become too warm for their existence.

If the climate, since the Glacial period, has ever been in any degree warmer than at present (as some geologists in the United States believe to have been the case, chiefly from the distribution of the fossil Gnathodon), then the arctic and temperate productions will at a very late period have marched a little further north, and subsequently have retreated to their present homes; but I have met with no satisfactory evidence with respect to this intercalated slightly warmer period, since the Glacial period.

The arctic forms, during their long southern migration and re-migration northward, will have been exposed to nearly the same climate, and, as is especially to be noticed, they will have kept in a body together; consequently their mutual relations will not have been much disturbed, and, in accordance with the principles inculcated in this volume, they will not have been liable to much modification. But with our Alpine productions, left isolated from the moment of the returning warmth, first at the bases and ultimately on the summits of the mountains, the case will have been somewhat different; for it is not likely that all the same arctic species will have been left on mountain-ranges distant from each other, and have survived there ever since; they will, also, in all probability have become mingled with ancient Alpine species, which must have existed on the mountains before the commencement of the Glacial epoch, and which during its coldest period will have been temporarily driven down to the plains; they will, also, have been exposed to somewhat different climatal influences. Their mutual relations will thus have been in some degree disturbed; consequently they will have been liable to modification; and this we find has been the case; for if we compare the present Alpine plants and animals of the several great European mountain-ranges, though very many of the species are identically the same, some present varieties, some are ranked as doubtful forms, and some few are distinct yet closely allied or representative species.

In illustrating what, as I believe, actually took place during the Glacial period, I assumed that at its commencement the arctic productions were as uniform round the polar regions as they are at the present day. But the foregoing remarks on distribution apply not only to strictly arctic forms, but also to many sub-arctic and to some few northern temperate forms, for some of these are the same on the lower mountains and on the plains of North America and Europe; and it may be reasonably asked how I account for the necessary degree of uniformity of the sub-arctic and northern temperate forms round the world, at the commencement of the Glacial period.

At the present day, the sub-arctic and northern temperate productions of the Old and New Worlds are separated from each other by the Atlantic Ocean and by the extreme northern part of the Pacific. During the Glacial period, when the inhabitants of the Old and New Worlds lived further southwards than at present, they must have been still more completely separated by wider spaces of ocean. I believe the above difficulty may be surmounted by looking to still earlier changes of climate of an opposite nature. We have good reason to believe that during the newer Pliocene period, before the Glacial epoch, and whilst the majority of the inhabitants of the world were specifically the same as now, the climate was warmer than at the present day. Hence we may suppose that the organisms now living under the climate of latitude 60°, during the Pliocene period lived further north under the Polar Circle, in latitude 66°–67°; and that the strictly arctic productions then lived on the broken land still nearer to the pole. Now if we look at a globe, we shall see that under the Polar Circle there is almost continuous land from western Europe, through Siberia, to eastern America. And to this continuity of the circumpolar land, and to the consequent freedom for intermigration under a more favourable climate, I attribute the necessary amount of uniformity in the sub-arctic and northern temperate productions of the Old and New Worlds, at a period anterior to the Glacial epoch.

Believing, from reasons before alluded to, that our continents have long remained in nearly the same relative position, though subjected to large, but partial oscillations of level, I am strongly inclined to extend the above view, and to infer that during some earlier and still warmer period, such as the older Pliocene period, a large number of the same plants and animals inhabited the almost continuous circumpolar land; and that these plants and animals, both in the Old and New Worlds, began slowly to migrate southwards as the climate became less warm, long before the commencement of the Glacial period. We now see, as I believe, their descendants, mostly in a modified condition, in the central parts of Europe and the United States. On this view we can understand the relationship, with very little identity, between the productions of North America and Europe,—a relationship which is most remarkable, considering the distance of the two areas, and their separation by the Atlantic Ocean. We can further understand the singular fact remarked on by several observers, that the productions of Europe and America during the later tertiary stages were more closely related to each other than they are at the present time; for during these warmer periods the northern parts of the Old and New Worlds will have been almost continuously united by land, serving as a bridge, since rendered impassable by cold, for the intermigration of their inhabitants.

During the slowly decreasing warmth of the Pliocene period, as soon as the species in common, which inhabited the New and Old Worlds, migrated south of the Polar Circle, they must have been completely cut off from each other. This separation, as far as the more temperate productions are concerned, took place long ages ago. And as the plants and animals migrated southward, they will have become mingled in the one great region with the native American productions, and have had to compete with them; and in the other great region, with those of the Old World. Consequently we have here everything favourable for much modification,—for far more modification than with the Alpine productions, left isolated, within a much more recent period, on the several mountain-ranges and on the arctic lands of the two Worlds. Hence it has come, that when we compare the now living productions of the temperate regions of the New and Old Worlds, we find very few identical species (though Asa Gray has lately shown that more plants are identical than was formerly supposed), but we find in every great class many forms, which some naturalists rank as geographical races, and others as distinct species; and a host of closely allied or representative forms which are ranked by all naturalists as specifically distinct.

As on the land, so in the waters of the sea, a slow southern migration of a marine fauna, which during the Pliocene or even a somewhat earlier period, was nearly uniform along the continuous shores of the Polar Circle, will account, on the theory of modification, for many closely allied forms now living in areas completely sundered. Thus, I think, we can understand the presence of many existing and tertiary representative forms on the eastern and western shores of temperate North America; and the still more striking case of many closely allied crustaceans (as described in Dana's admirable work), of some fish and other marine animals, in the Mediterranean and in the seas of Japan,—areas now separated by a continent and by nearly a hemisphere of equatorial ocean.

These cases of relationship, without identity, of the inhabitants of seas now disjoined, and likewise of the past and present inhabitants of the temperate lands of North America and Europe, are inexplicable on the theory of creation. We cannot say that they have been created alike, in correspondence with the nearly similar physical conditions of the areas; for if we compare, for instance, certain parts of South America with the southern continents of the Old World, we see countries closely corresponding in all their physical conditions, but with their inhabitants utterly dissimilar.

But we must return to our more immediate subject, the Glacial period. I am convinced that Forbes's view may be largely extended. In Europe we have the plainest evidence of the cold period, from the western shores of Britain to the Oural range, and southward to the Pyrenees. We may infer from the frozen mammals and nature of the mountain vegetation, that Siberia was similarly affected. Along the Himalaya, at points 900 miles apart, glaciers have left the marks of their former low descent; and in Sikkim, Dr Hooker saw maize growing on gigantic ancient moraines. South of the equator, we have some direct evidence of former glacial action in New Zealand; and the same plants, found on widely separated mountains in that island, tell the same story. If one account which has been published can be trusted, we have direct evidence of glacial action in the south-eastern corner of Australia.

Looking to America; in the northern half, ice-borne fragments of rock have been observed on the eastern side as far south as lat. 36°–37°, and on the shores of the Pacific, where the climate is now so different, as far south as lat. 46°; erratic boulders have, also, been noticed on the Rocky Mountains. In the Cordillera of Equatorial South America, glaciers once extended far below their present level. In central Chili I was astonished at the structure of a vast mound of detritus, about 800 feet in height, crossing a valley of the Andes; and this I now feel convinced was a gigantic moraine, left far below any existing glacier. Further south on both sides of the continent, from lat. 41° to the southernmost extremity, we have the clearest evidence of former glacial action, in huge boulders transported far from their parent source.

We do not know that the Glacial epoch was strictly simultaneous at these several far distant points on opposite sides of the world. But we have good evidence in almost every case, that the epoch was included within the latest geological period. We have, also, excellent evidence, that it endured for an enormous time, as measured by years, at each point. The cold may have come on, or have ceased, earlier at one point of the globe than at another, but seeing that it endured for long at each, and that it was contemporaneous in a geological sense, it seems to me probable that it was, during a part at least of the period, actually simultaneous throughout the world. Without some distinct evidence to the contrary, we may at least admit as probable that the glacial action was simultaneous on the eastern and western sides of North America, in the Cordillera under the equator and under the warmer temperate zones, and on both sides of the southern extremity of the continent. If this be admitted, it is difficult to avoid believing that the temperature of the whole world was at this period simultaneously cooler. But it would suffice for my purpose, if the temperature was at the same time lower along certain broad belts of longitude.

On this view of the whole world, or at least of broad longitudinal belts, having been simultaneously colder from pole to pole, much light can be thrown on the present distribution of identical and allied species. In America, Dr Hooker has shown that between forty and fifty of the flowering plants of Tierra del Fuego, forming no inconsiderable part of its scanty flora, are common to Europe, enormously remote as these two points are; and there are many closely allied species. On the lofty mountains of equatorial America a host of peculiar species belonging to European genera occur. On the highest mountains of Brazil, some few European genera were found by

Gardner, which do not exist in the wide intervening hot countries. So on the Silla of Caraccas the illustrious Humboldt long ago found species belonging to genera characteristic of the Cordillera. On the mountains of Abyssinia, several European forms and some few representatives of the peculiar flora of the Cape of Good Hope occur. At the Cape of Good Hope a very few European species, believed not to have been introduced by man, and on the mountains, some few representative European forms are found, which have not been discovered in the intertropical parts of Africa. On the Himalaya, and on the isolated mountain-ranges of the peninsula of India, on the heights of Ceylon, and on the volcanic cones of Java, many plants occur, either identically the same or representing each other, and at the same time representing plants of Europe, not found in the intervening hot lowlands. A list of the genera collected on the loftier peaks of Java raises a picture of a collection made on a hill in Europe! Still more striking is the fact that southern Australian forms are clearly represented by plants growing on the summits of the mountains of Borneo. Some of these Australian forms, as I hear from Dr Hooker, extend along the heights of the peninsula of Malacca, and are thinly scattered, on the one hand over India and on the other as far north as Japan.

On the southern mountains of Australia, Dr F. Müller has discovered several European species; other species, not introduced by man, occur on the lowlands; and a long list can be given, as I am informed by Dr Hooker, of European genera, found in Australia, but not in the intermediate torrid regions. In the admirable *Introduction to the Flora of New Zealand*, by Dr Hooker, analogous and striking facts are given in regard to the plants of that large island. Hence we see that throughout the world, the plants growing on the more lofty mountains, and on the temperate lowlands of the northern and southern hemispheres, are sometimes identically the same; but they are much oftener specifically distinct, though related to each other in a most remarkable manner.

This brief abstract applies to plants alone: some strictly analogous facts could be given on the distribution of terrestrial animals. In marine productions, similar cases occur; as an example, I may quote a remark by the highest authority, Prof. Dana, that 'it is certainly a wonderful fact that New Zealand should have a closer resemblance in its crustacea to Great Britain, its antipode, than to any other part of the world.' Sir J. Richardson, also, speaks of the reappearance on the shores of New Zealand, Tasmania, etc., of northern forms of fish. Dr Hooker informs me that twenty-five species of Algæ are common to New Zealand and to Europe, but have not been found in the intermediate tropical seas.

It should be observed that the northern species and forms found in the southern parts of the southern hemisphere, and on the mountain-ranges of the intertropical regions, are not arctic, but belong to the northern temperate zones. As Mr H. C. Watson has recently remarked, 'In receding from polar towards equatorial latitudes, the Alpine or mountain floras really become less and less arctic.' Many of the forms living on the mountains of the warmer regions of the earth and in the southern hemisphere are of doubtful value, being ranked by some naturalists as specifically distinct, by others as varieties; but some are certainly identical, and many, though closely related to northern forms, must be ranked as distinct species.

Now let us see what light can be thrown on the foregoing facts, on the belief, supported as it is by a large body of geological evidence, that the whole world, or a large part of it, was during the Glacial period simultaneously much colder than at present. The Glacial period, as measured by years, must have been very long; and when we remember over what vast spaces some naturalised plants and animals have spread within a few centuries, this period will have been ample for any amount of migration. As the cold came slowly on, all the tropical plants and other productions will have retreated from both sides towards the equator, followed in the rear by the temperate productions, and these by the arctic; but with the latter we are not now concerned. The tropical plants probably suffered much extinction; how much no one can say; perhaps formerly the tropics supported as many species as we see at the present day crowded together at the Cape of Good Hope, and in parts of temperate Australia. As we know that many tropical plants and animals can withstand a considerable amount of cold, many might have escaped extermination during a moderate fall of temperature, more especially by escaping into the lowest, most protected, and warmest districts. But the great fact to bear in mind is, that all tropical productions will have suffered to a certain extent. On the other hand, the temperate productions, after migrating nearer to the equator, though they will have been placed under somewhat new conditions, will have suffered less. And it is certain that many temperate plants, if protected from the inroads of competitors, can withstand a much warmer climate than their own. Hence, it seems to me possible, bearing in mind that the tropical productions were in a suffering state and could not have presented a firm front against intruders, that a certain number of the more vigorous and dominant temperate forms might have penetrated the native ranks and have reached or even crossed the equator. The invasion would, of course, have been greatly favoured by high land, and perhaps by a dry climate; for Dr Falconer informs me that it is the damp with the heat of the tropics which is so destructive to perennial plants from a temperate climate. On the other hand, the most humid and hottest districts will have afforded an asylum to the tropical natives. The mountain-ranges north-west of the Himalaya, and the long line of the Cordillera, seem to have afforded two great lines of invasion: and it is a striking fact, lately communicated to me by Dr Hooker, that all the flowering plants, about forty-six in number, common to Tierra del Fuego and to Europe still exist in North America, which must have lain on the line of march. But I do not doubt that some temperate productions entered and crossed even the *lowlands* of the tropics at the period when the cold was most intense,—when arctic forms had migrated some twenty-five degrees of latitude from their native country and covered the land at the foot of the Pyrenees. At this period of extreme cold, I believe that the climate under the equator at the level of the sea was about the same with that now felt there at the height of six or seven thousand feet. During this the coldest period, I suppose that large spaces of the tropical lowlands were clothed with a mingled tropical and temperate vegetation, like that now growing with strange luxuriance at the base of the Himalaya, as graphically described by Hooker.

Thus, as I believe, a considerable number of plants, a few terrestrial animals, and some marine productions, migrated during the Glacial period from the northern and southern temperate zones into the intertropical regions, and some even crossed the equator. As the warmth returned, these temperate forms would naturally ascend the higher mountains, being exterminated on the lowlands; those which had not reached the equator would re-migrate northward or southward towards their former homes; but the forms, chiefly northern, which had crossed the equator, would travel still further from their homes into the more temperate latitudes of the opposite hemisphere. Although we have reason to believe from geological evidence that the whole body of arctic shells underwent scarcely any modification during their long southern migration and re-migration northward, the case may have been wholly different with those intruding forms which settled themselves on the intertropical mountains, and in the southern hemisphere. These being surrounded by strangers will have had to compete with many new forms of life; and it is probable that selected modifications in their structure, habits, and constitutions will have profited them. Thus many of these wanderers, though still plainly related by inheritance to their brethren of the northern or southern hemispheres, now exist in their new homes as well-marked varieties or as distinct species.

It is a remarkable fact, strongly insisted on by Hooker in regard to America, and by Alph. de Candolle in regard to Australia, that many more identical plants and allied forms have apparently migrated from the north to the south, than in a reversed direction. We see, however, a few southern vegetable forms on the mountains of Borneo and Abyssinia. I suspect that this preponderant migration from north to south is due to the greater extent of land in the north, and to the northern forms having existed in their own homes in greater numbers, and having consequently been advanced through natural selection and competition to a higher state of perfection or dominating power, than the southern forms. And thus, when they became comming led during the Glacial period, the northern forms were enabled to beat the less powerful southern forms. Just in the same manner as we see at the present day, that very many European productions cover the ground in La Plata, and in a lesser degree in Australia, and have to a certain extent beaten the natives; whereas extremely few southern forms have become naturalised in any part of Europe, though hides, wool, and other objects likely to carry seeds have been largely imported into Europe during the last two or three centuries from La Plata, and during the last thirty or forty years from Australia. Something of the same kind must have occurred on the intertropical mountains: no doubt before the Glacial period they were stocked with endemic Alpine forms; but these have almost everywhere largely yielded to the more dominant forms, generated in the larger areas and more efficient workshops of the north. In many islands the native productions are nearly equalled or even outnumbered by the naturalised; and if the natives have not been actually exterminated, their numbers have been greatly reduced, and this is the first stage towards extinction. A mountain is an island on the land; and the intertropical mountains before the Glacial period must have been completely isolated; and I believe that the productions of these islands on the land yielded to those produced within the larger areas of the north, just in the same way as the productions of real islands have everywhere lately yielded to continental forms, naturalised by man's agency.

I am far from supposing that all difficulties are removed on the view here given in regard to the range and affinities of the allied species which live in the northern and southern temperate zones and on the mountains of the intertropical regions. Very many difficulties remain to be solved. I do not pretend to indicate the exact lines and means of migration, or the reason why certain species and not others have migrated; why certain species have been modified and have given rise to new groups of forms, and others have remained unaltered. We cannot hope to explain such facts, until we can say why one species and not another becomes naturalised by man's agency in a foreign land; why one ranges twice or thrice as far, and is twice or thrice as common, as another species within their own homes.

I have said that many difficulties remain to be solved: some of the most remarkable are stated with admirable clearness by Dr Hooker in his botanical works on the antarctic regions. These cannot be here discussed. I will only say that as far as regards the occurrence of identical species at points so enormously remote as Kerguelen Land, New Zealand, and Fuegia, I believe that towards the close of the Glacial period, icebergs, as suggested by Lyell, have been largely concerned in their dispersal. But the existence of several quite distinct species, belonging to genera exclusively confined to the south, at these and other distant points of the southern hemisphere, is, on my theory of descent with modification, a far more remarkable case of difficulty. For some of these species are so distinct, that we cannot suppose that there has been time since the commencement of the Glacial period for their migration, and for their subsequent modification to the necessary degree. The facts seem to me to indicate that peculiar and very distinct species have migrated in radiating lines from some common centre; and I am inclined to look in the southern, as in the northern hemisphere, to a former and warmer period, before the commencement of the Glacial period, when the antarctic lands, now covered with ice, supported a highly peculiar and isolated flora. I suspect that before this flora was exterminated by the Glacial epoch, a few forms were widely dispersed to various points of the southern hemisphere by occasional means of transport, and by the aid, as halting-places, of existing and now sunken islands. By these means, as I believe, the southern shores of America, Australia, New Zealand, have become slightly tinted by the same peculiar forms of vegetable life.

Sir C. Lyell in a striking passage has speculated, in language almost identical with mine, on the effects of great alternations of climate on geographical distribution. I believe that the world has recently felt one of his great cycles of change; and that on this view, combined with modification through natural selection, a multitude of facts in the present distribution both of the same and of allied forms of life can be explained. The living waters may be said to have flowed during one short period from the north and from the south, and to have crossed at the equator; but to have flowed with greater force from the north so as to have freely inundated the south. As the tide leaves its drift in horizontal lines, though rising higher on the shores where the tide rises highest, so have the living waters left their living drift on our mountain-summits, in a line gently rising from the arctic lowlands to a great height under the equator. The various beings thus left stranded may be compared with savage races of man, driven up and surviving in the mountain-fastnesses of almost every land, which serve as a record, full of interest to us, of the former inhabitants of the surrounding lowlands.

CHAPTER XII

GEOGRAPHICAL DISTRIBUTION—*continued*

Distribution of fresh-water productions—On the inhabitants of oceanic islands—Absence of Batrachians and of terrestrial Mammals—On the relation of the inhabitants of islands to those of the nearest mainland—On colonisation from the nearest source with subsequent modification—Summary of the last and present chapters.

As lakes and river-systems are separated from each other by barriers of land, it might have been thought that fresh-water productions would not have ranged widely within the same country, and as the sea is apparently a still more impassable barrier, that they never would have extended to distant countries. But the case is exactly the reverse. Not only have many fresh-water species, belonging to quite different classes, an enormous range, but allied species prevail in a remarkable manner throughout the world. I well remember, when first collecting in the fresh waters of Brazil, feeling much surprise at the similarity of the fresh-water insects, shells, etc., and at the dissimilarity of the surrounding terrestrial beings, compared with those of Britain.

But this power in fresh-water productions of ranging widely, though so unexpected, can, I think, in most cases be explained by their having become fitted, in a manner highly useful to them, for short and frequent migrations from pond to pond, or from stream to stream; and liability to wide dispersal would follow from this capacity as an almost necessary consequence. We can here consider only a few cases. In regard to fish, I believe that the same species never occur in the fresh waters of distant continents. But on the same continent the species often range widely and almost capriciously; for two river-systems will have some fish in common and some different. A few facts seem to favour the possibility of their occasional transport by accidental means; like that of the live fish not rarely dropped by whirlwinds in India, and the vitality of their ova when removed from the water. But I am inclined to attribute the dispersal of fresh-water fish mainly to slight changes within the recent period in the level of the land, having caused rivers to flow into each other. Instances, also, could be given of this having occurred during floods, without any change of level. We have evidence in the loess of the Rhine of considerable changes of level in the land within a very recent geological period, and when the surface was peopled by existing land and fresh-water shells. The wide difference of the fish on opposite sides of continuous mountain-ranges, which from an early period must have parted river-systems and completely prevented their inosculation, seems to lead to this same conclusion. With respect to allied fresh-water fish occurring at very distant points of the world, no doubt there are many cases which cannot at present be explained: but some fresh-water fish belong to very ancient forms, and in such cases there will have been ample time for great geographical changes, and consequently time and means for much migration. In the second place, salt-water fish can with care be slowly accustomed to live in fresh water; and, according to Valenciennes, there is hardly a single group of fishes confined exclusively to fresh water, so that we may imagine that a marine member of a fresh-water group might travel far along the shores of the sea, and subsequently become modified and adapted to the fresh waters of a distant land.

Some species of fresh-water shells have a very wide range, and allied species, which, on my theory, are descended from a common parent and must have proceeded from a single source, prevail throughout the world. Their distribution at first perplexed me much, as their ova are not likely to be transported by birds, and they are immediately killed by sea-water, as are the adults. I could not even understand how some naturalised species have rapidly spread throughout the same country. But two facts, which I have observed—and no doubt many others remain to be observed—throw some light on this subject. When a duck suddenly emerges from a pond covered with duck-weed, I have twice seen these little plants adhering to its back; and it has happened to me, in removing a little duck-weed from one aquarium to another, that I have quite unintentionally stocked the one with fresh-water shells from the other. But another agency is perhaps more effectual: I suspended a duck's feet, which might represent those of a bird sleeping in a natural pond, in an aquarium, where many ova of fresh-water shells were hatching; and I found that numbers of the extremely minute and just-hatched shells crawled on the feet, and clung to them so firmly that when taken out of the water they could not be jarred off, though at a somewhat more advanced age they would voluntarily drop off. These just hatched molluscs, though aquatic in their nature, survived on the duck's feet, in damp air, from twelve to twenty hours; and in this length of time a duck or heron might fly at least six or seven hundred miles, and would be sure to alight on a pool or rivulet, if blown across sea to an oceanic island or to any other distant point. Sir Charles Lyell also informs me that a Dyticus has been caught with an Ancylus (a fresh-water shell like a limpet) firmly adhering to it; and a water-beetle of the same family, a Colymbetes, once flew on board the *Beagle* when forty-five miles distant from the nearest land: how much farther it might have flown with a favouring gale no one can tell.

With respect to plants, it has long been known what enormous ranges many fresh-water and even marsh-species have, both over continents and to the most remote oceanic islands. This is strikingly shown, as remarked by Alph. de Candolle, in large groups of terrestrial plants, which have only a very few aquatic members; for these latter seem immediately to acquire, as if in consequence, a very wide range. I think favourable means of dispersal explain this fact. I have before mentioned that earth occasionally, though rarely, adheres in some quantity to the feet and beaks of birds. Wading birds, which frequent the muddy edges of ponds, if suddenly flushed, would be the most likely to have muddy feet. Birds of this order I can show are the greatest wanderers, and are occasionally found on the most remote and barren islands in the open ocean; they would not be likely to alight on the surface of the sea, so that the dirt would not be washed off their feet; when making land, they would be sure to fly to their natural fresh-water haunts. I do not believe that botanists are aware how charged the mud of ponds is with seeds: I have tried several little experiments, but will here give only the most striking case: I took in February three table-spoonfuls of mud from three different points, beneath water, on the edge of a little pond; this mud when dry weighed only 6 3/4 ounces; I kept it covered up in my study for six months, pulling up and counting each plant as it grew; the plants were of many kinds, and were altogether 537 in number; and yet the viscid mud was all contained in a breakfast cup! Considering these facts, I think it would be an inexplicable circumstance if water-birds did not transport the seeds of fresh-water plants to vast distances, and if consequently the range of these plants was not very great. The same agency may have come into play with the eggs of some of the smaller fresh-water animals.

Other and unknown agencies probably have also played a part. I have stated that fresh-water fish eat some kinds of seeds, though they reject many other kinds after having swallowed them; even small fish swallow seeds of moderate size, as of the yellow water-lily and Potamogeton. Herons and other birds, century after century, have gone on daily devouring fish; they then take flight and go to other waters, or are blown across the sea; and we have seen that seeds retain their power of germination, when rejected in pellets or in excrement, many hours afterwards. When I saw the great size of the seeds of that fine water-lily, the Nelumbium, and remembered Alph. de Candolle's remarks on this plant, I thought that its distribution must remain quite inexplicable; but Audubon states that he found the seeds of the great southern water-lily (probably, according to Dr Hooker, the Nelumbium luteum) in a heron's stomach; although I do not know the fact, yet analogy makes me believe that a heron flying to another pond and getting a hearty meal of fish, would probably reject from its stomach a pellet containing the seeds of the Nelumbium undigested; or the seeds might be dropped by the bird whilst feeding its young, in the same way as fish are known sometimes to be dropped.

In considering these several means of distribution, it should be remembered that when a pond or stream is first formed, for instance, on a rising islet, it will be unoccupied; and a single seed or egg will have a good chance of succeeding. Although there will always be a struggle for life between the individuals of the species, however few, already occupying any pond, yet as the number of kinds is small, compared with those on the land, the competition will probably be less severe between aquatic than between terrestrial species; consequently an intruder from the waters of a foreign country, would have a better chance of seizing on a place, than in the case of terrestrial colonists. We should, also, remember that some, perhaps many, fresh-water productions are low in the scale of nature, and that we have reason to believe that such low beings change or become modified less quickly than the high; and this will give longer time than the average for the migration of the same aquatic species. We should not forget the probability of many species having formerly ranged as continuously as fresh-water productions ever can range, over immense areas, and having subsequently become extinct in intermediate regions. But the wide distribution of fresh-water plants and of the lower animals, whether retaining the same identical form or in some degree modified, I believe mainly depends on the wide dispersal of their seeds and eggs by animals, more especially by fresh-water birds, which have large powers of flight, and naturally travel from one to another and often distant piece of water. Nature, like a careful gardener, thus takes her seeds from a bed of a particular nature, and drops them in another equally well fitted for them.

On the Inhabitants of Oceanic Islands.—We now come to the last of the three classes of facts, which I have selected as presenting the greatest amount of difficulty, on the view that all the individuals both of the same and of allied species have descended from a single parent; and therefore have all proceeded from a common birthplace, notwithstanding that in the course of time they have come to inhabit distant points of the globe. I have already stated that I cannot honestly admit Forbes's view on continental extensions, which, if legitimately followed out, would lead to the belief that within the recent period all existing islands have been nearly or quite joined to some continent. This view would remove many difficulties, but it would not, I think, explain all the facts in regard to insular productions. In the following remarks I shall not confine myself to the mere question of dispersal; but shall consider some other facts, which bear on the truth of the two theories of independent creation and of descent with modification.

The species of all kinds which inhabit oceanic islands are few in number compared with those on equal continental areas: Alph. de Candolle admits this for plants, and Wollaston for insects. If we look to the large size and varied stations of New Zealand, extending over 780 miles of latitude, and compare its flowering plants, only 750 in number, with those on an equal area at the Cape of Good Hope or in Australia, we must, I think, admit that something quite independently of any difference in physical conditions has caused so great a difference in number. Even the uniform county of Cambridge has 847 plants, and the little island of Anglesea 764, but a few ferns and a few introduced plants are included in these numbers, and the comparison in some other respects is not quite fair. We have evidence that the barren island of Ascension aboriginally possessed under half a dozen flowering plants; yet many have become naturalised on it, as they have on New Zealand and on every other oceanic island which can be named. In St. Helena there is reason to believe that the naturalised plants and animals have nearly or quite exterminated many native productions. He who admits the doctrine of the creation of each separate species, will have to admit, that a sufficient number of the best adapted plants and animals have not been created on oceanic islands; for man has unintentionally stocked them from various sources far more fully and perfectly than has nature.

Although in oceanic islands the number of kinds of inhabitants is scanty, the proportion of endemic species (*i.e.* those found nowhere else in the world) is often extremely large. If we compare, for instance, the number of the endemic land-shells in Madeira, or of the endemic birds in the Galapagos Archipelago, with the number found on any continent, and then compare the area of the islands with that of the continent, we shall see that this is true. This fact might have been expected on my theory, for, as already explained, species occasionally arriving after long intervals in a new and isolated district, and having to compete with new associates, will be eminently liable to modification, and will often produce groups of modified descendants. But it by no means follows, that, because in an island nearly all the species of one class are peculiar, those of another class, or of another section of the same class, are peculiar; and this difference seems to depend partly on the species which do not become modified having immigrated with facility and in a body, so that their mutual relations have not been much disturbed; and partly on the frequent arrival of unmodified immigrants from the mother-country, and the consequent intercrossing with them. With respect to the effects of this intercrossing, it should be remembered that the offspring of such crosses would almost certainly gain in vigour; so that even an occasional cross would produce more effect than might at first have been anticipated. To give a few examples: in the Galapagos Islands nearly every land-bird, but only two out of the eleven marine birds, are peculiar; and it is obvious that marine birds could arrive at these islands more easily than land-birds. Bermuda, on the other hand, which lies at about the same distance from North America as the Galapagos Islands do from South America, and which has a very peculiar soil, does not possess one endemic land-bird; and we know from Mr J. M. Jones's admirable account of Bermuda, that very many North American birds, during their great annual migrations, visit either periodically or occasionally this island. Madeira does not possess one peculiar bird, and many European and African birds are almost every year blown there, as I am informed by Mr E. V. Harcourt. So that these two islands of Bermuda and Madeira have been stocked by birds, which for long ages have struggled together in their former homes, and have become mutually adapted to each other; and when settled in their new homes, each kind will have been kept by the others to their proper places and habits, and will consequently have been little liable to modification. Any tendency to modification will, also, have been checked by intercrossing with the unmodified immigrants from the mother-country. Madeira, again, is inhabited by a wonderful number of peculiar land-shells, whereas not one species of sea-shell is confined to its shores: now, though we do not know how sea-shells are dispersed, yet we can see that their eggs or larvæ, perhaps attached to seaweed or floating timber, or to the feet of wading-birds, might be transported far more easily than land-shells, across three or four hundred miles of open sea. The different orders of insects in Madeira apparently present analogous facts.

Oceanic islands are sometimes deficient in certain classes, and their places are apparently occupied by the other inhabitants; in the Galapagos Islands reptiles, and in New Zealand gigantic wingless birds, take the place of mammals. In the plants of the Galapagos Islands, Dr Hooker has shown that the proportional numbers of the different orders are very different from what they are elsewhere. Such cases are generally accounted for by the physical conditions of the islands; but this explanation seems to me not a little doubtful. Facility of immigration, I believe, has been at least as important as the nature of the conditions.

Many remarkable little facts could be given with respect to the inhabitants of remote islands. For instance, in certain islands not tenanted by mammals, some of the endemic plants have beautifully hooked seeds; yet few relations are more striking than the adaptation of hooked seeds for transportal by the wool and fur of quadrupeds. This case presents no difficulty on my view, for a hooked seed might be transported to an island by some other means; and the plant then becoming slightly modified, but still retaining its hooked seeds, would form an endemic species, having as useless an appendage as any rudimentary organ,—for instance, as the shrivelled wings under the soldered elytra of many insular beetles. Again, islands often possess trees or bushes belonging to orders which elsewhere include only herbaceous species; now trees, as Alph. de Candolle has shown, generally have, whatever the cause may be, confined ranges. Hence trees would be little likely to reach distant oceanic islands; and an herbaceous plant, though it would have no chance of successfully competing in stature with a fully developed tree, when established on an island and having to compete with herbaceous plants alone, might readily gain an advantage by growing taller and taller and overtopping the other plants. If so, natural selection would often tend to add to the stature of herbaceous plants when growing on an oceanic island, to whatever order they belonged, and thus convert them first into bushes and ultimately into trees.

With respect to the absence of whole orders on oceanic islands, Bory St Vincent long ago remarked that Batrachians (frogs, toads, newts) have never been found on any of the many islands with which the great oceans are studded. I have taken pains to verify this assertion, and I have found it strictly true. I have, however, been assured that a frog exists on the mountains of the great island of New Zealand; but I suspect that this exception (if the information be correct) may be explained through glacial agency. This general absence of frogs, toads, and newts on so many oceanic islands cannot be accounted for by their physical conditions; indeed it seems that islands are peculiarly well fitted for these animals; for frogs have been introduced into Madeira, the Azores, and Mauritius, and have multiplied so as to become a nuisance. But as these animals and their spawn are known to be immediately killed by sea-water, on my view we can see that there would be great difficulty in their transportal across the sea, and therefore why they do not exist on any oceanic island. But why, on the theory of creation, they should not have been created there, it would be very difficult to explain.

Mammals offer another and similar case. I have carefully searched the oldest voyages, but have not finished my search; as yet I have not found a single instance, free from doubt, of a terrestrial mammal (excluding domesticated animals kept by the natives) inhabiting an island situated above 300 miles from a continent or great continental island; and many islands situated at a much less distance are equally barren. The Falkland Islands, which are inhabited by a wolf-like fox, come nearest to an exception; but this group cannot be considered as oceanic, as it lies on a bank connected with the mainland; moreover, icebergs formerly brought boulders to its western shores, and they may have formerly transported foxes, as so frequently now happens in the arctic regions. Yet it cannot be said that small islands will not support small mammals, for they occur in many parts of the world on very small islands, if close to a continent; and hardly an island can be named on which our smaller quadrupeds have not become naturalised and greatly multiplied. It cannot be said, on the ordinary view of creation, that there has not been time for the creation of mammals; many volcanic islands are sufficiently ancient, as shown by the stupendous degradation which they have suffered and by their tertiary strata: there has also been time for the production of endemic species belonging to other classes; and on continents it is thought that mammals appear and disappear at a quicker rate than other and lower animals. Though terrestrial mammals do not occur on oceanic islands, aërial mammals do occur on almost every island. New Zealand possesses two bats found nowhere else in the world: Norfolk Island, the Viti Archipelago, the Bonin Islands, the Caroline and Marianne Archipelagoes, and Mauritius, all possess their peculiar bats. Why, it may be asked, has the supposed creative force produced bats and no other mammals on remote islands? On my view this question can easily be answered; for no terrestrial mammal can be transported across a wide space of sea, but bats can fly across. Bats have been seen wandering by day far over the Atlantic Ocean; and two North American species either regularly or occasionally visit Bermuda, at the distance of 600 miles from the mainland. I hear from Mr Tomes, who has specially studied this family, that many of the same species have enormous ranges, and are found on continents and on far distant islands. Hence we have only to suppose that such wandering species have been modified through natural selection in their new homes in relation to their new position, and we can understand the presence of endemic bats on islands, with the absence of all terrestrial mammals.

Besides the absence of terrestrial mammals in relation to the remoteness of islands from continents, there is also a relation, to a certain extent independent of distance, between the depth of the sea separating an island from the neighbouring mainland, and the presence in both of the same mammiferous species or of allied species in a more or less modified condition. Mr Windsor Earl has made some striking observations on this head in regard to the great Malay Archipelago, which is traversed near Celebes by a space of deep ocean; and this space separates two widely distinct mammalian faunas. On either side the islands are situated on moderately deep submarine banks, and they are inhabited by closely allied or identical quadrupeds. No doubt some few anomalies occur in this great archipelago, and there is much difficulty in forming a judgment in some cases owing to the probable naturalisation of certain mammals through man's agency; but we shall soon have much light thrown on the natural history of this archipelago by the admirable zeal and researches of Mr Wallace. I have not as yet had time to follow up this subject in all other quarters of the world; but as far as I have gone, the relation generally holds good. We see Britain separated by a shallow channel from Europe, and the mammals are the same on both sides; we meet with analogous facts on many islands separated by similar channels from Australia. The West Indian Islands stand on a deeply submerged bank, nearly 1000 fathoms in depth, and here we find American forms, but the species and even the genera are distinct. As the amount of modification in all cases depends to a certain degree on the lapse of time, and as during changes of level it is obvious that islands separated by shallow channels are more likely to have been continuously united within a recent period to the mainland than islands separated by deeper channels, we can understand the frequent relation between the depth of the sea and the degree of affinity of the mammalian inhabitants of islands with those of a neighbouring contient,—an inexplicable relation on the view of independent acts of creation.

All the foregoing remarks on the inhabitants of oceanic islands,—namely, the scarcity of kinds—the richness in endemic forms in particular classes or sections of classes,—the absence of whole groups, as of batrachians, and of terrestrial mammals notwithstanding the presence of aërial bats,—the singular proportions of certain orders of plants,—herbaceous forms having been developed into trees, etc.,—seem to me to accord better with the view of occasional means of transport having been largely efficient in the long course of time, than with the view of all our oceanic islands having been formerly connected by continuous land with the nearest continent; for on this latter view the migration would probably have been more complete; and if modification be admitted, all the forms of life would have been more equally modified, in accordance with the paramount importance of the relation of organism to organism.

I do not deny that there are many and grave difficulties in understanding how several of the inhabitants of the more remote islands, whether still retaining the same specific form or modified since their arrival, could have reached their present homes. But the probability of many islands having existed as halting-places, of which not a wreck now remains, must not be overlooked. I will here give a single instance of one of the cases of difficulty. Almost all oceanic islands, even the most isolated and smallest, are inhabited by land-shells, generally by endemic species, but sometimes by species found elsewhere. Dr Aug. A. Gould has given several interesting cases in regard to the land-shells of the islands of the Pacific. Now it is notorious that land-shells are very easily killed by salt; their eggs, at least such as I have tried, sink in sea-water and are killed by it. Yet there must be, on my view, some unknown, but highly efficient means for their transportal. Would the just-hatched young occasionally crawl on and adhere to the feet of birds roosting on the ground, and thus get transported? It occurred to me that land-shells, when hibernating and having a membranous diaphragm over the mouth of the shell, might be floated in chinks of drifted timber across moderately wide arms of the sea. And I found that several species did in this state withstand uninjured an immersion in sea-water during seven days: one of these shells was the Helix pomatia, and after it had again hibernated I put it in sea-water for twenty days, and it perfectly recovered. As this species has a thick calcareous operculum, I removed it, and when it had formed a new membranous one, I immersed it for fourteen days in sea-water, and it recovered and crawled away: but more experiments are wanted on this head.

The most striking and important fact for us in regard to the inhabitants of islands, is their affinity to those of the nearest mainland, without being actually the same species. Numerous instances could be given of this fact. I will give only one, that of the Galapagos Archipelago, situated under the equator, between 500 and 600 miles from the shores of South America. Here almost every product of the land and water bears the unmistakable stamp of the American continent. There are twenty-six land-birds and twenty-five of these are ranked by Mr Gould as distinct species, supposed to have been created here; yet the close affinity of most of these birds to American species in every character, in their habits, gestures, and tones of voice, was manifest. So it is with the other animals, and with nearly all the plants, as shown by Dr Hooker in his admirable memoir on the Flora of this archipelago. The naturalist, looking at the inhabitants of these volcanic islands in the Pacific, distant several hundred miles from the continent, yet feels that he is standing on American land. Why should this be so? why should the species which are supposed to have been created in the Galapagos Archipelago, and nowhere else, bear so plain a stamp of affinity to those created in America? There is nothing in the conditions of life, in the geological nature of the islands, in their height or climate, or in the proportions in which the several classes are associated together, which resembles closely the conditions of the South American coast: in fact there is a considerable dissimilarity in all these respects. On the other hand, there is a considerable degree of resemblance in the volcanic nature of the soil, in climate, height, and size of the islands, between the Galapagos and Cape de Verde Archipelagos: but what an entire and absolute difference in their inhabitants! The inhabitants of the Cape de Verde Islands are related to those of Africa, like those of the Galapagos to America. I believe this grand fact can receive no sort of explanation on the ordinary view of independent creation; whereas on the view here maintained, it is obvious that the Galapagos Islands would be likely to receive colonists, whether by occasional means of transport or by formerly continuous land, from America; and the Cape de Verde Islands from Africa; and that such colonists would be liable to modification;—the principle of inheritance still betraying their original birthplace.

Many analogous facts could be given: indeed it is an almost universal rule that the endemic productions of islands are related to those of the nearest continent, or of other near islands. The exceptions are few, and most of them can be explained. Thus the plants of Kerguelen Land, though standing nearer to Africa than to America, are related, and that very closely, as we know from Dr Hooker's account, to those of America: but on the view that this island has been mainly stocked by seeds brought with earth and stones on icebergs, drifted by the prevailing currents, this anomaly disappears. New Zealand in its endemic plants is much more closely related to Australia, the nearest mainland, than to any other region: and this is what might have been expected; but it is also plainly related to South America, which, although the next nearest continent, is so enormously remote, that the fact becomes an anomaly. But this difficulty almost disappears on the view that both New Zealand, South America, and other southern lands were long ago partially stocked from a nearly intermediate though distant point, namely from the antarctic islands, when they were clothed with vegetation, before the commencement of the Glacial period. The affinity, which, though feeble, I am assured by Dr Hooker is real, between the flora of the south-western corner of Australia and of the Cape of Good Hope, is a far more remarkable case, and is at present inexplicable: but this affinity is confined to the plants, and will, I do not doubt, be some day explained.

The law which causes the inhabitants of an archipelago, though specifically distinct, to be closely allied to those of the nearest continent, we sometimes see displayed on a small scale, yet in a most interesting manner, within the limits of the same archipelago. Thus the several islands of the Galapagos Archipelago are tenanted, as I have elsewhere shown, in a quite marvellous manner, by very closely related species; so that the inhabitants of each separate island, though mostly distinct, are related in an incomparably closer degree to each other than to the inhabitants of any other part of the world. And this is just what might have been expected on my view, for the islands are situated so near each other that they would almost certainly receive immigrants from the same original source, or from each other. But this dissimilarity between the endemic inhabitants of the islands may be used as an argument against my views; for it may be asked, how has it happened in the several islands situated within sight of each other, having the same geological nature, the same height, climate, etc., that many of the immigrants should have been differently modified, though only in a small degree. This long appeared to me a great difficulty: but it arises in chief part from the deeply-seated error of considering the physical conditions of a country as the most important for its inhabitants; whereas it cannot, I think, be disputed that the nature of the other inhabitants, with which each has to compete, is at least as important, and generally a far more important element of success. Now if we look to those inhabitants of the Galapagos Archipelago which are found in other parts of the world (laying on one side for the moment the endemic species, which cannot be here fairly included, as we are considering how they have come to be modified since their arrival), we find a considerable amount of difference in the several islands. This difference might indeed have been expected on the view of the islands having been stocked by occasional means of transport—a seed, for instance, of one plant having been brought to one island, and that of another plant to another island. Hence when in former times an immigrant settled on any one or more of the islands, or when it subsequently spread from one island to another, it would undoubtedly be exposed to different conditions of life in the different islands, for it would have to compete with different sets of organisms: a plant for instance, would find the best-fitted ground more perfectly occupied by distinct plants in one island than in another, and it would be exposed to the attacks of somewhat different enemies. If then it varied, natural selection would probably favour different varieties in the different islands. Some species, however, might spread and yet retain the same character throughout the group, just as we see on continents some species spreading widely and remaining the same.

The really surprising fact in this case of the Galapagos Archipelago, and in a lesser degree in some analogous instances, is that the new species formed in the separate islands have not quickly spread to the other islands. But the islands, though in sight of each other, are separated by deep arms of the sea, in most cases wider than the British Channel, and there is no reason to suppose that they have at any former period been continuously united. The currents of the sea are rapid and sweep across the archipelago, and gales of wind are extraordinarily rare; so that the islands are far more effectually separated from each other than they appear to be on a map. Nevertheless a good many species, both those found in other parts of the world and those confined to the archipelago, are common to the several islands, and we may infer from certain facts that these have probably spread from some one island to the others. But we often take, I think, an erroneous view of the probability of closely-allied species invading each other's territory, when put into free intercommunication. Undoubtedly if one species has any advantage whatever over another, it will in a very brief time wholly or in part supplant it; but if both are equally well fitted for their own places in nature, both probably will hold their own places and keep separate for almost any length of time. Being familiar with the fact that many species, naturalised through man's agency, have spread with astonishing rapidity over new countries, we are apt to infer that most species would thus spread; but we should remember that the forms which become naturalised in new countries are not generally closely allied to the aboriginal inhabitants, but are very distinct species, belonging in a large proportion of cases, as shown by Alph. de Candolle, to distinct genera. In the Galapagos Archipelago, many even of the birds, though so well adapted for flying from island to island, are distinct on each; thus there are three closely-allied species of mocking-thrush, each confined to its own island. Now let us suppose the mocking-thrush of Chatham Island to be blown to Charles Island, which has its own mocking-thrush: why should it succeed in establishing itself there? We may safely infer that Charles Island is well stocked with its own species, for annually more eggs are laid there than can possibly be reared; and we may infer that the mocking-thrush peculiar to Charles Island is at least as well fitted for its home as is the species peculiar to Chatham Island. Sir C Lyell and Mr Wollaston have communicated to me a remarkable fact bearing on this subject; namely, that Madeira and the adjoining islet of Porto Santo possess many distinct but representative land-shells, some of which live in crevices of stone; and although large quantities of stone are annually transported from Porto Santo to Madeira, yet this latter island has not become colonised by the Porto Santo species: nevertheless both islands have been colonised by some European land-shells, which no doubt had some advantage over the indigenous species. From these considerations I think we need not greatly marvel at the endemic and representative species, which inhabit the several islands of the Galapagos Archipelago, not having universally spread from island to island. In many other instances, as in the several districts of the same continent, pre-occupation has probably played an important part in checking the commingling of species under the same conditions of life. Thus, the south-east and south-west corners of Australia have nearly the same physical conditions, and are united by continuous land, yet they are inhabited by a vast number of distinct mammals, birds, and plants.

The principle which determines the general character of the fauna and flora of oceanic islands, namely, that the inhabitants, when not identically the same, yet are plainly related to the inhabitants of that region whence colonists could most readily have been derived,—the colonists having been subsequently modified and better fitted to their new homes,—is of the widest application throughout nature. We see this on every mountain, in every lake and marsh. For Alpine species, excepting in so far as the same forms, chiefly of plants, have spread widely throughout the world during the recent Glacial epoch, are related to those of the surrounding lowlands;—thus we have in South America, Alpine humming-birds, Alpine rodents, Alpine plants, etc., all of strictly American forms, and it is obvious that a mountain, as it became slowly upheaved, would naturally be colonised from the surrounding lowlands. So it is with the inhabitants of lakes and marshes, excepting in so far as great facility of transport has given the same general forms to the whole world. We see this same principle in the blind animals inhabiting the caves of America and of Europe. Other analogous facts could be given. And it will, I believe, be universally found to be true, that wherever in two regions, let them be ever so distant, many closely-allied or representative species occur, there will likewise be found some identical species, showing, in accordance with the foregoing view, that at some former period there has been intercommunication or migration between the two regions. And wherever many closely-allied species occur, there will be found many forms which some naturalists rank as distinct species, and some as varieties; these doubtful forms showing us the steps in the process of modification.

This relation between the power and extent of migration of a species, either at the present time or at some former period under different physical conditions, and the existence at remote points of the world of other species allied to it, is shown in another and more general way. Mr Gould remarked to me long ago, that in those genera of birds which range over the world, many of the species have very wide ranges. I can hardly doubt that this rule is generally true, though it would be difficult to prove it. Amongst mammals, we see it strikingly displayed in Bats, and in a lesser degree in the Felidæ and Canidæ. We see it, if we compare the distribution of butterflies and beetles. So it is with most fresh-water productions, in which so many genera range over the world, and many individual species have enormous ranges. It is not meant that in world-ranging genera all the species have a wide range, or even that they have on an *average* a wide range; but only that some of the species range very widely; for the facility with which widely-ranging species vary and give rise to new forms will largely determine their average range. For instance, two varieties of the same species inhabit America and Europe, and the species thus has an immense range; but, if the variation had been a little greater, the two varieties would have been ranked as distinct species, and the common range would have been greatly reduced. Still less is it meant, that a species which apparently has the capacity of crossing barriers and ranging widely, as in the case of certain powerfully-winged birds, will necessarily range widely; for we should never forget that to range widely implies not only the power of crossing barriers, but the more important power of being victorious in distant lands in the struggle for life with foreign associates. But on the view of all the species of a genus having descended from a single parent, though now distributed to the most remote points of the world, we ought to find, and I believe as a general rule we do find, that some at least of the species range very widely; for it is necessary that the unmodified parent should range widely, undergoing modification during its diffusion, and should place itself under diverse conditions favourable for the conversion of its offspring, firstly into new varieties and ultimately into new species.

In considering the wide distribution of certain genera, we should bear in mind that some are extremely ancient, and must have branched off from a common parent at a remote epoch; so that in such cases there will have been ample time for great climatal and geographical changes and for accidents of transport; and consequently for the migration of some of the species into all quarters of the world, where they may have become slightly modified in relation to their new conditions. There is, also, some reason to believe from geological evidence that organisms low in the scale within each great class, generally change at a slower rate than the higher forms; and consequently the lower forms will have had a better chance of ranging widely and of still retaining the same specific character. This fact, together with the seeds and eggs of many low forms being very minute and better fitted for distant transportation, probably accounts for a law which has long been observed, and which has lately been admirably discussed by Alph. de Candolle in regard to plants, namely, that the lower any group of organisms is, the more widely it is apt to range.

The relations just discussed,—namely, low and slowly-changing organisms ranging more widely than the high,—some of the species of widely-ranging genera themselves ranging widely,—such facts, as alpine, lacustrine, and marsh productions being related (with the exceptions before specified) to those on the surrounding low lands and dry lands, though these stations are so different,—the very close relation of the distinct species which inhabit the islets of the same archipelago,—and especially the striking relation of the inhabitants of each whole archipelago or island to those of the nearest mainland,—are, I think, utterly inexplicable on the ordinary view of the independent creation of each species, but are explicable on the view of colonisation from the nearest or readiest source, together with the subsequent modification and better adaptation of the colonists to their new homes.

Summary of last and present Chapters.—In these chapters I have endeavoured to show, that if we make due allowance for our ignorance of the full effects of all the changes of climate and of the level of the land, which have certainly occurred within the recent period, and of other similar changes which may have occurred within the same period; if we remember how profoundly ignorant we are with respect to the many and curious means of occasional transport,—a subject which has hardly ever been properly experimentised on; if we bear in mind how often a species may have ranged continously over a wide area, and then have become extinct in the intermediate tracts, I think the difficulties in believing that all the individuals of the same species, wherever located, have descended from the same parents, are not insuperable. And we are led to this conclusion, which has been arrived at by many naturalists under the designation of single centres of creation, by some general considerations, more especially from the importance of barriers and from the analogical distribution of sub-genera, genera, and families.

With respect to the distinct species of the same genus, which on my theory must have spread from one parent-source; if we make the same allowances as before for our ignorance, and remember that some forms of life change most slowly, enormous periods of time being thus granted for their migration, I do not think that the difficulties are insuperable; though they often are in this case, and in that of the individuals of the same species, extremely great.

As exemplifying the effects of climatal changes on distribution, I have attempted to show how important has been the influence of the modern Glacial period, which I am fully convinced simultaneously affected the whole world, or at least great meridional belts. As showing how diversified are the means of occasional transport, I have discussed at some little length the means of dispersal of fresh-water productions.

If the difficulties be not insuperable in admitting that in the long course of time the individuals of the same species, and likewise of allied species, have proceeded from some one source; then I think all the grand leading facts of geographical distribution are explicable on the theory of migration (generally of the more dominant forms of life), together with subsequent modification and the multiplication of new forms. We can thus understand the high importance of barriers, whether of land or water, which separate our several zoological and botanical provinces. We can thus understand the localisation of sub-genera, genera, and families; and how it is that under different latitudes, for instance in South America, the inhabitants of the plains and mountains, of the forests, marshes, and deserts, are in so mysterious a manner linked together by affinity, and are likewise linked to the extinct beings which formerly inhabited the same continent. Bearing in mind that the mutual relation of organism to organism is of the highest importance, we can see why two areas having nearly the same physical conditions should often be inhabited by very different forms of life; for according to the length of time which has elapsed since new inhabitants entered one region; according to the nature of the communication which allowed certain forms and not others to enter, either in greater or lesser numbers; according or not, as those which entered happened to come in more or less direct competition with each other and with the aborigines; and according as the immigrants were capable of varying more or less rapidly, there would ensue in different regions, independently of their physical conditions, infinitely diversified conditions of life,—there would be an almost endless amount of organic action and reaction,—and we should find, as we do find, some groups of beings greatly, and some only slightly modified,—some developed in great force, some existing in scanty numbers—in the different great geographical provinces of the world.

On these same principles, we can understand, as I have endeavoured to show, why oceanic islands should have few inhabitants, but of these a great number should be endemic or peculiar; and why, in relation to the means of migration, one group of beings, even within the same class, should have all its species endemic, and another group should have all its species common to other quarters of the world. We can see why whole groups of organisms, as batrachians and terrestrial mammals, should be absent from oceanic islands, whilst the most isolated islands possess their own peculiar species of aërial mammals or bats. We can see why there should be some relation between the presence of mammals, in a more or less modified condition, and the depth of the sea between an island and the mainland. We can clearly see why all the inhabitants of an archipelago, though specifically distinct on the several islets, should be closely related to each other, and likewise be related, but less closely, to those of the nearest continent or other source whence immigrants were probably derived. We can see why in two areas, however distant from each other, there should be a correlation, in the presence of identical species, of varieties, of doubtful species, and of distinct but representative species.

As the late Edward Forbes often insisted, there is a striking parallelism in the laws of life throughout time and space: the laws governing the succession of forms in past time being nearly the same with those governing at the present time the differences in different areas. We see this in many facts. The endurance of each species and group of species is continuous in time; for the exceptions to the rule are so few, that they may fairly be attributed to our not having as yet discovered in an intermediate deposit the forms which are therein absent, but which occur above and below: so in space, it certainly is the general rule that the area inhabited by a single species, or by a group of species, is continuous; and the exceptions, which are not rare, may, as I have attempted to show, be accounted for by migration at some former period under different conditions or by occasional means of transport, and by the species having become extinct in the intermediate tracts. Both in time and space, species and groups of species have their points of maximum development. Groups of species, belonging either to a certain period of time, or to a certain area, are often characterised by trifling characters in common, as of sculpture or colour. In looking to the long succession of ages, as in now looking to distant provinces throughout the world, we find that some organisms differ little, whilst others belonging to a different class, or to a different order, or even only to a different family of the same order, differ greatly. In both time and space the lower members of each class generally change less than the higher; but there are in both cases marked exceptions to the rule. On my theory these several relations throughout time and space are intelligible; for whether we look to the forms of life which have changed during successive ages within the same quarter of the world, or to those which have changed after having migrated into distant quarters, in both cases the forms within each class have been connected by the same bond of ordinary generation; and the more nearly any two forms are related in blood, the nearer they will generally stand to each other in time and space; in both cases the laws of variation have been the same, and modifications have been accumulated by the same power of natural selection.

CHAPTER XIII

MUTUAL AFFINITIES OF ORGANIC BEINGS: MORPHOLOGY: EMBRYOLOGY: RUDIMENTARY ORGANS

CLASSIFICATION, groups subordinate to groups—Natural system—Rules and difficulties in classification, explained on the theory of descent with modification—Classification of varieties—Descent always used in classification—Analogical or adaptive characters—Affinities, general, complex and radiating—Extinction separates and defines groups—MORPHOLOGY, between members of the same class, between parts of the same individual—EMBRYOLOGY, laws of, explained by variations not supervening at an early age, and being inherited at a corresponding age—RUDIMENTARY ORGANS; their origin explained—Summary.

FROM the first dawn of life, all organic beings are found to resemble each other in descending degrees, so that they can be classed in groups under groups. This classification is evidently not arbitrary like the grouping of the stars in constellations. The existence of groups would have been of simple signification, if one group had been exclusively fitted to inhabit the land, and another the water; one to feed on flesh, another on vegetable matter, and so on; but the case is widely different in nature; for it is notorious how commonly members of even the same sub-group have different habits. In our second and fourth chapters, on Variation and on Natural Selection, I have attempted to show that it is the widely ranging, the much diffused and common, that is the dominant species belonging to the larger genera, which vary most. The varieties, or incipient species, thus produced ultimately become converted, as I believe, into new and distinct species; and these, on the principle of inheritance, tend to produce other new and dominant species. Consequently the groups which are now large, and which generally include many dominant species, tend to go on increasing indefinitely in size. I further attempted to show that from the varying descendants of each species trying to occupy as many and as different places as possible in the economy of nature, there is a constant tendency in their characters to diverge. This conclusion was supported by looking at the great diversity of the forms of life which, in any small area, come into the closest competition, and by looking to certain facts in naturalisation.

I attempted also to show that there is a constant tendency in the forms which are increasing in number and diverging in character, to supplant and exterminate the less divergent, the less improved, and preceding forms. I request the reader to turn to the diagram illustrating the action, as formerly explained, of these several principles; and he will see that the inevitable result is that the modified descendants proceeding from one progenitor become broken up into groups subordinate to groups. In the diagram each letter on the uppermost line may represent a genus including several species; and all the genera on this line form together one class, for all have descended from one ancient but unseen parent, and, consequently, have inherited something in common. But the three genera on the left hand have, on this same principle, much in common, and form a sub-family, distinct from that including the next two genera on the right hand, which diverged from a common parent at the fifth stage of descent. These five genera have also much, though less, in common; and they form a family distinct from that including the three genera still further to the right hand, which diverged at a still earlier period. And all these genera, descended from (A), form an order distinct from the genera descended from (I). So that we here have many species descended from a single progenitor grouped into genera; and the genera are included in, or subordinate to, sub-families, families, and orders, all united into one class. Thus, the grand fact in natural history of the subordination of group under group, which, from its familiarity, does not always sufficiently strike us, is in my judgment explained.

Naturalists try to arrange the species, genera, and families in each class, on what is called the Natural System. But what is meant by this system? Some authors look at it merely as a scheme for arranging together those living objects which are most alike, and for separating those which are most unlike; or as an artificial means for enunciating, as briefly as possible, general propositions,—that is, by one sentence to give the characters common, for instance, to all mammals, by another those common to all carnivora, by another those common to the dog-genus, and then by adding a single sentence, a full description is given of each kind of dog. The ingenuity and utility of this system are indisputable. But many naturalists think that something more is meant by the Natural System; they believe that it reveals the plan of the Creator; but unless it be specified whether order in time or space, or what else is meant by the plan of the Creator, it seems to me that nothing is thus added to our knowledge. Such expressions as that famous one of Linnæus, and which we often meet with in a more or less concealed form, that the characters do not make the genus, but that the genus gives the characters, seem to imply that something more is included in our classification, than mere resemblance. I believe that something more is included; and that propinquity of descent,—the only known cause of the similarity of organic beings,—is the bond, hidden as it is by various degrees of modification, which is partially revealed to us by our classifications.

Let us now consider the rules followed in classification, and the difficulties which are encountered on the view that classification either gives some unknown plan of creation, or is simply a scheme for enunciating general propositions and of placing together the forms most like each other. It might have been thought (and was in ancient times thought) that those parts of the structure which determined the habits of life, and the general place of each being in the economy of nature, would be of very high importance in classification. Nothing can be more false. No one regards the external similarity of a mouse to a shrew, of a dugong to a whale, of a whale to a fish, as of any importance. These resemblances, though so intimately connected with the whole life of the being, are ranked as merely 'adaptive or analogical characters'; but to the consideration of these resemblances we shall have to recur. It may even be given as a general rule, that the less any part of the organisation is concerned with special habits, the more important it becomes for classification. As an instance: Owen, in speaking of the dugong, says, 'The generative organs being those which are most remotely related to the habits and food of an animal, I have always regarded as affording very clear indications of its true affinities. We are least likely in the modifications of these organs to mistake a merely adaptive for an essential character.' So with plants, how remarkable it is that the organs of vegetation, on which their whole life depends, are of little signification, excepting in the first main divisions; whereas the organs of reproduction, with their product the seed, are of paramount importance!

We must not, therefore, in classifying, trust to resemblances in parts of the organisation, however important they may be for the welfare of the being in relation to the outer world. Perhaps from this cause it has partly arisen, that almost all naturalists lay the greatest stress on resemblances in organs of high vital or physiological importance. No doubt this view of the classificatory importance of organs which are important is generally, but by no means always, true. But their importance for classification, I believe, depends on their greater constancy throughout large groups of species; and this constancy depends on such organs having generally been subjected to less change in the adaptation of the species to their conditions of life. That the mere physiological importance of an organ does not determine its classificatory value, is almost shown by the one fact, that in allied groups, in which the same organ, as we have every reason to suppose, has nearly the same physiological value, its classificatory value is widely different. No naturalist can have worked at any group without being struck with this fact; and it has been fully acknowledged in the writings of almost every author. It will suffice to quote the highest authority, Robert Brown, who in speaking of certain organs in the Proteaceæ, says their generic importance, 'like that of all their parts, not only in this but, as I apprehend, in every natural family, is very unequal, and in some cases seems to be entirely lost.' Again in another work he says, the genera of the Connaraceæ 'differ in having one or more ovaria, in the existence or absence of albumen, in the imbricate or valvular æstivation. Any one of these characters singly is frequently of more than generic importance, though here even when all taken together they appear insufficient to separate Cnestis from Connarus.' To give an example amongst insects, in one great division of the Hymenoptera, the antennæ, as Westwood has remarked, are most constant in structure; in another division they differ much, and the differences are of quite subordinate value in classification; yet no one probably will say that the antennæ in these two divisions of the same order are of unequal physiological importance. Any number of instances could be given of the varying importance for classification of the same important organ within the same group of beings.

Again, no one will say that rudimentary or atrophied organs are of high physiological or vital importance; yet, undoubtedly, organs in this condition are often of high value in classification. No one will dispute that the rudimentary teeth in the upper jaws of young ruminants, and certain rudimentary bones of the leg, are highly serviceable in exhibiting the close affinity between Ruminants and Pachyderms. Robert Brown has strongly insisted on the fact that the rudimentary florets are of the highest importance in the classification of the Grasses.

Numerous instances could be given of characters derived from parts which must be considered of very trifling physiological importance, but which are universally admitted as highly serviceable in the definition of whole groups. For instance, whether or not there is an open passage from the nostrils to the mouth, the only character, according to Owen, which absolutely distinguishes fishes and reptiles—the inflection of the angle of the jaws in Marsupials—the manner in which the wings of insects are folded—mere colour in certain Algæ—mere pubescence on parts of the flower in grasses—the nature of the dermal covering, as hair or feathers, in the Vertebrata. If the Ornithorhynchus had been covered with feathers instead of hair, this external and trifling character would, I think, have been considered by naturalists as important an aid in determining the degree of affinity of this strange creature to birds and reptiles, as an approach in structure in any one internal and important organ.

The importance, for classification, of trifling characters, mainly depends on their being correlated with several other characters of more or less importance. The value indeed of an aggregate of characters is very evident in natural history. Hence, as has often been remarked, a species may depart from its allies in several characters, both of high physiological importance and of almost universal prevalence, and yet leave us in no doubt where it should be ranked. Hence, also, it has been found, that a classification founded on any single character, however important that may be, has always failed; for no part of the organisation is universally constant. The importance of an aggregate of characters, even when none are important, alone explains, I think, that saying of Linnæus, that the characters do not give the genus, but the genus gives the characters; for this saying seems founded on an appreciation of many trifling points of resemblance, too slight to be defined. Certain plants, belonging to the Malpighiaceæ, bear perfect and degraded flowers; in the latter, as A. de Jussieu has remarked, 'the greater number of the characters proper to the species, to the genus, to the family, to the class, disappear, and thus laugh at our classification.' But when Aspicarpa produced in France, during several years, only degraded flowers, departing so wonderfully in a number of the most important points of structure from the proper type of the order, yet M. Richard sagaciously saw, as Jussieu observes, that this genus should still be retained amongst the Malpighiaceæ. This case seems to me well to illustrate the spirit with which our classifications are sometimes necessarily founded.

Practically when naturalists are at work, they do not trouble themselves about the physiological value of the characters which they use in defining a group, or in allocating any particular species. If they find a character nearly uniform, and common to a great number of forms, and not common to others, they use it as one of high value; if common to some lesser number, they use it as of subordinate value. This principle has been broadly confessed by some naturalists to be the true one; and by none more clearly than by that excellent botanist, Aug. St. Hilaire. If certain characters are always found correlated with others, though no apparent bond of connection can be discovered between them, especial value is set on them. As in most groups of animals, important organs, such as those for propelling the blood, or for aërating it, or those for propagating the race, are found nearly uniform, they are considered as highly serviceable in classification; but in some groups of animals all these, the most important vital organs, are found to offer characters of quite subordinate value.

We can see why characters derived from the embryo should be of equal importance with those derived from the adult, for our classifications of course include all ages of each species. But it is by no means obvious, on the ordinary view, why the structure of the embryo should be more important for this purpose than that of the adult, which alone plays its full part in the economy of nature. Yet it has been strongly urged by those great naturalists, Milne Edwards and Agassiz, that embryonic characters are the most important of any in the classification of animals; and this doctrine has very generally been admitted as true. The same fact holds good with flowering plants, of which the two main divisions have been founded on characters derived from the embryo,—on the number and position of the embryonic leaves or cotyledons, and on the mode of development of the plumule and radicle. In our discussion on embryology, we shall see why such characters are so valuable, on the view of classification tacitly including the idea of descent.

Our classifications are often plainly influenced by chains of affinities. Nothing can be easier than to define a number of characters common to all birds; but in the case of crustaceans, such definition has hitherto been found impossible. There are crustaceans at the opposite ends of the series, which have hardly a character in common; yet the species at both ends, from being plainly allied to others, and these to others, and so onwards, can be recognised as unequivocally belonging to this, and to no other class of the Articulata.

Geographical distribution has often been used, though perhaps not quite logically, in classification, more especially in very large groups of closely allied forms. Temminck insists on the utility or even necessity of this practice in certain groups of birds; and it has been followed by several entomologists and botanists.

Finally, with respect to the comparative value of the various groups of species, such as orders, sub-orders, families, sub-families, and genera, they seem to be, at least at present, almost arbitrary. Several of the best botanists, such as Mr Bentham and others, have strongly insisted on their arbitrary value. Instances could be given amongst plants and insects, of a group of forms, first ranked by practised naturalists as only a genus, and then raised to the rank of a sub-family or family; and this has been done, not because further research has detected important structural differences, at first overlooked, but because numerous allied species, with slightly different grades of difference, have been subsequently discovered.

All the foregoing rules and aids and difficulties in classification are explained, if I do not greatly deceive myself, on the view that the natural system is founded on descent with modification; that the characters which naturalists consider as showing true affinity between any two or more species, are those which have been inherited from a common parent, and, in so far, all true classification is genealogical; that community of descent is the hidden bond which naturalists have been unconsciously seeking, and not some unknown plan of creation, or the enunciation of general propositions, and the mere putting together and separating objects more or less alike.

But I must explain my meaning more fully. I believe that the *arrangement* of the groups within each class, in due subordination and relation to the other groups, must be strictly genealogical in order to be natural; but that the *amount* of difference in the several branches or groups, though allied in the same degree in blood to their common progenitor, may differ greatly, being due to the different degrees of modification which they have undergone; and this is expressed by the forms being ranked under different genera, families, sections, or orders. The reader will best understand what is meant, if he will take the trouble of referring to the diagram in the preliminary. We will suppose the letters A to L to represent allied genera, which lived during the Silurian epoch, and these have descended from a species which existed at an unknown anterior period. Species of three of these genera (A, F, and I) have transmitted modified descendants to the present day, represented by the fifteen genera (a^{14} to z^{14}) on the uppermost horizontal line. Now all these modified descendants from a single species, are represented as related in blood or descent to the same degree; they may metaphorically be called cousins to the same millionth degree; yet they differ widely and in different degrees from each other. The forms descended from A, now broken up into two or three families, constitute a distinct order from those descended from I, also broken up into two families. Nor can the existing species, descended from A, be ranked in the same genus with the parent A; or those from I, with the parent I. But the existing genus F^{14} may be supposed to have been but slightly modified; and it will then rank with the parent-genus F; just as some few still living organic beings belong to Silurian genera. So that the amount or value of the differences between organic beings all related to each other in the same degree in blood, has come to be widely different. Nevertheless their genealogical *arrangement* remains strictly true, not only at the present time, but at each successive period of descent. All the modified descendants from A will have inherited something in common from their common parent, as will all the descendants from I; so will it be with each subordinate branch of descendants, at each successive period. If, however, we choose to suppose that any of the descendants of A or of I have been so much modified as to have more or less completely lost traces of their parentage, in this case, their places in a natural classification will have been more or less completely lost,—as sometimes seems to have occurred with existing organisms. All the descendants of the genus F, along its whole line of descent, are supposed to have been but little modified, and they yet form a single genus. But this genus, though much isolated, will still occupy its proper intermediate position; for F originally was intermediate in character between A and I, and the several genera descended from these two genera will have inherited to a certain extent their characters. This natural arrangement is shown, as far as is possible on paper, in the diagram, but in much too simple a manner. If a branching diagram had not been used, and only the names of the groups had been written in a linear series, it would have been still less possible to have given a natural arrangement; and it is notoriously not possible to represent in a series, on a flat surface, the affinities which we discover in nature amongst the beings of the same group. Thus, on the view which I hold, the natural system is genealogical in its arrangement, like a pedigree; but the degrees of modification which the different groups have undergone, have to be expressed by ranking them under different so-called genera, sub-families, families, sections, orders, and classes.

It may be worth while to illustrate this view of classification, by taking the case of languages. If we possessed a perfect pedigree of mankind, a genealogical arrangement of the races of man would afford the best classification of the various languages now spoken throughout the world; and if all extinct languages, and all intermediate and slowly changing dialects, had to be included, such an arrangement would, I think, be the only possible one. Yet it might be that some very ancient language had altered little, and had given rise to few new languages, whilst others (owing to the spreading and subsequent isolation and states of civilisation of the several races, descended from a common race) had altered much, and had given rise to many new languages and dialects. The various degrees of difference in the languages from the same stock, would have to be expressed by groups subordinate to groups; but the proper or even only possible arrangement would still be genealogical; and this would be strictly natural, as it would connect together all languages, extinct and modern, by the closest affinities, and would give the filiation and origin of each tongue.

In confirmation of this view, let us glance at the classification of varieties, which are believed or known to have descended from one species. These are grouped under species, with sub-varieties under varieties; and with our domestic productions, several other grades of difference are requisite, as we have seen with pigeons. The origin of the existence of groups subordinate to groups, is the same with varieties as with species, namely, closeness of descent with various degrees of modification. Nearly the same rules are followed in classifying varieties, as with species. Authors have insisted on the necessity of classing varieties on a natural instead of an artificial system; we are cautioned, for instance, not to class two varieties of the pine-apple together, merely because their fruit, though the most important part, happens to be nearly identical; no one puts the swedish and common turnips together, though the esculent and thickened stems are so similar. Whatever part is found to be most constant, is used in classing varieties: thus the great agriculturist Marshall says the horns are very useful for this purpose with cattle, because they are less variable than the shape or colour of the body, etc.; whereas with sheep the horns are much less serviceable, because less constant. In classing varieties, I apprehend if we had a real pedigree, a genealogical classification would be universally preferred; and it has been attempted by some authors. For we might feel sure, whether there had been more or less modification, the principle of inheritance would keep the forms together which were allied in the greatest number of points. In tumbler pigeons, though some sub-varieties differ from the others in the important character of having a longer beak, yet all are kept together from having the common habit of tumbling; but the short-faced breed has nearly or quite lost this habit; nevertheless, without any reasoning or thinking on the subject, these tumblers are kept in the same group, because allied in blood and alike in some other respects. If it could be proved that the Hottentot had descended from the Negro, I think he would be classed under the Negro group, however much he might differ in colour and other important characters from negroes.

With species in a state of nature, every naturalist has in fact brought descent into his classification; for he includes in his lowest grade, or that of a species, the two sexes; and how enormously these sometimes differ in the most important characters, is known to every naturalist: scarcely a single fact can be predicated in common of the males and hermaphrodites of certain cirripedes, when adult, and yet no one dreams of separating them. The naturalist includes as one species the several larval stages of the same individual, however much they may differ from each other and from the adult; as he likewise includes the so-called alternate generations of Steenstrup, which can only in a technical sense be considered as the same individual. He includes monsters; he includes varieties, not solely because they closely resemble the parent-form, but because they are descended from it. He who believes that the cowslip is descended from the primrose, or conversely, ranks them together as a single species, and gives a single definition. As soon as three Orchidean forms (Monochanthus, Myanthus, and Catasetum), which had previously been ranked as three distinct genera, were known to be sometimes produced on the same spike, they were immediately included as a single species.

As descent has universally been used in classing together the individuals of the same species, though the males and females and larvæ are sometimes extremely different; and as it has been used in classing varieties which have undergone a certain, and sometimes a considerable amount of modification, may not this same element of descent have been unconsciously used in grouping species under genera, and genera under higher groups, though in these cases the modification has been greater in degree, and has taken a longer time to complete? I believe it has thus been unconsciously used; and only thus can I understand the several rules and guides which have been followed by our best systematists. We have no written pedigrees; we have to make out community of descent by resemblances of any kind. Therefore we choose those characters which, as far as we can judge, are the least likely to have been modified in relation to the conditions of life to which each species has been recently exposed. Rudimentary structures on this view are as good as, or even sometimes better than, other parts of the organisation. We care not how trifling a character may be—let it be the mere inflection of the angle of the jaw, the manner in which an insect's wing is folded, whether the skin be covered by hair or feathers—if it prevail throughout many and different species, especially those having very different habits of life, it assumes high value; for we can account for its presence in so many forms with such different habits, only by its inheritance from a common parent. We may err in this respect in regard to single points of structure, but when several characters, let them be ever so trifling, occur together throughout a large group of beings having different habits, we may feel almost sure, on the theory of descent, that these characters have been inherited from a common ancestor. And we know that such correlated or aggregated characters have especial value in classification.

We can understand why a species or a group of species may depart, in several of its most important characteristics, from its allies, and yet be safely classed with them. This may be safely done, and is often done, as long as a sufficient number of characters, let them be ever so unimportant, betrays the hidden bond of community of descent. Let two forms have not a single character in common, yet if these extreme forms are connected together by a chain of intermediate groups, we may at once infer their community of descent, and we put them all into the same class. As we find organs of high physiological importance—those which serve to preserve life under the most diverse conditions of existence—are generally the most constant, we attach especial value to them; but if these same organs, in another group or section of a group, are found to differ much, we at once value them less in our classification. We shall hereafter, I think, clearly see why embryological characters are of such high classificatory importance. Geographical distribution may sometimes be brought usefully into play in classing large and widely-distributed genera, because all the species of the same genus, inhabiting any distinct and isolated region, have in all probability descended from the same parents.

We can understand, on these views, the very important distinction between real affinities and analogical or adaptive resemblances. Lamarck first called attention to this distinction, and he has been ably followed by Macleay and others. The resemblance, in the shape of the body and in the fin-like anterior limbs, between the dugong, which is a pachydermatous animal, and the whale, and between both these mammals and fishes, is analogical. Amongst insects there are innumerable instances: thus Linnæus, misled by external appearances, actually classed an homopterous insect as a moth. We see something of the same kind even in our domestic varieties, as in the thickened stems of the common and swedish turnip. The resemblance of the greyhound and racehorse is hardly more fanciful than the analogies which have been drawn by some authors between very distinct animals. On my view of characters being of real importance for classification, only in so far as they reveal descent, we can clearly understand why analogical or adaptive character, although of the utmost importance to the welfare of the being, are almost valueless to the systematist. For animals, belonging to two most distinct lines of descent, may readily become adapted to similar conditions, and thus assume a close external resemblance; but such resemblances will not reveal—will rather tend to conceal their blood-relationship to their proper lines of descent. We can also understand the apparent paradox, that the very same characters are analogical when one class or order is compared with another, but give true affinities when the members of the same class or order are compared one with another: thus the shape of the body and fin-like limbs are only analogical when whales are compared with fishes, being adaptations in both classes for swimming through the water; but the shape of the body and fin-like limbs serve as characters exhibiting true affinity between the several members of the whale family; for these cetaceans agree in so many characters, great and small, that we cannot doubt that they have inherited their general shape of body and structure of limbs from a common ancestor. So it is with fishes.

As members of distinct classes have often been adapted by successive slight modifications to live under nearly similar circumstances,—to inhabit for instance the three elements of land, air, and water,—we can perhaps understand how it is that a numerical parallelism has sometimes been observed between the sub-groups in distinct classes. A naturalist, struck by a parallelism of this nature in any one class, by arbitrarily raising or sinking the value of the groups in other classes (and all our experience shows that this valuation has hitherto been arbitrary), could easily extend the parallelism over a wide range; and thus the septenary, quinary, quaternary, and ternary classifications have probably arisen.

As the modified descendants of dominant species, belonging to the larger genera, tend to inherit the advantages, which made the groups to which they belong large and their parents dominant, they are almost sure to spread widely, and to seize on more and more places in the economy of nature. The larger and more dominant groups thus tend to go on increasing in size; and they consequently supplant many smaller and feebler groups. Thus we can account for the fact that all organisms, recent and extinct, are included under a few great orders, under still fewer classes, and all in one great natural system. As showing how few the higher groups are in number, and how widely spread they are throughout the world, the fact is striking, that the discovery of Australia has not added a single insect belonging to a new class; and that in the vegetable kingdom, as I learn from Dr Hooker, it has added only two or three orders of small size.

In the chapter on geological succession I attempted to show, on the principle of each group having generally diverged much in character during the long-continued process of modification, how it is that the more ancient forms of life often present characters in some slight degree intermediate between existing groups. A few old and intermediate parent-forms having occasionally transmitted to the present day descendants but little modified, will give to us our so-called osculant or aberrant groups. The more aberrant any form is, the greater must be the number of connecting forms which on my theory have been exterminated and utterly lost. And we have some evidence of aberrant forms having suffered severely from extinction, for they are generally represented by extremely few species; and such species as do occur are generally very distinct from each other, which again implies extinction. The genera Ornithorhynchus and Lepidosiren, for example, would not have been less aberrant had each been represented by a dozen species instead of by a single one; but such richness in species, as I find after some investigation, does not commonly fall to the lot of aberrant genera. We can, I think, account for this fact only by looking at aberrant forms as failing groups conquered by more successful competitors, with a few members preserved by some unusual coincidence of favourable circumstances.

Mr Waterhouse has remarked that, when a member belonging to one group of animals exhibits an affinity to a quite distinct group, this affinity in most cases is general and not special: thus, according to Mr Waterhouse, of all Rodents, the bizcacha is most nearly related to Marsupials; but in the points in which it approaches this order, its relations are general, and not to any one marsupial species more than to another. As the points of affinity of the bizcacha to Marsupials are believed to be real and not merely adaptive, they are due on my theory to inheritance in common. Therefore we must suppose either that all Rodents, including the bizcacha, branched off from some very ancient Marsupial, which will have had a character in some degree intermediate with respect to all existing Marsupials; or that both Rodents and Marsupials branched off from a common progenitor, and that both groups have since undergone much modification in divergent directions. On either view we may suppose that the bizcacha has retained, by inheritance, more of the character of its ancient progenitor than have other Rodents; and therefore it will not be specially related to any one existing Marsupial, but indirectly to all or nearly all Marsupials, from having partially retained the character of their common progenitor, or of an early member of the group. On the other hand, of all Marsupials, as Mr Waterhouse has remarked, the phascolomys resembles most nearly, not any one species, but the general order of Rodents. In this case, however, it may be strongly suspected that the resemblance is only analogical, owing to the phascolomys having become adapted to habits like those of a Rodent. The elder De Candolle has made nearly similar observations on the general nature of the affinities of distinct orders of plants.

On the principle of the multiplication and gradual divergence in character of the species descended from a common parent, together with their retention by inheritance of some characters in common, we can understand the excessively complex and radiating affinities by which all the members of the same family or higher group are connected together. For the common parent of a whole family of species, now broken up by extinction into distinct groups and sub-groups, will have transmitted some of its characters, modified in various ways and degrees, to all; and the several species will consequently be related to each other by circuitous lines of affinity of various lengths (as may be seen in the diagram so often referred to), mounting up through many predecessors. As it is difficult to show the blood-relationship between the numerous kindred of any ancient and noble family, even by the aid of a genealogical tree, and almost impossible to do this without this aid, we can understand the extraordinary difficulty which naturalists have experienced in describing, without the aid of a diagram, the various affinities which they perceive between the many living and extinct members of the same great natural class.

Extinction, as we have seen in the fourth chapter, has played an important part in defining and widening the intervals between the several groups in each class. We may thus account even for the distinctness of whole classes from each other—for instance, of birds from all other vertebrate animals—by the belief that many ancient forms of life have been utterly lost, through which the early progenitors of birds were formerly connected with the early progenitors of the other vertebrate classes. There has been less entire extinction of the forms of life which once connected fishes with batrachians. There has been still less in some other classes, as in that of the Crustacea, for here the most wonderfully diverse forms are still tied together by a long, but broken, chain of affinities. Extinction has only separated groups: it has by no means made them; for if every form which has ever lived on this earth were suddenly to reappear, though it would be quite impossible to give definitions by which each group could be distinguished from other groups, as all would blend together by steps as fine as those between the finest existing varieties, nevertheless a natural classification, or at least a natural arrangement, would be possible. We shall see this by turning to the diagram: the letters, A to L, may represent eleven Silurian genera, some of which have produced large groups of modified descendants. Every intermediate link between these eleven genera and their primordial parent, and every intermediate link in each branch and sub-branch of their descendants, may be supposed to be still alive; and the links to be as fine as those between the finest varieties. In this case it would be quite impossible to give any definition by which the several members of the several groups could be distinguished from their more immediate parents; or these parents from their ancient and unknown progenitor. Yet the natural arrangement in the diagram would still hold good;

and, on the principle of inheritance, all the forms descended from A, or from I, would have something in common. In a tree we can specify this or that branch, though at the actual fork the two unite and blend together. We could not, as I have said, define the several groups; but we could pick out types, or forms, representing most of the characters of each group, whether large or small, and thus give a general idea of the value of the differences between them. This is what we should be driven to, if we were ever to succeed in collecting all the forms in any class which have lived throughout all time and space. We shall certainly never succeed in making so perfect a collection: nevertheless, in certain classes, we are tending in this direction; and Milne Edwards has lately insisted, in an able paper, on the high importance of looking to types, whether or not we can separate and define the groups to which such types belong.

Finally, we have seen that natural selection, which results from the struggle for existence, and which almost inevitably induces extinction and divergence of character in the many descendants from one dominant parent-species, explains that great and universal feature in the affinities of all organic beings, namely, their subordination in group under group. We use the element of descent in classing the individuals of both sexes and of all ages, although having few characters in common, under one species; we use descent in classing acknowledged varieties, however different they may be from their parent; and I believe this element of descent is the hidden bond of connection which naturalists have sought under the term of the Natural-System. On this idea of the natural system being, in so far as it has been perfected, genealogical in its arrangement, with the grades of difference between the descendants from a common parent, expressed by the terms genera, families, orders, etc., we can understand the rules which we are compelled to follow in our classification. We can understand why we value certain resemblances far more than others; why we are permitted to use rudimentary and useless organs, or others of trifling physiological importance; why, in comparing one group with a distinct group, we summarily reject analogical or adaptive characters, and yet use these same characters within the limits of the same group. We can clearly see how it is that all living and extinct forms can be grouped together in one great system; and how the several members of each class are connected together by the most complex and radiating lines of affinities. We shall never, probably, disentangle the inextricable web of affinities between the members of any one class; but when we have a distinct object in view, and do not look to some unknown plan of creation, we may hope to make sure but slow progress.

Morphology.—We have seen that the members of the same class, independently of their habits of life, resemble each other in the general plan of their organisation. This resemblance is often expressed by the term 'unity of type;' or by saying that the several parts and organs in the different species of the class are homologous. The whole subject is included under the general name of Morphology. This is the most interesting department of natural history, and may be said to be its very soul. What can be more curious than that the hand of a man, formed for grasping, that of a mole for digging, the leg of the horse, the paddle of the porpoise, and the wing of the bat, should all be constructed on the same pattern, and should include similar bones, in the same relative positions? Geoffroy St Hilaire has insisted strongly on the high importance of relative connection in homologous organs: the parts may change to almost any extent in form and size, and yet they always remain connected together in the same order. We never find, for instance, the bones of the arm and forearm, or of the thigh and leg, transposed. Hence the same names can be given to the homologous bones in widely different animals. We see the same great law in the construction of the mouths of insects: what can be more different than the immensely long spiral proboscis of a sphinx-moth, the curious folded one of a bee or bug, and the great jaws of a beetle?—yet all these organs, serving for such different purposes, are formed by infinitely numerous modifications of an upper lip, mandibles, and two pairs of maxillæ. Analogous laws govern the construction of the mouths and limbs of crustaceans. So it is with the flowers of plants.

Nothing can be more hopeless than to attempt to explain this similarity of pattern in members of the same class, by utility or by the doctrine of final causes. The hopelessness of the attempt has been expressly admitted by Owen in his most interesting work on the 'Nature of Limbs.' On the ordinary view of the independent creation of each being, we can only say that so it is;—that it has so pleased the Creator to construct each animal and plant.

The explanation is manifest on the theory of the natural selection of successive slight modifications,—each modification being profitable in some way to the modified form, but often affecting by correlation of growth other parts of the organisation. In changes of this nature, there will be little or no tendency to modify the original pattern, or to transpose parts. The bones of a limb might be shortened and widened to any extent, and become gradually enveloped in thick membrane, so as to serve as a fin; or a webbed foot might have all its bones, or certain bones, lengthened to any extent, and the membrane connecting them increased to any extent, so as to serve as a wing: yet in all this great amount of modification there will be no tendency to alter the framework of bones or the relative connection of the several parts. If we suppose that the ancient progenitor, the archetype as it may be called, of all mammals, had its limbs constructed on the existing general pattern, for whatever purpose they served, we can at once perceive the plain signification of the homologous construction of the limbs throughout the whole class. So with the mouths of insects, we have only to suppose that their common progenitor had an upper lip, mandibles, and two pairs of maxillæ, these parts being perhaps very simple in form; and then natural selection, acting on some originally created form, will account for the infinite diversity in structure and function of the mouths of insects. Nevertheless, it is conceivable that the general pattern of an organ might become so much obscured as to be finally lost, by the atrophy and ultimately by the complete abortion of certain parts, by the soldering together of other parts, and by the doubling or multiplication of others,—variations which we know to be within the limits of possibility. In the paddles of the extinct gigantic sea-lizards, and in the mouths of certain suctorial crustaceans, the general pattern seems to have been thus to a certain extent obscured.

There is another and equally curious branch of the present subject; namely, the comparison not of the same part in different members of a class, but of the different parts or organs in the same individual. Most physiologists believe that the bones of the skull are homologous with—that is correspond in number and in relative connection with—the elemental parts of a certain number of vertebræ. The anterior and posterior limbs in each member of the vertebrate and articulate classes are plainly homologous. We see the same law in comparing the wonderfully complex jaws and legs in crustaceans. It is familiar to almost every one, that in a flower the relative position of the sepals, petals, stamens, and pistils, as well as their intimate structure, are intelligible on the view that they consist of metamorphosed leaves, arranged in a spire. In monstrous plants, we often get direct evidence of the possibility of one organ being transformed into another; and we can actually see in embryonic crustaceans and in many other animals, and in flowers, that organs, which when mature become extremely different, are at an early stage of growth exactly alike.

How inexplicable are these facts on the ordinary view of creation! Why should the brain be enclosed in a box composed of such numerous and such extraordinary shaped pieces of bone? As Owen has remarked, the benefit derived from the yielding of the separate pieces in the act of parturition of mammals, will by no means explain the same construction in the skulls of birds. Why should similar bones have been created in the formation of the wing and leg of a bat, used as they are for such totally different purposes? Why should one crustacean, which has an extremely complex mouth formed of many parts, consequently always have fewer legs; or conversely, those with many legs have simpler mouths? Why should the sepals, petals, stamens, and pistils in any individual flower, though fitted for such widely different purposes, be all constructed on the same pattern?

On the theory of natural selection, we can satisfactorily answer these questions. In the vertebrata, we see a series of internal vertebræ bearing certain processes and appendages; in the articulata, we see the body divided into a series of segments, bearing external appendages; and in flowering plants, we see a series of successive spiral whorls of leaves. An indefinite repetition of the same part or organ is the common characteristic (as Owen has observed) of all low or little-modified forms; therefore we may readily believe that the unknown progenitor of the vertebrata possessed many vertebræ; the unknown progenitor of the articulata, many segments; and the unknown progenitor of flowering plants, many spiral whorls of leaves. We have formerly seen that parts many times repeated are eminently liable to vary in number and structure; consequently it is quite probable that natural selection, during a long-continued course of modification, should have seized on a certain number of the primordially similar elements, many times repeated, and have adapted them to the most diverse purposes. And as the whole amount of modification will have been effected by slight successive steps, we need not wonder at discovering in such parts or organs, a certain degree of fundamental resemblance, retained by the strong principle of inheritance.

In the great class of molluscs, though we can homologise the parts of one species with those of other and distinct species, we can indicate but few serial homologies; that is, we are seldom enabled to say that one part or organ is homologous with another in the same individual. And we can understand this fact; for in molluscs, even in the lowest members of the class, we do not find nearly so much indefinite repetition of any one part, as we find in the other great classes of the animal and vegetable kingdoms.

Naturalists frequently speak of the skull as formed of metamorphosed vertebræ; the jaws of crabs as metamorphosed legs; the stamens and pistils of flowers as metamorphosed leaves; but it would in these cases probably be more correct, as Professor Huxley has remarked, to speak of both skull and vertebræ, both jaws and legs, etc.,—as having been metamorphosed, not one from the other, but from some common element. Naturalists, however, use such language only in a metaphorical sense: they are far from meaning that during a long course of descent, primordial organs of any kind—vertebræ in the one case and legs in the other—have actually been modified into skulls or jaws. Yet so strong is the appearance of a modification of this nature having occurred, that naturalists can hardly avoid employing language having this plain signification. On my view these terms may be used literally; and the wonderful fact of the jaws, for instance, of a crab retaining numerous characters, which they would probably have retained through inheritance, if they had really been metamorphosed during a long course of descent from true legs, or from some simple appendage, is explained.

Embryology.—It has already been casually remarked that certain organs in the individual, which when mature become widely different and serve for different purposes, are in the embryo exactly alike. The embryos, also, of distinct animals within the same class are often strikingly similar: a better proof of this cannot be given, than a circumstance mentioned by Agassiz, namely, that having forgotten to ticket the embryo of some vertebrate animal, he cannot now tell whether it be that of a mammal, bird, or reptile. The vermiform larvæ of moths, flies, beetles, etc., resemble each other much more closely than do the mature insects; but in the case of larvæ, the embryos are active, and have been adapted for special lines of life. A trace of the law of embryonic resemblance, sometimes lasts till a rather late age: thus birds of the same genus, and of closely allied genera, often resemble each other in their first and second plumage; as we see in the spotted feathers in the thrush group. In the cat tribe, most of the species are striped or spotted in lines; and stripes can be plainly distinguished in the whelp of the lion. We occasionally though rarely see something of this kind in plants: thus the embryonic leaves of the ulex or furze, and the first leaves of the phyllodineous acaceas, are pinnate or divided like the ordinary leaves of the leguminosæ.

The points of structure, in which the embryos of widely different animals of the same class resemble each other, often have no direct relation to their conditions of existence. We cannot, for instance, suppose that in the embryos of the vertebrata the peculiar loop-like course of the arteries near the branchial slits are related to similar conditions,—in the young mammal which is nourished in the womb of its mother, in the egg of the bird which is hatched in a nest, and in the spawn of a frog under water. We have no more reason to believe in such a relation, than we have to believe that the same bones in the hand of a man, wing of a bat, and fin of a porpoise, are related to similar conditions of life. No one will suppose that the stripes on the whelp of a lion, or the spots on the young blackbird, are of any use to these animals, or are related to the conditions to which they are exposed.

The case, however, is different when an animal during any part of its embryonic career is active, and has to provide for itself. The period of activity may come on earlier or later in life; but whenever it comes on, the adaptation of the larva to its conditions of life is just as perfect and as beautiful as in the adult animal. From such special adaptations, the similarity of the larvæ or active embryos of allied animals is sometimes much obscured; and cases could be given of the larvæ of two species, or of two groups of species, differing quite as much, or even more, from each other than do their adult parents. In most cases, however, the larvæ, though active, still obey, more or less closely, the law of common embryonic resemblance. Cirripedes afford a good instance of this: even the illustrious Cuvier did not perceive that a barnacle was, as it certainly is, a crustacean; but a glance at the larva shows this to be the case in an unmistakable manner. So again the two main divisions of cirripedes, the pedunculated and sessile, which differ widely in external appearance, have larvæ in all their stages barely distinguishable.

The embryo in the course of development generally rises in organisation: I use this expression, though I am aware that it is hardly possible to define clearly what is meant by the organisation being higher or lower. But no one probably will dispute that the butterfly is higher than the caterpillar. In some cases, however, the mature animal is generally considered as lower in the scale than the larva, as with certain parasitic crustaceans. To refer once again to cirripedes: the larvæ in the first stage have three pairs of legs, a very simple single eye, and a probosciformed mouth, with which they feed largely, for they increase much in size. In the second stage, answering to the chrysalis stage of butterflies, they have six pairs of beautifully constructed natatory legs, a pair of magnificent compound eyes, and extremely complex antennæ; but they have a closed and imperfect mouth, and cannot feed: their function at this stage is, to search by their well-developed organs of sense, and to reach by their active powers of swimming, a proper place on which to become attached and to undergo their final metamorphosis. When this is completed they are fixed for life: their legs are now converted into prehensile organs; they again obtain a well-constructed mouth; but they have no antennæ, and their two eyes are now reconverted into a minute, single, and very simple eye-spot. In this last and complete state, cirripedes may be considered as either more highly or more lowly organised than they were in the larval condition. But in some genera the larvæ become developed either into hermaphrodites having the ordinary structure, or into what I have called complemental males: and in the latter, the development has assuredly been retrograde; for the male is a mere sack, which lives for a short time, and is destitute of mouth, stomach, or other organ of importance, excepting for reproduction.

We are so much accustomed to see differences in structure between the embryo and the adult, and likewise a close similarity in the embryos of widely different animals within the same class, that we might be led to look at these facts as necessarily contingent in some manner on growth. But there is no obvious reason why, for instance, the wing of a bat, or the fin of a porpoise, should not have been sketched out with all the parts in proper proportion, as soon as any structure became visible in the embryo. And in some whole groups of animals and in certain members of other groups, the embryo does not at any period differ widely from the adult: thus Owen has remarked in regard to cuttle-fish, 'there is no metamorphosis; the cephalopodic character is manifested long before the parts of the embryo are completed'; and again in spiders, 'there is nothing worthy to be called a metamorphosis.' The larvæ of insects, whether adapted to the most diverse and active habits, or quite inactive, being fed by their parents or placed in the midst of proper nutriment, yet nearly all pass through a similar worm-like stage of development; but in some few cases, as in that of Aphis, if we look to the admirable drawings by Professor Huxley of the development of this insect, we see no trace of the vermiform stage.

How, then, can we explain these several facts in embryology,—namely the very general, but not universal difference in structure between the embryo and the adult;—of parts in the same individual embryo, which ultimately become very unlike and serve for diverse purposes, being at this early period of growth alike;—of embryos of different species within the same class, generally, but not universally, resembling each other;—of the structure of the embryo not being closely related to its conditions of existence, except when the embryo becomes at any period of life active and has to provide for itself;—of the embryo apparently having sometimes a higher organisation than the mature animal, into which it is developed? I believe that all these facts can be explained, as follows, on the view of descent with modification.

It is commonly assumed, perhaps from monstrosities often affecting the embryos at a very early period, that slight variations necessarily appear at an equally early period. But we have little evidence on this head—indeed the evidence rather points the other way; for it is notorious that breeders of cattle, horses, and various fancy animals, cannot positively tell, until some time after the animal has been born, what its merits or form will ultimately turn out. We see this plainly in our own children; we cannot always tell whether the child will be tall or short, or what its precise features will be. The question is not, at what period of life any variation has been caused, but at what period it is fully displayed. The cause may have acted, and I believe generally has acted, even before the embryo is formed; and the variation may be due to the male and female sexual elements having been affected by the conditions to which either parent, or their ancestors, have been exposed. Nevertheless an effect thus caused at a very early period, even before the formation of the embryo, may appear late in life; as when an hereditary disease, which appears in old age alone, has been communicated to the offspring from the reproductive element of one parent. Or again, as when the horns of cross-bred cattle have been affected by the shape of the horns of either parent. For the welfare of a very young animal, as long as it remains in its mother's womb, or in the egg, or as long as it is nourished and protected by its parent, it must be quite unimportant whether most of its characters are fully acquired a little earlier or later in life. It would not signify, for instance, to a bird which obtained its food best by having a long beak, whether or not it assumed a beak of this particular length, as long as it was fed by its parents. Hence, I conclude, that it is quite possible that each of the many successive modifications, by which each species has acquired its present structure, may have supervened at a not very early period of life; and some direct evidence from our domestic animals supports this view. But in other cases it is quite possible that each successive modification, or most of them, may have appeared at an extremely early period.

I have stated in the first chapter, that there is some evidence to render it probable, that at whatever age any variation first appears in the parent, it tends to reappear at a corresponding age in the offspring. Certain variations can only appear at corresponding ages, for instance, peculiarities in the caterpillar, cocoon, or imago states of the silk-moth; or, again, in the horns of almost full-grown cattle. But further than this, variations which, for all that we can see, might have appeared earlier or later in life, tend to appear at a corresponding age in the offspring and parent. I am far from meaning that this is invariably the case; and I could give a good many cases of variations (taking the word in the largest sense) which have supervened at an earlier age in the child than in the parent.

These two principles, if their truth be admitted, will, I believe, explain all the above specified leading facts in embryology. But first let us look at a few analogous cases in domestic varieties. Some authors who have written on Dogs, maintain that the greyhound and bull-dog, though appearing so different, are really varieties most closely allied, and have probably descended from the same wild stock; hence I was curious to see how far their puppies differed from each other: I was told by breeders that they differed just as much as their parents, and this, judging by the eye, seemed almost to be the case; but on actually measuring the old dogs and their six-days old puppies, I found that the puppies had not nearly acquired their full amount of proportional difference. So, again, I was told, that the foals of cart and race-horses differed as much as the full-grown animals; and this surprised me greatly, as I think it probable that the difference between these two breeds has been wholly caused by selection under domestication; but having had careful measurements made of the dam and of a three-days old colt of a race and heavy cart-horse, I find that the colts have by no means acquired their full amount of proportional difference.

As the evidence appears to me conclusive, that the several domestic breeds of Pigeon have descended from one wild species, I compared young pigeons of various breeds, within twelve hours after being hatched; I carefully measured the proportions (but will not here give details) of the beak, width of mouth, length of nostril and of eyelid, size of feet and length of leg, in the wild stock, in pouters, fantails, runts, barbs, dragons, carriers, and tumblers. Now some of these birds, when mature, differ so extraordinarily in length and form of beak, that they would, I cannot doubt, be ranked in distinct genera, had they been natural productions. But when the nestling birds of these several breeds were placed in a row, though most of them could be distinguished from each other, yet their proportional differences in the above specified several points were incomparably less than in the full-grown birds. Some characteristic points of difference—for instance, that of the width of mouth—could hardly be detected in the young. But there was one remarkable exception to this rule, for the young of the short-faced tumbler differed from the young of the wild rock-pigeon and of the other breeds, in all its proportions, almost exactly as much as in the adult state.

The two principles above given seem to me to explain these facts in regard to the later embryonic stages of our domestic varieties. Fanciers select their horses, dogs, and pigeons, for breeding, when they are nearly grown up: they are indifferent whether the desired qualities and structures have been acquired earlier or later in life, if the full-grown animal possesses them. And the cases just given, more especially that of pigeons, seem to show that the characteristic differences which give value to each breed, and which have been accumulated by man's selection, have not generally first appeared at an early period of life, and have been inherited by the offspring at a corresponding not early period. But the case of the short-faced tumbler, which when twelve hours old had acquired its proper proportions, proves that this is not the universal rule; for here the characteristic differences must either have appeared at an earlier period than usual, or, if not so, the differences must have been inherited, not at the corresponding, but at an earlier age.

Now let us apply these facts and the above two principles—which latter, through not proved true, can be shown to be in some degree probable—to species in a state of nature. Let us take a genus of birds, descended on my theory from some one parent-species, and of which the several new species have become modified through natural selection in accordance with their diverse habits. Then, from the many slight successive steps of variation having supervened at a rather late age, and having been inherited at a corresponding age, the young of the new species of our supposed genus will manifestly tend to resemble each other much more closely than do the adults, just as we have seen in the case of pigeons. We may extend this view to whole families or even classes. The fore-limbs, for instance, which served as legs in the parent-species, may have become, by a long course of modification, adapted in one descendant to act as hands, in another as paddles, in another as wings; and on the above two principles—namely of each successive modification supervening at a rather late age, and being inherited at a corresponding late age—the fore-limbs in the embryos of the several descendants of the parent-species will still resemble each other closely, for they will not have been modified. But in each of our new species, the embryonic fore-limbs will differ greatly from the fore-limbs in the mature animal; the limbs in the latter having undergone much modification at a rather later period of life, and having thus been converted into hands, or paddles, or wings. Whatever influence long-continued exercise or use on the one hand, and disuse on the other, may have in modifying an organ, such influence will mainly affect the mature animal, which has come to its full powers of activity and has to gain its own living; and the effects thus produced will be inherited at a corresponding mature age. Whereas the young will remain unmodified, or be modified in a lesser degree, by the effects of use and disuse.

In certain cases the successive steps of variation might supervene, from causes of which we are wholly ignorant, at a very early period of life, or each step might be inherited at an earlier period than that at which it first appeared. In either case (as with the short-faced tumbler) the young or embryo would closely resemble the mature parent-form. We have seen that this is the rule of development in certain whole groups of animals, as with cuttle-fish and spiders, and with a few members of the great class of insects, as with Aphis. With respect to the final cause of the young in these cases not undergoing any metamorphosis, or closely resembling their parents from their earliest age, we can see that this would result from the two following contingencies: firstly, from the young, during a course of modification carried on for many generations, having to provide for their own wants at a very early stage of development, and secondly, from their following exactly the same habits of life with their parents; for in this case, it would be indispensable for the existence of the species, that the child should be modified at a very early age in the same manner with its parents, in accordance with their similar habits. Some further explanation, however, of the embryo not undergoing any metamorphosis is perhaps requisite. If, on the other hand, it profited the young to follow habits of life in any degree different from those of their parent, and consequently to be constructed in a slightly different manner, then, on the principle of inheritance at corresponding ages, the active young or larvæ might easily be rendered by natural selection different to any conceivable extent from their parents. Such differences might, also, become correlated with successive stages of development; so that the larvæ, in the first stage, might differ greatly from the larvæ in the second stage, as we have seen to be the case with cirripedes. The adult might become fitted for sites or habits, in which organs of locomotion or of the senses, etc., would be useless; and in this case the final metamorphosis would be said to be retrograde.

As all the organic beings, extinct and recent, which have ever lived on this earth have to be classed together, and as all have been connected by the finest gradations, the best, or indeed, if our collections were nearly perfect, the only possible arrangement, would be genealogical. Descent being on my view the hidden bond of connection which naturalists have been seeking under the term of the natural system. On this view we can understand how it is that, in the eyes of most naturalists, the structure of the embryo is even more important for classification than that of the adult. For the embryo is the animal in its less modified state; and in so far it reveals the structure of its progenitor. In two groups of animals, however much they may at present differ from each other in structure and habits, if they pass through the same or similar embryonic stages, we may feel assured that they have both descended from the same or nearly similar parents, and are therefore in that degree closely related. Thus, community in embryonic structure reveals community of descent. It will reveal this community of descent, however much the structure of the adult may have been modified and obscured; we have seen, for instance, that cirripedes can at once be recognised by their larvæ as belonging to the great class of crustaceans. As the embryonic state of each species and group of species partially shows us the structure of their less modified ancient progenitors, we can clearly see why ancient and extinct forms of life should resemble the embryos of their descendants,—our existing species. Agassiz believes this to be a law of nature; but I am bound to confess that I only hope to see the law hereafter proved true. It can be proved true in those cases alone in which the ancient state, now supposed to be represented in existing embryos, has not been obliterated, either by the successive variations in a long course of modification having supervened at a very early age, or by the variations having been inherited at an earlier period than that at which they first appeared. It should also be borne in mind, that the supposed law of resemblance of ancient forms of life to the embryonic stages of recent forms, may be true, but yet, owing to the geological record not extending far enough back in time, may remain for a long period, or for ever, incapable of demonstration.

Thus, as it seems to me, the leading facts in embryology, which are second in importance to none in natural history, are explained on the principle of slight modifications not appearing, in the many descendants from some one ancient progenitor, at a very early period in the life of each, though perhaps caused at the earliest, and being inherited at a corresponding not early period. Embryology rises greatly in interest, when we thus look at the embryo as a picture, more or less obscured, of the common parent-form of each great class of animals.

Rudimentary, atrophied, or aborted Organs.—Organs or parts in this strange condition, bearing the stamp of inutility, are extremely common throughout nature. For instance, rudimentary mammæ are very general in the males of mammals: I presume that the 'bastard-wing' in birds may be safely considered as a digit in a rudimentary state: in very many snakes one lobe of the lungs is rudimentary; in other snakes there are rudiments of the pelvis and hind limbs. Some of the cases of rudimentary organs are extremely curious; for instance, the presence of teeth in fœtal whales, which when grown up have not a tooth in their heads; and the presence of teeth, which never cut through the gums, in the upper jaws of our unborn calves. It has even been stated on good authority that rudiments of teeth can be detected in the beaks of certain embryonic birds. Nothing can be plainer than that wings are formed for flight, yet in how many insects do we see wings so reduced in size as to be utterly incapable of flight, and not rarely lying under wing-cases, firmly soldered together!

The meaning of rudimentary organs is often quite unmistakable: for instance there are beetles of the same genus (and even of the same species) resembling each other most closely in all respects, one of which will have full-sized wings, and another mere rudiments of membrane; and here it is impossible to doubt, that the rudiments represent wings. Rudimentary organs sometimes retain their potentiality, and are merely not developed: this seems to be the case with the mammæ of male mammals, for many instances are on record of these organs having become well developed in full-grown males, and having secreted milk. So again there are normally four developed and two rudimentary teats in the udders of the genus Bos, but in our domestic cows the two sometimes become developed and give milk. In plants of the same species the petals sometimes occur as mere rudiments, and sometimes in a well-developed state. In plants with separated sexes, the male flowers often have a rudiment of a pistil; and Kölreuter found that by crossing such male plants with an hermaphrodite species, the rudiment of the pistil in the hybrid offspring was much increased in size; and this shows that the rudiment and the perfect pistil are essentially alike in nature.

An organ serving for two purposes, may become rudimentary or utterly aborted for one, even the more important purpose; and remain perfectly efficient for the other. Thus in plants, the office of the pistil is to allow the pollen-tubes to reach the ovules protected in the ovarium at its base. The pistil consists of a stigma supported on the style; but in some Compositæ, the male florets, which of course cannot be fecundated, have a pistil, which is in a rudimentary state, for it is not crowned with a stigma; but the style remains well developed, and is clothed with hairs as in other Compositæ, for the purpose of brushing the pollen out of the surrounding anthers. Again, an organ may become rudimentary for its proper purpose, and be used for a distinct object: in certain fish the swim-bladder seems to be nearly rudimentary for its proper function of giving buoyancy, but has become converted into a nascent breathing organ or lung. Other similar instances could be given.

Organs, however little developed, if of use, should not be called rudimentary; they cannot properly be said to be in an atrophied condition; they may be called nascent, and may hereafter be developed to any extent by natural selection. Rudimentary organs, on the other hand, are essentially useless, as teeth which never cut through the gums; in a still less developed condition, they would be of still less use. They cannot, therefore, under their present condition, have been formed by natural selection, which acts solely by the preservation of useful modifications; they have been retained, as we shall see, by inheritance, and relate to a former condition of their possessor. It is difficult to know what are nascent organs;

looking to the future, we cannot of course tell how any part will be developed, and whether it is now nascent; looking to the past, creatures with an organ in a nascent condition will generally have been supplanted and exterminated by their successors with the organ in a more perfect and developed condition. The wing of the penguin is of high service, and acts as a fin; it may, therefore, represent the nascent state of the wings of birds; not that I believe this to be the case, it is more probably a reduced organ, modified for a new function: the wing of the Apteryx is useless, and is truly rudimentary. The mammary glands of the Ornithorhynchus may, perhaps, be considered, in comparison with the udder of a cow, as in a nascent state. The ovigerous frena of certain cirripedes, which are only slightly developed and which have ceased to give attachment to the ova, are nascent branchiæ.

Rudimentary organs in the individuals of the same species are very liable to vary in degree of development and in other respects. Moreover, in closely allied species, the degree to which the same organ has been rendered rudimentary occasionally differs much. This latter fact is well exemplified in the state of the wings of the female moths in certain groups. Rudimentary organs may be utterly aborted; and this implies, that we find in an animal or plant no trace of an organ, which analogy would lead us to expect to find, and which is occasionally found in monstrous individuals of the species. Thus in the snapdragon (antirrhinum) we generally do not find a rudiment of a fifth stamen; but this may sometimes be seen. In tracing the homologies of the same part in different members of a class, nothing is more common, or more necessary, than the use and discovery of rudiments. This is well shown in the drawings given by Owen of the bones of the leg of the horse, ox, and rhinoceros.

It is an important fact that rudimentary organs, such as teeth in the upper jaws of whales and ruminants, can often be detected in the embryo, but afterwards wholly disappear. It is also, I believe, a universal rule, that a rudimentary part or organ is of greater size relatively to the adjoining parts in the embryo, than in the adult; so that the organ at this early age is less rudimentary, or even cannot be said to be in any degree rudimentary. Hence, also, a rudimentary organ in the adult is often said to have retained its embryonic condition.

I have now given the leading facts with respect to rudimentary organs. In reflecting on them, every one must be struck with astonishment: for the same reasoning power which tells us plainly that most parts and organs are exquisitely adapted for certain purposes, tells us with equal plainness that these rudimentary or atrophied organs, are imperfect and useless. In works on natural history rudimentary organs are generally said to have been created 'for the sake of symmetry', or in order 'to complete the scheme of nature'; but this seems to me no explanation, merely a re-statement of the fact. Would it be thought sufficient to say that because planets revolve in elliptic courses round the sun, satellites follow the same course round the planets, for the sake of symmetry, and to complete the scheme of nature? An eminent physiologist accounts for the presence of rudimentary organs, by supposing that they serve to excrete matter in excess, or injurious to the system; but can we suppose that the minute papilla, which often represents the pistil in male flowers, and which is formed merely of cellular tissue, can thus act? Can we suppose that the formation of rudimentary teeth, which are subsequently absorbed, can be of any service to the rapidly growing embryonic calf by the excretion of precious phosphate of lime? When a man's fingers have been amputated, imperfect nails sometimes appear on the stumps: I could as soon believe that these vestiges of nails have appeared, not from unknown laws of growth, but in order to excrete horny matter, as that the rudimentary nails on the fin of the manatee were formed for this purpose.

On my view of descent with modification, the origin of rudimentary organs is simple. We have plenty of cases of rudimentary organs in our domestic productions,—as the stump of a tail in tailless breeds,—the vestige of an ear in earless breeds,—the reappearance of minute dangling horns in hornless breeds of cattle, more especially, according to Youatt, in young animals,—and the state of the whole flower in the cauliflower. We often see rudiments of various parts in monsters. But I doubt whether any of these cases throw light on the origin of rudimentary organs in a state of nature, further than by showing that rudiments can be produced; for I doubt whether species under nature ever undergo abrupt changes. I believe that disuse has been the main agency; that it has led in successive generations to the gradual reduction of various organs, until they have become rudimentary,—as in the case of the eyes of animals inhabiting dark caverns, and of the wings of birds inhabiting oceanic islands, which have seldom been forced to take flight, and have ultimately lost the power of flying. Again, an organ useful under certain conditions, might become injurious under others, as with the wings of beetles living on small and exposed islands; and in this case natural selection would continue slowly to reduce the organ, until it was rendered harmless and rudimentary.

Any change in function, which can be effected by insensibly small steps, is within the power of natural selection; so that an organ rendered, during changed habits of life, useless or injurious for one purpose, might be modified and used for another purpose. Or an organ might be retained for one alone of its former functions. An organ, when rendered useless, may well be variable, for its variations cannot be checked by natural selection. At whatever period of life disuse or selection reduces an organ, and this will generally be when the being has come to maturity and to its full powers of action, the principle of inheritance at corresponding ages will reproduce the organ in its reduced state at the same age, and consequently will seldom affect or reduce it in the embryo. Thus we can understand the greater relative size of rudimentary organs in the embryo, and their lesser relative size in the adult. But if each step of the process of reduction were to be inherited, not at the corresponding age, but at an extremely early period of life (as we have good reason to believe to be possible), the rudimentary part would tend to be wholly lost, and we should have a case of complete abortion. The principle, also, of economy, explained in a former chapter, by which the materials forming any part or structure, if not useful to the possessor, will be saved as far as is possible, will probably often come into play; and this will tend to cause the entire obliteration of a rudimentary organ.

As the presence of rudimentary organs is thus due to the tendency in every part of the organisation, which has long existed, to be inherited—we can understand, on the genealogical view of classification, how it is that systematists have found rudimentary parts as useful as, or even sometimes more useful than, parts of high physiological importance. Rudimentary organs may be compared with the letters in a word, still retained in the spelling, but become useless in the pronunciation, but which serve as a clue in seeking for its derivation. On the view of descent with modification, we may conclude that the existence of organs in a rudimentary, imperfect, and useless condition, or quite aborted, far from presenting a strange difficulty, as they assuredly do on the ordinary doctrine of creation, might even have been anticipated, and can be accounted for by the laws of inheritance.

Summary.—In this chapter I have attempted to show, that the subordination of group to group in all organisms throughout all time; that the nature of the relationship, by which all living and extinct beings are united by complex, radiating, and circuitous lines of affinities into one grand system; the rules followed and the difficulties encountered by naturalists in their classifications; the value set upon characters, if constant and prevalent, whether of high vital importance, or of the most trifling importance, or, as in rudimentary organs, of no importance; the wide opposition in value between analogical or adaptive characters, and characters of true affinity; and other such rules;—all naturally follow on the view of the common parentage of those forms which are considered by naturalists as allied, together with their modification through natural selection, with its contingencies of extinction and divergence of character. In considering this view of classification, it should be borne in mind that the element of descent has been universally used in ranking together the sexes, ages, and acknowledged varieties of the same species, however different they may be in structure. If we extend the use of this element of descent,—the only certainly known cause of similarity in organic beings,—we shall understand what is meant by the natural system: it is genealogical in its attempted arrangement, with the grades of acquired difference marked by the terms varieties, species, genera, families, orders, and classes.

On this same view of descent with modification, all the great facts in Morphology become intelligible,—whether we look to the same pattern displayed in the homologous organs, to whatever purpose applied, of the different species of a class; or to the homologous parts constructed on the same pattern in each individual animal and plant.

On the principle of successive slight variations, not necessarily or generally supervening at a very early period of life, and being inherited at a corresponding period, we can understand the great leading facts in Embryology; namely, the resemblance in an individual embryo of the homologous parts, which when matured will become widely different from each other in structure and function; and the resemblance in different species of a class of the homologous parts or organs, though fitted in the adult members for purposes as different as possible. Larvæ are active embryos, which have become specially modified in relation to their habits of life, through the principle of modifications being inherited at corresponding ages. On this same principle—and bearing in mind, that when organs are reduced in size, either from disuse or selection, it will generally be at that period of life when the being has to provide for its own wants, and bearing in mind how strong is the principle of inheritance—the occurrence of rudimentary organs and their final abortion, present to us no inexplicable difficulties; on the contrary, their presence might have been even anticipated. The importance of embryological characters and of rudimentary organs in classification is intelligible, on the view that an arrangement is only so far natural as it is genealogical.

Finally, the several classes of facts which have been considered in this chapter, seem to me to proclaim so plainly, that the innumerable species, genera, and families of organic beings, with which this world is peopled, have all descended, each within its own class or group, from common parents, and have all been modified in the course of descent, that I should without hesitation adopt this view, even if it were unsupported by other facts or arguments.

CHAPTER XIV

RECAPITULATION AND CONCLUSION

Recapitulation of the difficulties on the theory of Natural Selection—Recapitulation of the general and special circumstances in its favour—Causes of the general belief in the immutability of species—How far the theory of natural selection may be extended—Effects of its adoption on the study of natural history—Concluding remarks.

As this whole volume is one long argument, it may be convenient to the reader to have the leading facts and inferences briefly recapitulated.

That many and serious objections may be advanced against the theory of descent with modification through natural selection, I do not deny. I have endeavoured to give to them their full force. Nothing at first can appear more difficult to believe than that the more complex organs and instincts should have been perfected, not by means superior to, though analogous with, human reason, but by the accumulation of innumerable slight variations, each good for the individual possessor. Nevertheless, this difficulty, though appearing to our imagination insuperably great, cannot be considered real if we admit the following propositions, namely,—that gradations in the perfection of any organ or instinct which we may consider, either do now exist or could have existed, each good of its kind,—that all organs and instincts are, in ever so slight a degree, variable,—and, lastly, that there is a struggle for existence leading to the preservation of each profitable deviation of structure or instinct. The truth of these propositions cannot, I think, be disputed.

It is, no doubt, extremely difficult even to conjecture by what gradations many structures have been perfected, more especially amongst broken and failing groups of organic beings; but we see so many strange gradations in nature, that we ought to be extremely cautious in saying that any organ or instinct, or any whole being, could not have arrived at its present state by many graduated steps. There are, it must be admitted, cases of special difficulty on the theory of natural selection; and one of the most curious of these is the existence of two or three defined castes of workers or sterile females in the same community of ants; but I have attempted to show how this difficulty can be mastered.

With respect to the almost universal sterility of species when first crossed, which forms so remarkable a contrast with the almost universal fertility of varieties when crossed, I must refer the reader to the recapitulation of the facts given at the end of the eighth chapter, which seem to me conclusively to show that this sterility is no more a special endowment than is the incapacity of two trees to be grafted together; but that it is incidental on constitutional differences in the reproductive systems of the intercrossed species. We see the truth of this conclusion in the vast difference in the result, when the same two species are crossed reciprocally; that is, when one species is first used as the father and then as the mother.

The fertility of varieties when intercrossed and of their mongrel offspring cannot be considered as universal; nor is their very general fertility surprising when we remember that it is not likely that either their constitutions or their reproductive systems should have been profoundly modified. Moreover, most of the varieties which have been experimentised on have been produced under domestication; and as domestication (I do not mean mere confinement) apparently tends to eliminate sterility, we ought not to expect it also to produce sterility.

The sterility of hybrids is a very different case from that of first crosses, for their reproductive organs are more or less functionally impotent; whereas in first crosses the organs on both sides are in a perfect condition. As we continually see that organisms of all kinds are rendered in some degree sterile from their constitutions having been disturbed by slightly different and new conditions of life, we need not feel surprise at hybrids being in some degree sterile, for their constitutions can hardly fail to have been disturbed from being compounded of two distinct organisations. This parallelism is supported by another parallel, but directly opposite, class of facts; namely, that the vigour and fertility of all organic beings are increased by slight changes in their conditions of life, and that the offspring of slightly modified forms or varieties acquire from being crossed increased vigour and fertility. So that, on the one hand, considerable changes in the conditions of life and crosses between greatly modified forms, lessen fertility; and on the other hand, lesser changes in the conditions of life and crosses between less modified forms, increase fertility.

Turning to geographical distribution, the difficulties encountered on the theory of descent with modification are grave enough. All the individuals of the same species, and all the species of the same genus, or even higher group, must have descended from common parents; and therefore, in however distant and isolated parts of the world they are now found, they must in the course of successive generations have passed from some one part to the others. We are often wholly unable even to conjecture how this could have been effected. Yet, as we have reason to believe that some species have retained the same specific form for very long periods, enormously long as measured by years, too much stress ought not to be laid on the occasional wide diffusion of the same species; for during very long periods of time there will always have been a good chance for wide migration by many means. A broken or interrupted range may often be accounted for by the extinction of the species in the intermediate regions. It cannot be denied that we are as yet very ignorant of the full extent of the various climatal and geographical changes which have affected the earth during modern periods; and such changes will obviously have greatly facilitated migration. As an example, I have attempted to show how potent has been the influence of the Glacial period on the distribution both of the same and of representative species throughout the world. We are as yet profoundly ignorant of the many occasional means of transport. With respect to distinct species of the same genus inhabiting very distant and isolated regions, as the process of modification has necessarily been slow, all the means of migration will have been possible during a very long period; and consequently the difficulty of the wide diffusion of species of the same genus is in some degree lessened.

As on the theory of natural selection an interminable number of intermediate forms must have existed, linking together all the species in each group by gradations as fine as our present varieties, it may be asked, Why do we not see these linking forms all around us? Why are not all organic beings blended together in an inextricable chaos? With respect to existing forms, we should remember that we have no right to expect (excepting in rare cases) to discover *directly* connecting links between them, but only between each and some extinct and supplanted form. Even on a wide area, which has during a long period remained continuous, and of which the climate and other conditions of life change insensibly in going from a district occupied by one species into another district occupied by a closely allied species, we have no just right to expect often to find intermediate varieties in the intermediate zone. For we have reason to believe that only a few species are undergoing change at any one period; and all changes are slowly effected. I have also shown that the intermediate varieties which will at first probably exist in the intermediate zones, will be liable to be supplanted by the allied forms on either hand; and the latter, from existing in greater numbers, will generally be modified and improved at a quicker rate than the intermediate varieties, which exist in lesser numbers; so that the intermediate varieties will, in the long run, be supplanted and exterminated.

On this doctrine of the extermination of an infinitude of connecting links, between the living and extinct inhabitants of the world, and at each successive period between the extinct and still older species, why is not every geological formation charged with such links? Why does not every collection of fossil remains afford plain evidence of the gradation and mutation of the forms of life? We meet with no such evidence, and this is the most obvious and forcible of the many objections which may be urged against my theory. Why, again, do whole groups of allied species appear, though certainly they often falsely appear, to have come in suddenly on the several geological stages? Why do we not find great piles of strata beneath the Silurian system, stored with the remains of the progenitors of the Silurian groups of fossils? For certainly on my theory such strata must somewhere have been deposited at these ancient and utterly unknown epochs in the world's history.

I can answer these questions and grave objections only on the supposition that the geological record is far more imperfect than most geologists believe. It cannot be objected that there has not been time sufficient for any amount of organic change; for the lapse of time has been so great as to be utterly inappreciable by the human intellect. The number of specimens in all our museums is absolutely as nothing compared with the countless generations of countless species which certainly have existed. We should not be able to recognise a species as the parent of any one or more species if we were to examine them ever so closely, unless we likewise possessed many of the intermediate links between their past or parent and present states; and these many links we could hardly ever expect to discover, owing to the imperfection of the geological record. Numerous existing doubtful forms could be named which are probably varieties; but who will pretend that in future ages so many fossil links will be discovered, that naturalists will be able to decide, on the common view, whether or not these doubtful forms are varieties? As long as most of the links between any two species are unknown, if any one link or intermediate variety be discovered, it will simply be classed as another and distinct species. Only a small portion of the world has been geologically explored. Only organic beings of certain classes can be preserved in a fossil condition, at least in any great number. Widely ranging species vary most, and varieties are often at first local,—both causes rendering the discovery of intermediate links less likely. Local varieties will not spread into other and distant regions until they are considerably modified and improved; and when they do spread, if discovered in a geological formation, they will appear as if suddenly created there, and will be simply classed as new species. Most formations have been intermittent in their accumulation; and their duration, I am inclined to believe, has been shorter than the average duration of specific forms. Successive formations are separated from each other by enormous blank intervals of time; for fossiliferous formations, thick enough to resist future degradation, can be accumulated only where much sediment is deposited on the subsiding bed of the sea. During the alternate periods of elevation and of stationary level the record will be blank. During these latter periods there will probably be more variability in the forms of life; during periods of subsidence, more extinction.

With respect to the absence of fossiliferous formations beneath the lowest Silurian strata, I can only recur to the hypothesis given in the ninth chapter. That the geological record is imperfect all will admit; but that it is imperfect to the degree which I require, few will be inclined to admit. If we look to long enough intervals of time, geology plainly declares that all species have changed; and they have changed in the manner which my theory requires, for they have changed slowly and in a graduated manner. We clearly see this in the fossil remains from consecutive formations invariably being much more closely related to each other, than are the fossils from formations distant from each other in time.

Such is the sum of the several chief objections and difficulties which may justly be urged against my theory; and I have now briefly recapitulated the answers and explanations which can be given to them. I have felt these difficulties far too heavily during many years to doubt their weight. But it deserves especial notice that the more important objections relate to questions on which we are confessedly ignorant; nor do we know how ignorant we are. We do not know all the possible transitional gradations between the simplest and the most perfect organs; it cannot be pretended that we know all the varied means of Distribution during the long lapse of years, or that we know how imperfect the Geological Record is. Grave as these several difficulties are, in my judgment they do not overthrow the theory of descent from a few created forms with subsequent modification.

Now let us turn to the other side of the argument. Under domestication we see much variability. This seems to be mainly due to the reproductive system being eminently susceptible to changes in the conditions of life; so that this system, when not rendered impotent, fails to reproduce offspring exactly like the parent-form. Variability is governed by many complex laws,—by correlation of growth, by use and disuse, and by the direct action of the physical conditions of life. There is much difficulty in ascertaining how much modification our domestic productions have undergone; but we may safely infer that the amount has been large, and that modifications can be inherited for long periods. As long as the conditions of life remain the same, we have reason to believe that a modification, which has already been inherited for many generations, may continue to be inherited for an almost infinite number of generations. On the other hand we have evidence that variability, when it has once come into play, does not wholly cease; for new varieties are still occasionally produced by our most anciently domesticated productions.

Man does not actually produce variability; he only unintentionally exposes organic beings to new conditions of life, and then nature acts on the organisation, and causes variability. But man can and does select the variations given to him by nature, and thus accumulate them in any desired manner. He thus adapts animals and plants for his own benefit or pleasure. He may do this methodically, or he may do it unconsciously by preserving the individuals most useful to him at the time, without any thought of altering the breed. It is certain that he can largely influence the character of a breed by selecting, in each successive generation, individual differences so slight as to be quite inappreciable by an uneducated eye. This process of selection has been the great agency in the production of the most distinct and useful domestic breeds. That many of the breeds produced by man have to a large extent the character of natural species, is shown by the inextricable doubts whether very many of them are varieties or aboriginal species.

There is no obvious reason why the principles which have acted so efficiently under domestication should not have acted under nature. In the preservation of favoured individuals and races, during the constantly-recurrent Struggle for Existence, we see the most powerful and ever-acting means of selection. The struggle for existence inevitably follows from the high geometrical ratio of increase which is common to all organic beings. This high rate of increase is proved by calculation,—by the rapid increase of many animals and plants during a succession of peculiar seasons, or when naturalised in a new country. More individuals are born than can possibly survive.

A grain in the balance will determine which individual shall live and which shall die,—which variety or species shall increase in number, and which shall decrease, or finally become extinct. As the individuals of the same species come in all respects into the closest competition with each other, the struggle will generally be most severe between them; it will be almost equally severe between the varieties of the same species, and next in severity between the species of the same genus. But the struggle will often be very severe between beings most remote in the scale of nature. The slightest advantage in one being, at any age or during any season, over those with which it comes into competition, or better adaptation in however slight a degree to the surrounding physical conditions, will turn the balance.

With animals having separated sexes there will in most cases be a struggle between the males for possession of the females. The most vigorous individuals, or those which have most successfully struggled with their conditions of life, will generally leave most progeny. But success will often depend on having special weapons or means of defence, or on the charms of the males; and the slightest advantage will lead to victory.

As geology plainly proclaims that each land has undergone great physical changes, we might have expected that organic beings would have varied under nature, in the same way as they generally have varied under the changed conditions of domestication. And if there be any variability under nature, it would be an unaccountable fact if natural selection had not come into play. It has often been asserted, but the assertion is quite incapable of proof, that the amount of variation under nature is a strictly limited quantity. Man, though acting on external characters alone and often capriciously, can produce within a short period a great result by adding up mere individual differences in his domestic productions; and every one admits that there are at least individual differences in species under nature. But, besides such differences, all naturalists have admitted the existence of varieties, which they think sufficiently distinct to be worthy of record in systematic works. No one can draw any clear distinction between individual differences and slight varieties; or between more plainly marked varieties and sub-species, and species. Let it be observed how naturalists differ in the rank which they assign to the many representative forms in Europe and North America.

If then we have under nature variability and a powerful agent always ready to act and select, why should we doubt that variations in any way useful to beings, under their excessively complex relations of life, would be preserved, accumulated, and inherited? Why, if man can by patience select variations most useful to himself, should nature fail in selecting variations useful, under changing conditions of life, to her living products? What limit can be put to this power, acting during long ages and rigidly scrutinising the whole constitution, structure, and habits of each creature,—favouring the good and rejecting the bad? I can see no limit to this power, in slowly and beautifully adapting each form to the most complex relations of life. The theory of natural selection, even if we looked no further than this, seems to me to be in itself probable. I have already recapitulated, as fairly as I could, the opposed difficulties and objections: now let us turn to the special facts and arguments in favour of the theory.

On the view that species are only strongly marked and permanent varieties, and that each species first existed as a variety, we can see why it is that no line of demarcation can be drawn between species, commonly supposed to have been produced by special acts of creation, and varieties which are acknowledged to have been produced by secondary laws. On this same view we can understand how it is that in each region where many species of a genus have been produced, and where they now flourish, these same species should present many varieties; for where the manufactory of species has been active, we might expect, as a general rule, to find it still in action; and this is the case if varieties be incipient species. Moreover, the species of the larger genera, which afford the greater number of varieties or incipient species, retain to a certain degree the character of varieties; for they differ from each other by a less amount of difference than do the species of smaller genera. The closely allied species also of the larger genera apparently have restricted ranges, and in their affinities they are clustered in little groups round other species—in which respects they resemble varieties. These are strange relations on the view of each species having been independently created, but are intelligible if all species first existed as varieties.

As each species tends by its geometrical ratio of reproduction to increase inordinately in number; and as the modified descendants of each species will be enabled to increase by so much the more as they become diversified in habits and structure, so as to be enabled to seize on many and widely different places in the economy of nature, there will be a constant tendency in natural selection to preserve the most divergent offspring of any one species. Hence during a long-continued course of modification, the slight differences, characteristic of varieties of the same species, tend to be augmented into the greater differences characteristic of species of the same genus. New and improved varieties will inevitably supplant and exterminate the older, less improved and intermediate varieties; and thus species are rendered to a large extent defined and distinct objects. Dominant species belonging to the larger groups tend to give birth to new and dominant forms; so that each large group tends to become still larger, and at the same time more divergent in character. But as all groups cannot thus succeed in increasing in size, for the world would not hold them, the more dominant groups beat the less dominant. This tendency in the large groups to go on increasing in size and diverging in character, together with the almost inevitable contingency of much extinction, explains the arrangement of all the forms of life, in groups subordinate to groups, all within a few great classes, which we now see everywhere around us, and which has prevailed throughout all time. This grand fact of the grouping of all organic beings seems to me utterly inexplicable on the theory of creation.

As natural selection acts solely by accumulating slight, successive, favourable variations, it can produce no great or sudden modification; it can act only by very short and slow steps. Hence the canon of 'Natura non facit saltum,' which every fresh addition to our knowledge tends to make truer, is on this theory simply intelligible. We can plainly see why nature is prodigal in variety, though niggard in innovation. But why this should be a law of nature if each species has been independently created, no man can explain.

Many other facts are, as it seems to me, explicable on this theory. How strange it is that a bird, under the form of woodpecker, should have been created to prey on insects on the ground; that upland geese, which never or rarely swim, should have been created with webbed feet; that a thrush should have been created to dive and feed on sub-aquatic insects; and that a petrel should have been created with habits and structure fitting it for the life of an auk or grebe! and so on in endless other cases. But on the view of each species constantly trying to increase in number, with natural selection always ready to adapt the slowly varying descendants of each to any unoccupied or ill-occupied place in nature, these facts cease to be strange, or perhaps might even have been anticipated.

As natural selection acts by competition, it adapts the inhabitants of each country only in relation to the degree of perfection of their associates; so that we need feel no surprise at the inhabitants of any one country, although on the ordinary view supposed to have been specially created and adapted for that country, being beaten and supplanted by the naturalised productions from another land. Nor ought we to marvel if all the contrivances in nature be not, as far as we can judge, absolutely perfect; and if some of them be abhorrent to our ideas of fitness. We need not marvel at the sting of the bee causing the bee's own death; at drones being produced in such vast numbers for one single act, with the great majority slaughtered by their sterile sisters; at the astonishing waste of pollen by our fir-trees; at the instinctive hatred of the queen bee for her own fertile daughters; at ichneumonidæ feeding within the live bodies of caterpillars; and at other such cases. The wonder indeed is, on the theory of natural selection, that more cases of the want of absolute perfection have not been observed.

The complex and little known laws governing variation are the same, as far as we can see, with the laws which have governed the production of so-called specific forms. In both cases physical conditions seem to have produced but little direct effect; yet when varieties enter any zone, they occasionally assume some of the characters of the species proper to that zone. In both varieties and species, use and disuse seem to have produced some effect; for it is difficult to resist this conclusion when we look, for instance, at the logger-headed duck, which has wings incapable of flight, in nearly the same condition as in the domestic duck; or when we look at the burrowing tucutucu, which is occasionally blind, and then at certain moles, which are habitually blind and have their eyes covered with skin; or when we look at the blind animals inhabiting the dark caves of America and Europe. In both varieties and species correlation of growth seems to have played a most important part, so that when one part has been modified other parts are necessarily modified. In both varieties and species reversions to long-lost characters occur. How inexplicable on the theory of creation is the occasional appearance of stripes on the shoulder and legs of the several species of the horse-genus and in their hybrids! How simply is this fact explained if we believe that these species have descended from a striped progenitor, in the same manner as the several domestic breeds of pigeon have descended from the blue and barred rock-pigeon!

On the ordinary view of each species having been independently created, why should the specific characters, or those by which the species of the same genus differ from each other, be more variable than the generic characters in which they all agree? Why, for instance, should the colour of a flower be more likely to vary in any one species of a genus, if the other species, supposed to have been created independently, have differently coloured flowers, than if all the species of the genus have the same coloured flowers? If species are only well-marked varieties, of which the characters have become in a high degree permanent, we can understand this fact; for they have already varied since they branched off from a common progenitor in certain characters, by which they have come to be specifically distinct from each other; and therefore these same characters would be more likely still to be variable than the generic characters which have been inherited without change for an enormous period. It is inexplicable on the theory of creation why a part developed in a very unusual manner in any one species of a genus, and therefore, as we may naturally infer, of great importance to the species, should be eminently liable to variation; but, on my view, this part has undergone, since the several species branched off from a common progenitor, an unusual amount of variability and modification, and therefore we might expect this part generally to be still variable. But a part may be developed in the most unusual manner, like the wing of a bat, and yet not be more variable than any other structure, if the part be common to many subordinate forms, that is, if it has been inherited for a very long period; for in this case it will have been rendered constant by long-continued natural selection.

Glancing at instincts, marvellous as some are, they offer no greater difficulty than does corporeal structure on the theory of the natural selection of successive, slight, but profitable modifications. We can thus understand why nature moves by graduated steps in endowing different animals of the same class with their several instincts. I have attempted to show how much light the principle of gradation throws on the admirable architectural powers of the hive-bee. Habit no doubt sometimes comes into play in modifying instincts; but it certainly is not indispensable, as we see, in the case of neuter insects, which leave no progeny to inherit the effects of long-continued habit. On the view of all the species of the same genus having descended from a common parent, and having inherited much in common, we can understand how it is that allied species, when placed under considerably different conditions of life, yet should follow nearly the same instincts; why the thrush of South America, for instance, lines her nest with mud like our British species. On the view of instincts having been slowly acquired through natural selection we need not marvel at some instincts being apparently not perfect and liable to mistakes, and at many instincts causing other animals to suffer.

If species be only well-marked and permanent varieties, we can at once see why their crossed offspring should follow the same complex laws in their degrees and kinds of resemblance to their parents,—in being absorbed into each other by successive crosses, and in other such points,—as do the crossed offspring of acknowledged varieties. On the other hand, these would be strange facts if species have been independently created, and varieties have been produced by secondary laws.

If we admit that the geological record is imperfect in an extreme degree, then such facts as the record gives, support the theory of descent with modification. New species have come on the stage slowly and at successive intervals; and the amount of change, after equal intervals of time, is widely different in different groups. The extinction of species and of whole groups of species, which has played so conspicuous a part in the history of the organic world, almost inevitably follows on the principle of natural selection; for old forms will be supplanted by new and improved forms. Neither single species nor groups of species reappear when the chain of ordinary generation has once been broken. The gradual diffusion of dominant forms, with the slow modification of their descendants, causes the forms of life, after long intervals of time, to appear as if they had changed simultaneously throughout the world. The fact of the fossil remains of each formation being in some degree intermediate in character between the fossils in the formations above and below, is simply explained by their intermediate position in the chain of descent. The grand fact that all extinct organic beings belong to the same system with recent beings, falling either into the same or into intermediate groups, follows from the living and the extinct being the offspring of common parents. As the groups which have descended from an ancient progenitor have generally diverged in character, the progenitor with its early descendants will often be intermediate in character in comparison with its later descendants; and thus we can see why the more ancient a fossil is, the oftener it stands in some degree intermediate between existing and allied groups. Recent forms are generally looked at as being, in some vague sense, higher than ancient and extinct forms; and they are in so far higher as the later and more improved forms have conquered the older and less improved organic beings in the struggle for life. Lastly, the law of the long endurance of allied forms on the same continent,—of marsupials in Australia, of edentata in America, and other such cases,—is intelligible, for within a confined country, the recent and the extinct will naturally be allied by descent.

Looking to geographical distribution, if we admit that there has been during the long course of ages much migration from one part of the world to another, owing to former climatal and geographical changes and to the many occasional and unknown means of dispersal, then we can understand, on the theory of descent with modification, most of the great leading facts in Distribution. We can see why there should be so striking a parallelism in the distribution of organic beings throughout space, and in their geological succession throughout time; for in both cases the beings have been connected by the bond of ordinary generation, and the means of modification have been the same. We see the full meaning of the wonderful fact, which must have struck every traveller, namely, that on the same continent, under the most diverse conditions, under heat and cold, on mountain and lowland, on deserts and marshes, most of the inhabitants within each great class are plainly related; for they will generally be descendants of the same progenitors and early colonists. On this same principle of former migration, combined in most cases with modification, we can understand, by the aid of the Glacial period, the identity of some few plants, and the close alliance of many others, on the most distant mountains, under the most different climates; and likewise the close alliance of some of the inhabitants of the sea in the northern and southern temperate zones, though separated by the whole intertropical ocean. Although two areas may present the same physical conditions of life, we need feel no surprise at their inhabitants being widely different, if they have been for a long period completely separated from each other; for as the relation of organism to organism is the most important of all relations, and as the two areas will have received colonists from some third source or from each other, at various periods and in different proportions, the course of modification in the two areas will inevitably be different.

On this view of migration, with subsequent modification, we can see why oceanic islands should be inhabited by few species, but of these, that many should be peculiar. We can clearly see why those animals which cannot cross wide spaces of ocean, as frogs and terrestrial mammals, should not inhabit oceanic islands; and why, on the other hand, new and peculiar species of bats, which can traverse the ocean, should so often be found on islands far distant from any continent. Such facts as the presence of peculiar species of bats, and the absence of all other mammals, on oceanic islands, are utterly inexplicable on the theory of independent acts of creation.

The existence of closely allied or representative species in any two areas, implies, on the theory of descent with modification, that the same parents formerly inhabited both areas; and we almost invariably find that wherever many closely allied species inhabit two areas, some identical species common to both still exist. Wherever many closely allied yet distinct species occur, many doubtful forms and varieties of the same species likewise occur. It is a rule of high generality that the inhabitants of each area are related to the inhabitants of the nearest source whence immigrants might have been derived. We see this in nearly all the plants and animals of the Galapagos Archipelago, of Juan Fernandez, and of the other American islands being related in the most striking manner to the plants and animals of the neighbouring American mainland; and those of the Cape de Verde Archipelago and other African islands to the African mainland. It must be admitted that these facts receive no explanation on the theory of creation.

The fact, as we have seen, that all past and present organic beings constitute one grand natural system, with group subordinate to group, and with extinct groups often falling in between recent groups, is intelligible on the theory of natural selection with its contingencies of extinction and divergence of character. On these same principles we see how it is, that the mutual affinities of the species and genera within each class are so complex and circuitous. We see why certain characters are far more serviceable than others for classification;—why adaptive characters, though of paramount importance to the being, are of hardly any importance in classification; why characters derived from rudimentary parts, though of no service to the being, are often of high classificatory value; and why embryological characters are the most valuable of all. The real affinities of all organic beings are due to inheritance or community of descent. The natural system is a genealogical arrangement, in which we have to discover the lines of descent by the most permanent characters, however slight their vital importance may be.

The framework of bones being the same in the hand of a man, wing of a bat, fin of the porpoise, and leg of the horse,—the same number of vertebræ forming the neck of the giraffe and of the elephant,—and innumerable other such facts, at once explain themselves on the theory of descent with slow and slight successive modifications. The similarity of pattern in the wing and leg of a bat, though used for such different purpose,—in the jaws and legs of a crab,—in the petals, stamens, and pistils of a flower, is likewise intelligible on the view of the gradual modification of parts or organs, which were alike in the early progenitor of each class. On the principle of successive variations not always supervening at an early age, and being inherited at a corresponding not early period of life, we can clearly see why the embryos of mammals, birds, reptiles, and fishes should be so closely alike, and should be so unlike the adult forms. We may cease marvelling at the embryo of an airbreathing mammal or bird having branchial slits and arteries running in loops, like those in a fish which has to breathe the air dissolved in water, by the aid of well-developed branchiæ.

Disuse, aided sometimes by natural selection, will often tend to reduce an organ, when it has become useless by changed habits or under changed conditions of life; and we can clearly understand on this view the meaning of rudimentary organs. But disuse and selection will generally act on each creature, when it has come to maturity and has to play its full part in the struggle for existence, and will thus have little power of acting on an organ during early life; hence the organ will not be much reduced or rendered rudimentary at this early age. The calf, for instance, has inherited teeth, which never cut through the gums of the upper jaw, from an early progenitor having well-developed teeth; and we may believe, that the teeth in the mature animal were reduced, during successive generations, by disuse or by the tongue and palate having been better fitted by natural selection to browse without their aid; whereas in the calf, the teeth have been left untouched by selection or disuse, and on the principle of inheritance at corresponding ages have been inherited from a remote period to the present day. On the view of each organic being and each separate organ having been specially created, how utterly inexplicable it is that parts, like the teeth in the embryonic calf or like the shrivelled wings under the soldered wing-covers of some beetles, should thus so frequently bear the plain stamp of inutility! Nature may be said to have taken pains to reveal, by rudimentary organs and by homologous structure, her scheme of modification, which it seems that we wilfully will not understand.

I have now recapitulated the chief facts and considerations which have thoroughly convinced me that species have been modified, during a long course of descent, by the preservation or the natural selection of many successive slight favourable variations. I cannot believe that a false theory would explain, as it seems to me that the theory of natural selection does explain, the several large classes of facts above specified. I see no good reason why the views given in this volume should shock the religious feelings of any one. A celebrated author and divine has written to me that 'he has gradually learnt to see that it is just as noble a conception of the Deity to believe that He created a few original forms capable of self-development into other and needful forms, as to believe that He required a fresh act of creation to supply the voids caused by the action of His laws.'

Why, it may be asked, have all the most eminent living naturalists and geologists rejected this view of the mutability of species? It cannot be asserted that organic beings in a state of nature are subject to no variation; it cannot be proved that the amount of variation in the course of long ages is a limited quantity; no clear distinction has been, or can be, drawn between species and well-marked varieties. It cannot be maintained that species when intercrossed are invariably sterile, and varieties invariably fertile; or that sterility is a special endowment and sign of creation. The belief that species were immutable productions was almost unavoidable as long as the history of the world was thought to be of short duration; and now that we have acquired some idea of the lapse of time, we are too apt to assume, without proof, that the geological record is so perfect that it would have afforded us plain evidence of the mutation of species, if they had undergone mutation.

But the chief cause of our natural unwillingness to admit that one species has given birth to other and distinct species, is that we are always slow in admitting any great change of which we do not see the intermediate steps. The difficulty is the same as that felt by so many geologists, when Lyell first insisted that long lines of inland cliffs had been formed, and great valleys excavated, by the slow action of the coast-waves. The mind cannot possibly grasp the full meaning of the term of a hundred million years; it cannot add up and perceive the full effects of many slight variations, accumulated during an almost infinite number of generations.

Although I am fully convinced of the truth of the views given in this volume under the form of an abstract, I by no means expect to convince experienced naturalists whose minds are stocked with a multitude of facts all viewed, during a long course of years, from a point of view directly opposite to mine. It is so easy to hide our ignorance under such expressions as the 'plan of creation', 'unity of design', etc., and to think that we give an explanation when we only restate a fact. Any one whose disposition leads him to attach more weight to unexplained difficulties than to the explanation of a certain number of facts will certainly reject my theory. A few naturalists, endowed with much flexibility of mind, and who have already begun to doubt on the immutability of species, may be influenced by this volume; but I look with confidence to the future, to young and rising naturalists, who will be able to view both sides of the question with impartiality. Whoever is led to believe that species are mutable will do good service by conscientiously expressing his conviction; for only thus can the load of prejudice by which this subject is overwhelmed be removed.

Several eminent naturalists have of late published their belief that a multitude of reputed species in each genus are not real species; but that other species are real, that is, have been independently created. This seems to me a strange conclusion to arrive at. They admit that a multitude of forms, which till lately they themselves thought were special creations, and which are still thus looked at by the majority of naturalists, and which consequently have every external characteristic feature of true species,—they admit that these have been produced by variation, but they refuse to extend the same view to other and very slightly different forms. Nevertheless they do not pretend that they can define, or even conjecture, which are the created forms of life, and which are those produced by secondary laws. They admit variation as a *vera causa* in one case, they arbitrarily reject it in another, without assigning any distinction in the two cases. The day will come when this will be given as a curious illustration of the blindness of preconceived opinion. These authors seem no more startled at a miraculous act of creation than at an ordinary birth. But do they really believe that at innumerable periods in the earth's history certain elemental atoms have been commanded suddenly to flash into living tissues? Do they believe that at each supposed act of creation one individual or many were produced? Were all the infinitely numerous kinds of animals and plants created as eggs or seed, or as full grown? and in the case of mammals, were they created bearing the false marks of nourishment from the mother's womb? Although naturalists very properly demand a full explanation of every difficulty from those who believe in the mutability of species, on their own side they ignore the whole subject of the first appearance of species in what they consider reverent silence.

It may be asked how far I extend the doctrine of the modification of species. The question is difficult to answer, because the more distinct the forms are which we may consider, by so much the arguments fall away in force. But some arguments of the greatest weight extend very far. All the members of whole classes can be connected together by chains of affinities, and all can be classified on the same principle, in groups subordinate to groups. Fossil remains sometimes tend to fill up very wide intervals between existing orders. Organs in a rudimentary condition plainly show that an early progenitor had the organ in a fully developed state; and this in some instances necessarily implies an enormous amount of modification in the descendants. Throughout whole classes various structures are formed on the same pattern, and at an embryonic age the species closely resemble each other. Therefore I cannot doubt that the theory of descent with modification embraces all the members of the same class. I believe that animals have descended from at most only four or five progenitors, and plants from an equal or lesser number.

Analogy would lead me one step further, namely, to the belief that all animals and plants have descended from some one prototype. But analogy may be a deceitful guide. Nevertheless all living things have much in common, in their chemical composition, their germinal vesicles, their cellular structure, and their laws of growth and reproduction. We see this even in so trifling a circumstance as that the same poison often similarly affects plants and animals; or that the poison secreted by the gall-fly produces monstrous growths on the wild rose or oak-tree. Therefore I should infer from analogy that probably all the organic beings which have ever lived on this earth have descended from some one primordial form, into which life was first breathed by the Creator.

When the views advanced by me in this volume, and by Mr Wallace in the *Linnean Journal*, or when analogous views on the origin of species are generally admitted, we can dimly foresee that there will be a considerable revolution in natural history. Systematists will be able to pursue their labours as at present; but they will not be incessantly haunted by the shadowy doubt whether this or that form be in essence a species. This I feel sure, and I speak after experience, will be no slight relief. The endless disputes whether or not some fifty species of British brambles are true species will cease. Systematists will have only to decide (not that this will be easy) whether any form be sufficiently constant and distinct from other forms, to be capable of definition; and if definable, whether the differences be sufficiently important to deserve a specific name. This latter point will become a far more essential consideration than it is at present; for differences, however slight, between any two forms, if not blended by intermediate gradations, are looked at by most naturalists as sufficient to raise both forms to the rank of species. Hereafter we shall be compelled to acknowledge that the only distinction between species and well-marked varieties is, that the latter are known, or believed, to be connected at the present day by intermediate gradations, whereas species were formerly thus connected. Hence, without rejecting the consideration of the present existence of intermediate gradations between any two forms, we shall be led to weigh more carefully and to value higher the actual amount of difference between them. It is quite possible that forms now generally acknowledged to be merely varieties may hereafter be thought worthy of specific names, as with the primrose and cowslip; and in this case scientific and common language will come into accordance. In short, we shall have to treat species in the same manner as those naturalists treat genera, who admit that genera are merely artificial combinations made for convenience. This may not be a cheering prospect; but we shall at least be freed from the vain search for the undiscovered and undiscoverable essence of the term species.

The other and more general departments of natural history will rise greatly in interest. The terms used by naturalists of affinity, relationship, community of type, paternity, morphology, adaptive characters, rudimentary and aborted organs, etc., will cease to be metaphorical, and will have a plain signification. When we no longer look at an organic being as a savage looks at a ship, as at something wholly beyond his comprehension; when we regard every production of nature as one which has had a history; when we contemplate every complex structure and instinct as the summing up of many contrivances, each useful to the possessor, nearly in the same way as when we look at any great mechanical invention as the summing up of the labour, the experience, the reason, and even the blunders of numerous workmen; when we thus view each organic being, how far more interesting, I speak from experience, will the study of natural history become!

A grand and almost untrodden field of inquiry will be opened, on the causes and laws of variation, on correlation of growth, on the effects of use and disuse, on the direct action of external conditions, and so forth. The study of domestic productions will rise immensely in value. A new variety raised by man will be a more important and interesting subject for study than one more species added to the infinitude of already recorded species. Our classifications will come to be, as far as they can be so made, genealogies; and will then truly give what may be called the plan of creation. The rules for classifying will no doubt become simpler when we have a definite object in view. We possess no pedigrees or armorial bearings; and we have to discover and trace the many diverging lines of descent in our natural genealogies, by characters of any kind which have long been inherited. Rudimentary organs will speak infallibly with respect to the nature of long-lost structures. Species and groups of species, which are called aberrant, and which may fancifully be called living fossils, will aid us in forming a picture of the ancient forms of life. Embryology will reveal to us the structure, in some degree obscured, of the prototypes of each great class.

When we can feel assured that all the individuals of the same species, and all the closely allied species of most genera, have within a not very remote period descended from one parent, and have migrated from some one birthplace; and when we better know the many means of migration, then, by the light which geology now throws, and will continue to throw, on former changes of climate and of the level of the land, we shall surely be enabled to trace in an admirable manner the former migrations of the inhabitants of the whole world. Even at present, by comparing the differences of the inhabitants of the sea on the opposite sides of a continent, and the nature of the various inhabitants of that continent in relation to their apparent means of immigration, some light can be thrown on ancient geography.

The noble science of Geology loses glory from the extreme imperfection of the record. The crust of the earth with its embedded remains must not be looked at as a well-filled museum, but as a poor collection made at hazard and at rare intervals. The accumulation of each great fossiliferous formation will be recognised as having depended on an unusual concurrence of circumstances, and the blank intervals between the successive stages as having been of vast duration. But we shall be able to gauge with some security the duration of these intervals by a comparison of the preceding and succeeding organic forms. We must be cautious in attempting to correlate as strictly contemporaneous two formations, which include few identical species, by the general succession of their forms of life. As species are produced and exterminated by slowly acting and still existing causes, and not by miraculous acts of creation and by catastrophes; and as the most important of all causes of organic change is one which is almost independent of altered and perhaps suddenly altered physical conditions, namely, the mutual relation of organism to organism,—the improvement of one being entailing the improvement or the extermination of others; it follows, that the amount of organic change in the fossils of consecutive formations probably serves as a fair measure of the lapse of actual time. A number of species, however, keeping in a body might remain for a long period unchanged, whilst within this same period, several of these species, by migrating into new countries and coming into competition with foreign associates, might become modified; so that we must not overrate the accuracy of organic change as a measure of time. During early periods of the earth's history, when the forms of life were probably slower; and at the first dawn of life, when very few forms of the simplest structure existed, the rate of change may have been slow in an extreme degree. The whole history of the world, as at present known, although of a length quite incomprehensible by us, will hereafter be recognised as a mere fragment of time, compared with the ages which have elapsed since the first creature, the progenitor of innumerable extinct and living descendants, was created.

In the distant future I see open fields for far more important researches. Psychology will be based on a new foundation, that of the necessary acquirement of each mental power and capacity by gradation. Light will be thrown on the origin of man and his history.

Authors of the highest eminence seem to be fully satisfied with the view that each species has been independently created. To my mind it accords better with what we know of the laws impressed on matter by the Creator, that the production and extinction of the past and present inhabitants of the world should have been due to secondary causes, like those determining the birth and death of the individual. When I view all beings not as special creations, but as the lineal descendants of some few beings which lived long before the first bed of the Silurian system was deposited, they seem to me to become ennobled. Judging from the past, we may safely infer that not one living species will transmit its unaltered likeness to a distant futurity. And of the species now living very few will transmit progeny of any kind to a far distant futurity; for the manner in which all organic beings are grouped, shows that the greater number of species of each genus, and all the species of many genera, have left no descendants, but have become utterly extinct. We can so far take a prophetic glance into futurity as to foretell that it will be the common and widely-spread species, belonging to the larger and dominant groups, which will ultimately prevail and procreate new and dominant species. As all the living forms of life are the lineal descendants of those which lived long before the Silurian epoch, we may feel certain that the ordinary succession by generation has never once been broken, and that no cataclysm has desolated the whole world. Hence we may look with some confidence to a secure future of equally inappreciable length. And as natural selection works solely by and for the good of each being, all corporeal and mental endowments will tend to progress towards perfection.

It is interesting to contemplate an entangled bank, clothed with many plants of many kinds, with birds singing on the bushes, with various insects flitting about, and with worms crawling through the damp earth, and to reflect that these elaborately constructed forms, so different from each other, and dependent on each other in so complex a manner, have all been produced by laws acting around us. These laws, taken in the largest sense, being Growth with Reproduction; Inheritance, which is almost implied by reproduction; Variability, from the indirect and direct action of the external conditions of life, and from use and disuse; a Ratio of Increase so high as to lead to a Struggle for Life, and as a consequence to Natural Selection, entailing Divergence of Character and the Extinction of less-improved forms. Thus, from the war of nature, from famine and death, the most exalted object which we are capable of conceiving, namely, the production of the higher animals, directly follows. There is grandeur in this view of life, with its several powers, having been originally breathed by the Creator into a few forms or into one; and that, whilst this planet has gone cycling on according to the fixed law of gravity, from so simple a beginning endless forms most beautiful and most wonderful have been, and are being, evolved.

APPENDIX I
REGISTER OF WRITERS REFERRED TO IN THE TEXT OF THE *ORIGIN*

Works by a considerable number of the people referred to are listed in Darwin's reading notebooks which he kept during the years he was preparing the *Origin*. See *The Correspondence of Charles Darwin*, eds. Frederick Burkhardt and Sydney Smith, 4 (Cambridge, 1988), pp. 434–573.

Some of the people to whose work Darwin refers in the *Origin* later took strong positions for or against his theories; where known, this is indicated below.

JEAN LOUIS AGASSIZ (1807–73). Swiss zoologist and geologist, later Professor at Harvard; he became one of Darwin's inveterate critics, continuing to hold a polygenist view (that races were separate creations) and that creation was divided by an ice-age: see 'A Period in the History of Our Planet', *Edinburgh New Philosophical Journal* (1843) and *An Essay on Classification* (London, 1859).

JOHN JAMES AUDUBON (1785–1851). American ornithologist and natural history artist; his observations are cited by Darwin both in *The Voyage of the Beagle* and the *Origin*. See his *Ornithological Biography, or an Account of the Habits of the Birds of the United States of America, and Interspersed with Delineations of American Scenery and Manners* (5 vols.; Edinburgh, 1831–49).

CHARLES C. BABINGTON (1808–94). Botanist, taxonomist, and beetle collector; worked in the Channel Islands; Darwin's contemporary at Cambridge, later Professor of Botany there.

ROBERT BAKEWELL (1725–95). By experimental breeding produced new and more profitable strains of sheep and cattle; his endeavours are a striking example of the power of 'artificial selection'. The attempt by Buckley and Burgess to keep Bakewell's strain pure and consistent over fifty years threw into prominence the natural propensity to diversification: important for Darwin's argument (see p. 31).

JOACHIM BARRANDE (1799–1883). French geologist; worked particularly on fossils in the Silurian system of Europe, *Système silurien du centre de la Bohême*, 29 parts (Prague and Paris, 1852–1911).

JOHN BARROW (1764–1848). Traveller and prolific author; his observations of South Africa, Europe, the Arctic, etc., recorded in *An autobiographical memoir of Sir John Barrow . . . including reflections, observations, and reminiscences at home and abroad . . .* (London, 1847).

GEORGE BENTHAM (1800–84). Systematic botanist and logician, whose knowledge of fourteen languages enhanced the perfection of his descriptions; President of the Linnaean Society for thirteen years from 1861, his known scientific caution made him a very valuable supporter of Darwin. His *Flora* was published in seven vols. from 1863 to 1878.

MILES BERKELEY (1803–89). Perpetual curate in Northamptonshire; described the *Beagle* fungi and assisted in Darwin's seed experiments.

EDWARD BLYTH(E) (1810–73). Curator of the Asiatic Society's Museum in Calcutta; Darwin's informant on Indian domestic animals and frequently referred to by Darwin in his later work, *The Variation of Animals and Plants Under Domestication* (1868).

JEAN BAPTISTE BORY DE ST VINCENT (1778–1846). French traveller and naturalist; *Voyage dans les quatres principales isles des mers d'Afrique* (3 vols; Paris, 1801) in Darwin's reading lists.

JOSEPHUS BOSQUET (1814–80). French geologist; described fossilized cirripedes as well as flora and fauna in works published in the 1850s.

BERNHARD BRENT (d. 1867). Pigeon fancier; studied and wrote about European pigeon breeding.

HEINRICH BRONN (1800–62). palaeontologist; Professor at Freiburg and at Heidelberg; translated the *Origin* into German, with some omissions and notes critical of the theory.

ROBERT BROWN (1773–1858). Botanist and microscopist; friend of Joseph Banks; Keeper of the Botanical Collection at the British Museum; described fungi and fossil wood from the *Beagle*; interested in classification and an innovative worker on the relations of the parts of plants to each other.

BUCKLEY. *See* BAKEWELL.

JAMES BUCKMAN (1816–84). Geologist and botanist; Professor at the Royal Agricultural College, Cirencester; studied botany from an agricultural point of view, e.g., his work on grasses.

BURGESS. *See* BAKEWELL.

PROBY THOMAS CAUTLEY (1802–71). Indian army colonel; projected and constructed the Ganges Canal, and was a well-known researcher in geology and palaeontology; together with Hugh Falconer (q.v.), who was in charge of the Botanical Garden at Saharanpur, he explored the geology of the Sivalik range and presented his immense collection of fossils to the British Museum. Published mainly in the journal *Asiatic Researches*.

ROBERT CHAMBERS (1802–71). Journalist and publisher; anonymous author of the immensely successful *Vestiges of the Natural History of Creation* (1844), which gives a 'transmutationist' account of the continuous generation of nature in the cosmos from nebulae into a progressive ordering of kinds and species. The mechanism Chambers supposed was the occasional power of a lower to breed a higher type. Responded to criticisms by writing *Explanations: A Sequel* (1845), and *Vestiges* ran through fourteen edns. Chambers later welcomed the *Origin*, but Darwin relegated him in his discussions to the role of amateur forerunner.

WILLIAM CLIFT (1775–1849). Assistant to the physician John Hunter and afterwards curator of his museum and collection. Much respected as an anatomist and student of fossils, deferred to by scientists such as Cuvier and Lyell. Father-in-law of Richard Owen.

GEORGES CUVIER (1769–1832). Distinguished French taxonomist who argued that animal organization is functional and that its perfection of adaptation to conditions produces the distinctiveness of species, which cannot therefore be studied on a progressive or an archetypal basis. See his *Le Règne animal* (Paris 1817).

JAMES DWIGHT DANA (1813–95). Geologist and naturalist; described zoophytes and crustacea from expedition to the Pacific 1838–42; like Darwin, worked on coral islands; later Professor of Geology at Yale.

ÉTIENNE VISCOUNT D'ARCHIAC (1802–68?). French geologist, wrote on the fossils of the Rhenish Provinces and prepared a tabular list of organic remains in the Devonian system (1842).

FELIX D'AZARA (1746–1821). Spanish zoologist and traveller in Paraguay and Brazil; his *Voyages* (1809) and *Quadrupeds* (1801) are listed in Darwin's Reading Notebooks as read more than once.

AUGUSTIN DE CANDOLLE (1778–1841). Classificatory botanist; Professor of Natural History at Geneva; cited by Darwin as 'the elder De Candolle'. His son, Alphonse De Candolle, Swiss biologist, worked on plant distribution: *Géographie Botanique Raisonnée* (2 vols.; Paris, 1855).

ALCIDE D'ORBIGNY (1802–57). French palaeontologist; worked on fossil remains in South America.

ANDREW DOWNING (1815–52). American architect and pomologist; wrote *The Fruits and Fruit Trees of America* (London, 1845).

GEORGE WINDSOR EARL. Author of *The Eastern Seas: Or, Voyages and Adventures in the Indian Archipelago, in 1832, 1833, 1834* (London, 1837). Darwin cites him on the Malay Archipelago.

W. W. EDWARDS. Informant on race-horses.

JEAN BAPTISTE ELIE DE BEAUMONT (1798–1874). French mining engineer and geologist; catastrophist; worked on mountain ranges. Disbelieved in universal catastrophe as cause of extinctions.

WALTER ELLIOT (1803–87). Civil servant, archaeologist, natural historian and numismologist of India.

THOMAS CAMPBELL EYTON (1809–80). Naturalist; wrote on cattle, fishes, birds, and particularly ducks. Friend and correspondent of Darwin, who nevertheless opposed his theories; Eyton was chagrined to find some of his observations on hybridism used in support of natural selection (e.g., *see* p. 205).

JEAN FABRÉ (1823–1915). French entomologist and popular scientific writer; worked on reproductive systems of beetles and insects in 1850s.

HUGH FALCONER (1808–65). Palaeontologist and botanist; with Cautley (q.v.), explored the Indian Sivalik hills and uncovered mastodons and other fossil mammals; established the tertiary geological age of the hills. Brought enormous collections of dried plants and fossil bones to England. His *Fauna Antiqua Silva*, parts 1–9 (London, 1845–9) was a valuable source for Darwin.

EDWARD FORBES (1815–54). Botanist, palaeontologist, and biogeographer; a formidable generalizer from empirical experiment and observation, his work on the distribution of plants and molluscs involved both the idea of sunken landbridges and continental drift. From Darwin's point of view his most important paper was 'On the Connection Between the Distribution of the existing Fauna and Flora of the British Isles and the Geological Changes which have affected their Area' (1846).

ELIAS FRIES (1794–1878). Swedish botanist; author of *Systema orbis vegetabilis* (Lund, 1825), listed in Darwin's reading notebooks.

GEORGE GARDNER (1812–49). Botanist; worked in Brazil in 1830s sending home 60,000 specimens and contributing to Hooker's *London Journal of Botany*. Later Superintendent of the Botanical Garden of Ceylon.

KARL FRIEDRICH VON GÄRTNER (1772–1850). German botanist, whose work on plant hybridization, along with that of the German botanist Joseph Kölreuter, was Darwin's principal source in Ch. 8, 'Hybridism'.

ETIENNE GEOFFROY SAINT-HILAIRE (1772–1844). 'The elder Geoffroy'; French zoologist much concerned with monsters, 'sports', and nature's balances. His son Isidore Geoffroy Saint-Hilaire (1805–61) continued his work on teratology.

LOUIS GIROU DE BUZAREINGUES (1773–1856). French physician; published a memoir on the distribution of the sexes in France, and a work on physiological, political and moral philosophy, both in 1828.

JOHANN GEORG GMELIN (1709–55). Explorer and natural historian; Professor of Chemistry at Academy of Sciences, St Petersburg; observed presence of same species in geographically remote areas.

ROBERT GODWIN-AUSTEN (1808–84). Geologist; studied the weald and the geology of southern England and the Channel; later published *Land and Freshwater Mollusca of India*, vols. 1–3 (London, 1882–).

JOHANN WOLFGANG VON GOETHE (1749–1832). Poet and scientist; alluded to by Darwin alongside Geoffroy Saint-Hilaire (q.v.).

AUGUSTUS A. GOULD (1805–66). American author, editor, and naturalist; studied shells, and edited important anthology *The Naturalist's Library*, 1849.

JOHN GOULD (1804–81). Ornithologist and artist; son of a working gardener and himself a gardener by training, Gould collected, drew, and stuffed birds from all over the world.

ASA GRAY (1810–88). American botanist; Professor of Natural History at Harvard. A friend of Darwin, his teleological reading of natural selection (which sought to reconcile evolutionary theory and theology) appealed to Darwin when it appeared in 1863.

E. W. HARCOURT. Author of *A Sketch of Madeira: Containing Information for the Traveller or Invalid Visitor* (London, 1851), which includes lists of birds and plants.

GEORGE HARTUNG (1822?–91). German geologist and traveller; correspondent of Lyell (q.v.).

SAMUEL HEARNE (1745–92). Traveller; wrote description of expedition in search of copper mines, *A journey from Prince of Wales's Fort in Hudson's Bay to the Northern Ocean* (London, 1795).

OSWALD HEER (1809–83). Swiss botanist, climatologist, and traveller to Arctic; Director of Zurich Botanical Gardens.

WILLIAM HERBERT (1778–1847). Dean of Manchester, and botanist, especially interested in bulbous plants; helped edit White's *Selborne* (1833).

ROBERT HERON (1765–1854). Politician and naturalist; kept 'curious animals' in his private menagerie; Darwin cites his observation of his peacocks' behaviour in his own writing on sexual selection.

KARL VON HEUSINGER (1792–1883). German physician and physiologist, specializing in animal husbandry.

JOSEPH HOOKER (1817–1911). Among the most important of Darwin's informants and counsellors; botanist and traveller; later, Director of Kew Gardens; analysed the plants collected in the Galapagos during the voyage of the *Beagle*; read the 1844 essay at that time; was convinced by the Darwin–Wallace theory and incorporated it in his own work in 1858: his 'Introductory Essay on the Flora of Tasmania' which advanced the study of the geographical distribution of plants.

LEONARD HORNER (1785–1864). Educational reformer and geologist; one of the founders of the Mechanics' Institutes, and commissioner enquiring into the employment of children in factories; catalogued Geological Society's collections.

PIERRE HUBER (1777–1840). French entomologist; his *The Natural History of Ants* published in translation (London, 1820).

ALEXANDER VON HUMBOLDT (1769–1859). Prussian naturalist and explorer, who worked in government mining service; travelled particularly in South America and Siberia. His *Personal Narrative* was an inspiration to Darwin and in *The Voyage of the Beagle* he is frequently cited as an authority.

JOHN HUNTER (d. 1809). Physician; cited by Darwin on secondary sexual characteristics.

THOMAS HENRY HUXLEY (1825–95). Palaeontologist, botanist and zoologist, lecturer; later of great importance to Darwin's career and renown by virtue of his zeal in promoting the influence of Darwin's theories through writings such as *Man's Place in Nature* (1863). Mentioned twice only in the *Origin.*

ALEXANDER JOHNSTON (1804–71). His *Physical Atlas . . . illustrating the geographical distribution of natural phenomena* (1848) was revised and enlarged 1856.

ANTOINE DE JUSSIEU (1748–1836). French botanist and taxonomist. His *Genera plantarum* (1789) is directly quoted by Darwin in Ch. XIII, 'Classification'. His son, ADRIEN DE JUSSIEU (1753–1853), succeeded his father as Professor of Botany at the Musée National d'Histoire Naturelle in Paris.

WILLIAM KIRBY (1759–1850). Clergyman and entomologist; described the insects brought home from Franklin's first two arctic expeditions; wrote the seventh Bridgewater Treatise, *On the History, Habits and Instincts of Animals* (1835).

JEAN BAPTISTE DE LAMARCK (1744–1829). French naturalist; his *Histoire naturelle des animaux sans vertèbres* (Paris, 1815–22) advanced an innovative transformist argument that argued for the transmission by descent of characteristics acquired during the course of life. Darwin emphasized his differences with Lamarck when establishing his theory, but in the revisions of the *Origin* during the 1860s moved back towards a more Lamarckian position.

CHARLES LE ROY (1723–89). French philosopher, wrote (often under the pseudonym 'The Naturalist of Nuremberg') on the question of animal intelligence and perfectability; his *Lettres philosophiques sur l'intelligence et la perfectibilité des animaux avec quelques lettres sur l'homme* (Paris, 1802) was translated as *The Intelligence and Perfectability of Animals from a Philosophic Point of View* (London, 1870).

LINNAEUS (CARL VON LINNÉ) (1708–78). Swedish botanist, zoologist, taxonomist; he reordered scientific classification and taxonomic terms. Darwin cites his observation of the potential hyperproductivity of plants if all seeds survive.

JOHN LUBBOCK (1834–1913). Entomologist; a neighbour of Darwin's at Down, and later an important anthropologist; created Baron Avebury. Cited here as 'this philosophical naturalist' for his observations of larvae muscles.

PROSPER LUCAS (1805–85). French systematist; his *Traité philosophique et physiologique de l'heredité naturelle* (2 vols.; Paris, 1847–50) is an important source for Darwin.

CHARLES LYELL (1797–1875). Geologist; one of Darwin's most influential advisors and sources; his *Principles of Geology* (1833), with its emphasis on uniform processes of change, greatly affected Darwin's thought. He, together with Hooker, arranged Darwin's issue from the dilemma created by Wallace's parallel formulation of the principle of natural selection.

ALEXANDER MACLEAY (1767–1848). Entomologist and colonial civil servant in Australia.

THOMAS ROBERT MALTHUS (1766–1834). Political economist. His *An Essay on the Principle of Population* (1798) opens with examples from natural history (e.g., the potential spread of fennel) which may have led Darwin and Wallace more readily to transfer his economic argument to the natural world. Malthus argued that draconian measures were required to prevent human population outrunning the supply of natural resources. He argued that all uncontrolled populations have a capacity for exponential growth which will inevitably exceed supply.

'a man whom I can implicitly trust'. Perhaps a servant? see p. 134

ANDREW MARSHALL (1742–1813). Anatomist and agriculturist.

MARTIN MARTENS (1797–1863). Belgian physician, Professor of Chemistry and Botany at Louvain.

WILLIAM C. MARTIN (1798–1864). Son of the naturalist William Martin; superintendent of the Museum of the National Zoological Society in the 1830s; particularly interested in birds.

CARLO MATTEUCCI (1811–68). Italian physiologist; wrote extensively on electro–physiological phenomena, e.g., his 10-part *Electro–Physiological Researches* (London, 1845–6).

HUGH MILLER (1802–56). Geologist, first apprenticed to a stone-mason; Scottish nationalist and committed Christian; admired by others such as Buckland; wrote *The Testimony of the Rocks* (1857), a work which combines geological theory and religious conviction.

WILLIAM MILLER (1801–80). Mineralogist; Professor of Mineralogy at Cambridge.

HENRI MILNE-EDWARDS (1800–85). French zoologist; worked on crustacea; drew on economic theory for his concept of physiological division of labour, a useful tool for Darwin's thought.

LORD MORETON, HENRY GEORGE FRANCIS (1802–53). Famous breeder of shorthorn cattle.

FELIX MÜLLER (b. 1818). Belgian botanist.

RODERICK MURCHISON (1792–1871). Geologist; first ally, then contender, with Sedgwick (q.v.), and much involved in the controversy over the Silurian system.

CHARLES MURRAY, Bart. (1806–95). Traveller and novelist; recounted his prolonged stay with Pawnee Indians, and visits to Cuba and the Azores, in *Travels in North America* (London, 1839). His novel concerning American Indian life, *The Prairie Bird* (1844), went through numerous editions.

'My son'. See p. 134.

RICHARD OWEN (1804–92). Anatomist, palaeontologist, natural historical writer, and superintendent of the Natural History Museum. A powerful figure in Victorian scientific circles, he became one of Darwin's most formidable opponents.

PETER PALLAS (1741–1811). German systematist and explorer; wrote on geographical distribution of plants and birds; travelled to Russia, Persia, Tibet, etc.

FRANÇOIS PICTET (1809–72). French palaeontologist; Darwin cites his four-vol. *Traité élémentaire de paléontologie* (Paris, 1844–5).

PLINY (AD 23–74), Darwin cites his *The historie of the world, commonly called the Naturall Historie of C. Plinius Secundus*, trans. Philemon Holland in 1601.

JOSEPH PRESTWICH (1812–96). British geologist, concerned with water-bearing strata and with fossils.

LOUIS RAMOND DE CARBONNIÈRES (1753–1827). French botanist. Darwin here cites from his articles on plants of the high Pyrenees, published in the *Bulletin des Sciences de la Societé Philomathique de Paris* (1799–1801).

ANDREW RAMSAY (1814–91). Geologist; director of the Geological Survey; friend and fellow Alpinist with John Tyndall.

JOHANN RENGGER (1795–1832). German zoologist; wrote on the mammals of Paraguay (1830) and on the revolution there.

JOHN RICHARDSON (1787–1865). Arctic explorer; surgeon and naturalist to Franklin's 1819 expedition; ichthyologist; taught Thomas Henry Huxley.

HORACE STEBBING ST JOHN (1832–88). Journalist who wrote on the history and geography of India. His wife worked on Audubon (q.v.).

AUGUSTIN SAGARET. French botanist; produced a book on the varieties of melon (1826).

JØRGEN SCHIØDTE (1815–84). Danish entomologist and natural historian, concerned particularly with creatures living underground.

HERMANN SCHLEGEL (1804–84). Dutch natural historian and prolific writer, especially on monkeys; his *Essay on the Physiognomy of Serpents* was translated by Thomas Traill in 1843.

JOHN SEBRIGHT (1767–1846). Landowner who campaigned for the amendment of the draconian game laws; animal breeder concerned with improvements, and interested in animal instinct.

ADAM SEDGWICK (1785–1873). Geologist; started seriously to study geology after being elected to the Woodwardian Professorship of Geology in Cambridge in 1818; instrumental in founding the scientific triposes in the university; distinguished for work within the British Isles; after the publication of the *Origin*, strongly opposed to Darwin's theories.

BENJAMIN SILLIMAN (1779–1864). American natural historian and geologist; Professor of Chemistry and Natural History at Yale. Darwin cites his observations on sight in animals that live in darkness.

FREDERICK SMITH (1806?–79). Of the British Museum; see his *Catalogue of British Hymenoptera in the British Museum*: Part 1, Apidae-Bees (1855). Smith catalogued Wallace's collection of hymenopterous insects.

JAMES SMITH, of Jordan Hill (1782–1867). Geologist and West India merchant, studied glacial geology.

LORD SOMERVILLE, JOHN (1765–1819). Agriculturalist; largest breeder and owner of merinos in Britain.

LORD SPENCER, JOHN CHARLES (1782–1845). Agriculturalist; wrote an essay on the gestation of cows.

CHRISTIAN C. [KONRAD] SPRENGEL (1750–1816). German botanist; studied fertilization of flowers by insects and self-pollination.

IGNAZ TAUSCH. Botanist; first vol. of his work *Hortus Canalius seu Plantarum rariorum quae in Horto botanico Jósephi Malabaila comitis de Canal coluntur icones et descriptiones* published 1823, no further vols. forthcoming.

WILLIAM TEGETMEIER (1816–1912). Pigeon fancier and naturalist; here cited on activity of bees; also wrote *The Poultry Book: Including pigeons and rabbits* (London, 1856–7).

COENRAAD TEMMINCK (1778–1858). Dutch ornithologist; wrote a three-vol. treatise on pigeons, as well as work on European, Japanese, and Indonesian mammals (all listed in Darwin's reading notebooks).

GEORGE THWAITES (1811–82). Accountant, botanist and entomologist; Superintendent of the Botanical Gardens in Ceylon; expert on lichen, fungus, and non-flowering plants, and excellent microscopist.

JOHN TOMES (1815–1895). Dental surgeon; worked on histology of bones and teeth.

JEAN BAPTISTE VAN MONS (1765–1842). Belgian botanist and physicist; author of *Arbres fruitières* (2 vols.; 1835–6).

ALFRED RUSSEL WALLACE (1823–1913). Mentioned twice in the argument of the *Origin*; independently formulated the concept of natural selection (see Introduction p. vii), and propelled Darwin into publishing his work. His studies and collections in the Malay Archipelago were the foundation of his work on species distribution. Later he involved himself in large social questions such as land nationalization; he was a socialist and became a spiritualist.

GEORGE ROBERT WATERHOUSE (1810–88). Zoologist; beetle enthusiast; later Keeper of the Department of Geology at the British Museum; he published descriptions of various species of mammals and insects collected by Darwin on the *Beagle* voyage.

HEWETT C. WATSON (1804–81). Botanist, especially interested in plant distribution; published accounts of his expedition to the Azores and, in 1845, an article on 'Progressive Development' in the *Phytologist.*

JOHN WESTWOOD (1805–93). Entomologist and palaeontologist; author of *An Introduction to the Modern Classification of Insects* (2 vols.; London, 1839–40); later Professor of Invertebrate Zoology at Oxford. He was not persuaded by Darwin's evolutionary theories.

THOMAS VERNON WOLLASTON (1822–78). Entomologist; insect specialist at the British Museum; a friend of Darwin's, he dedicated to him his *On the Variation of Species with special reference to the Insecta; followed by an enquiry into the nature of genera* (London, 1856).

Samuel Woodward (1790–1838). Geologist and local historian in Norfolk; the first to enumerate all known British fossils.

William Youatt (1776–1847). Veterinary surgeon; wrote a series of books on the management and breeding of farm animals as part of the 'Library of Useful Knowledge', and edited later editions of the eighteenth-century manual *The Complete Grazier*.

APPENDIX II

GLOSSARY OF PRINCIPAL SCIENTIFIC TERMS, ADDED IN THE FIFTH EDITION OF THE *ORIGIN**

ABERRANT. Forms or groups of animals or plants which deviate in important characters from their nearest allies, so as not to be easily included in the same group with them, are said to be aberrant.

ABERRATION (in optics). In the refraction of light by a convex lens the rays passing through different parts of the lens are brought to a focus at slightly different distances—this is called *spherical aberration*; at the same time the coloured rays are separated by the prismatic action of the lens and likewise brought to a focus at different distances—this is *chromatic aberration*.

ABNORMAL. Contrary to the general rule.

ABORTED. An organ is said to be aborted, when its development has been arrested at a very early stage.

ALBINISM. Albinos are animals in which the usual colouring matters characteristic of the species have not been produced in the skin and its appendages. Albinism is the state of being an albino.

ALGAE. A class of plants including the ordinary seaweeds and the filamentous freshwater weeds.

ALTERNATION OF GENERATIONS. This term is applied to a peculiar mode of reproduction which prevails among many of the lower animals, in which the egg produces a living form quite different from its parent, but from which the parent form is reproduced by a process of budding, or by the division of the substance of the first product of the egg.

* I am indebted to the kindness of Mr W. S. Dallas for this glossary, which has been given because several readers have complained to me that some of the terms used were unintelligible to them. Mr Dallas has endeavoured to give the explanation of the terms in as popular a form as possible. (*Darwin's note*)

AMMONITES. A group of fossil, spiral, chambered shells, allied to the existing pearly nautilus, but having the partitions between the chambers waved in complicated patterns at their junction with the outer wall of the shell.

ANALOGY. That resemblance of structures which depends upon similarity of function, as in the wings of insects and birds. Such structures are said to be *analogous* and to be *analogues* of each other.

ANIMALCULE. A minute animal: generally applied to those visible only by the microscope.

ANNELIDS. A class of worms in which the surface of the body exhibits a more or less distinct division into rings or segments, generally provided with appendages for locomotion and with gills. It includes the ordinary marine worms, the earthworms, and the leeches.

ANTENNAE. Jointed organs appended to the head in insects, crustacea and centipedes, and not belonging to the mouth.

ANTHERS. The summits of the stamens of flowers, in which the pollen or fertilizing dust is produced.

APLACENTALIA, APLACENTATA OR APLACENTAL MAMMALS. See *Mammalia*.

ARCHETYPAL. Of or belonging to the archetype, or ideal primitive form upon which all the beings of a group seem to be organized.

ARTICULATA. A great division of the animal kingdom characterized generally by having the surface of the body divided into rings, called segments, a greater or less number of which are furnished with jointed legs (such as insects, crustaceans and centipedes).

ASYMMETRICAL. Having the two sides unlike.

ATROPHIED. Arrested in development at a very early stage.

BALANUS. The genus including the common acorn-shells which live in abundance on the rocks of the sea-coast.

BATRACHIANS. A class of animals allied to the reptiles, but undergoing a peculiar metamorphosis, in which the young animal is generally aquatic and breathes by gills. (*Examples*, frogs, toads, and newts.)

BOULDERS. Large transported blocks of stone generally imbedded in clays or gravels.

BRACHIOPODA. A class of marine Mollusca, or soft-bodied animals, furnished with a bivalve shell, attached to submarine objects by a stalk which passes through an aperture in one of the valves, and furnished with fringed arms, by the action of which food is carried to the mouth.

BRANCHIAE. Gills or organs for respiration in water.

BRANCHIAL. Pertaining to gills or branchiae.

CAMBRIAN SYSTEM. A series of very ancient Palaeozoic rocks, between the Laurentian and the Silurian. Until recently these were regarded as the oldest fossiliferous rocks.

CANIDAE. The dog-family, including the dog, wolf, fox, jackal, etc.

CARAPACE. The shell enveloping the anterior part of the body in crustaceans generally; applied also to the hard shelly pieces of the cirripedes.

CARBONIFEROUS. This term is applied to the great formation which includes, among other rocks, the coal-measures. It belongs to the oldest, or Palaeozoic, system of formations.

CAUDAL. Of or belonging to the tail.

CEPHALOPODS. The highest class of the Mollusca, or soft-bodied animals, characterized by having the mouth surrounded by a greater or less number of fleshy arms or tentacles, which, in most living species, are furnished with sucking-cups. (*Examples*, cuttle-fish, nautilus.)

CETACEA. An order of Mammalia, including the whales, dolphins, etc., having the form of the body fish-like, the skin naked, and only the forelimbs developed.

CHELONIA. An order of reptiles including the turtles, tortoises, etc.

CIRRIPEDES. An order of crustaceans including the barnacles and acorn-shells. Their young resemble those of many other crustaceans in form; but when mature they are always attached to other objects, either directly or by means of a stalk, and their bodies are enclosed by a calcareous shell composed of several pieces, two of which can open to give issue to a bunch of curled, jointed tentacles, which represent the limbs.

COCCUS. The genus of insects, including the cochineal. In these the male is a minute, winged fly, and the female generally a motionless, berry-like mass.

COCOON. A case usually of silky material, in which insects are frequently enveloped during the second or resting-stage (pupa) of their existence. The term 'cocoon-stage' is here used as equivalent to 'pupa-stage'.

COELOSPERMOUS. A term applied to those fruits of the Umbelliferae which have the seed hollowed on the inner face.

COLEOPTERA. Beetles, an order of insects, having a biting mouth and the first pair of wings more or less horny, forming sheaths for the second pair, and usually meeting in a straight line down the middle of the back.

COLUMN. A peculiar organ in the flowers of orchids, in which the stamens, style and stigma (or the reproductive parts) are united.

COMPOSITAE or COMPOSITOUS PLANTS. Plants in which the inflorescence consists of numerous small flowers (florets) brought together into a dense head, the base of which is enclosed by a common envelope. (*Examples*, the daisy, dandelion, etc.)

CONFERVAE. The filamentous weeds of freshwater.

CONGLOMERATE. A rock made up of fragments of rock or pebbles, cemented together by some other material.

COROLLA. The second envelope of a flower, usually composed of coloured, leaf-like organs (petals), which may be united by their edges either in the basal part or throughout.

CORRELATION. The normal coincidence of one phenomenon, character, etc., with another.

CORYMB. A bunch of flowers in which those springing from the lower part of the flower stalk are supported on long stalks so as to be nearly on a level with the upper ones.

COTYLEDONS. The first or seed-leaves of plants.

CRUSTACEANS. A class of articulated animals, having the skin of the body generally more or less hardened by the deposition of calcareous matter, breathing by means of gills. (*Examples*, crab, lobster, shrimp, etc.)

CURCULIO. The old generic term for the beetles known as weevils, characterized by their four-jointed feet, and by the head being produced into a sort of beak, upon the sides of which the antennae are inserted.

CUTANEOUS. Of or belonging to the skin.

DEGRADATION. The wearing down of land by the action of the sea or of meteoric agencies.

DENUDATION. The wearing away of the surface of the land by water.

DEVONIAN SYSTEM or formation. A series of Palaeozoic rocks, including the Old Red Sandstone.

DICOTYLEDONS OF DICOTYLEDONOUS PLANTS. A class of plants characterized by having two seed-leaves, by the formation of new wood between the bark and the old wood (exogenous growth) and by the reticulation of the veins of the leaves. The parts of the flowers are generally in multiples of five.

DIFFERENTIATION. The separation or discrimination of parts or organs which in simpler forms of life are more or less united.

DIMORPHIC. Having two distinct forms. Dimorphism is the condition of the appearance of the same species under two dissimilar forms.

DIOECIOUS. Having the organs of the sexes upon distinct individuals.

DIORITE. A peculiar form of greenstone.

DORSAL. Of or belonging to the back.

EDENTATA. A peculiar order of quadrupeds, characterized by the absence of at least the middle incisor (front) teeth in both jaws. (*Examples*, the sloths and armadillos.)

ELYTRA. The hardened forewings of beetles, serving as sheaths for the membranous hind-wings, which constitute the true organs of flight.

EMBRYO. The young animal undergoing development within the egg or womb.

EMBRYOLOGY. The study of the development of the embryo.

ENDEMIC. Peculiar to a given locality.

ENTOMOSTRACA. A division of the class Crustacea, having all the segments of the body usually distinct, gills attached to the feet or organs of the mouth, and the feet fringed with fine hairs. They are generally of small size.

EOCENE. The earliest of the three divisions of the Tertiary epoch of geologists. Rocks of this age contain a small proportion of shells identical with species now living.

EPHEMEROUS INSECTS. Insects allied to the mayfly.

FAUNA. The totality of the animals naturally inhabiting a certain country or region, or which have lived during a given geological period.

FELIDAE. The cat-family.

FERAL. Having become wild from a state of cultivation or domestication.

FLORA. The totality of the plants growing naturally in a country, or during a given geological period.

FLORETS. Flowers imperfectly developed in some respects, and collected into a dense spike or head, as in the grasses, the dandelion, etc.

FOETAL. Of or belonging to the foetus, or embryo in course of development.

FORAMINIFERA. A class of animals of very low organization, and generally of small size, having a jelly-like body, from the surface of which delicate filaments can be given off and retracted for the prehension of external objects, and having a calcareous or sandy shell, usually divided into chambers, and perforated with small apertures.

FOSSILIFEROUS. Containing fossils.

FOSSORIAL. Having a faculty of digging. The fossorial Hymenoptera are a group of wasp-like insects, which burrow in sandy soil to make nests for their young.

FRENUM (pl. FRENA). A small band or fold of skin.

FUNGI (sing. FUNGUS). A class of cellular plants, of which mushrooms, toadstools, and moulds, are familiar examples.

FURCULA. The forked bone formed by the union of the collar-bones in many birds, such as the common fowl.

GALLINACEOUS BIRDS. An order of birds of which the common fowl, turkey, and pheasant, are well-known examples.

GALLUS. The genus of birds which includes the common fowl.

GANGLION. A swelling or knot from which nerves are given off as from a centre.

GANOID fishes. Fishes covered with peculiar enamelled bony scales. Most of them are extinct.

GERMINAL VESICLE. A minute vesicle in the eggs of animals, from which the development of the embryo proceeds.

GLACIAL PERIOD. A period of great cold and of enormous extension of ice upon the surface of the earth. It is believed that Glacial periods have occurred repeatedly during the geological history of the earth, but the term is generally applied to the close of the Tertiary epoch, when nearly the whole of Europe was subjected to an arctic climate.

GLAND. An organ which secretes or separates some peculiar product from the blood or sap of animals or plants.

GLOTTIS. The opening of the windpipe into the oesophagus or gullet.

GNEISS. A rock approaching granite in composition, but more or less laminated, and really produced by the alteration of a sedimentary deposit after its consolidation.

GRALLATORES. The so-called wading-birds (storks, cranes, snipes, etc.), which are generally furnished with long legs, bare of feathers above the heel, and have no membranes between the toes.

GRANITE. A rock consisting essentially of crystals of felspar and mica in a mass of quartz.

HABITAT. The locality in which a plant or animal naturally lives.

HEMIPTERA. An order or suborder of insects, characterized by the possession of a jointed beak or rostrum, and by having the forewings horny in the basal portion and membranous at the extremity, where they cross each other. This group includes the various species of bugs.

HERMAPHRODITE. Possessing the organs of both sexes.

HOMOLOGY. That relation between parts which results from their development from corresponding embryonic parts, either in different animals, as in the case of the arm of man, the foreleg of a quadruped, and the wing of a bird; or in the same individual, as in the case of the fore and hind legs in quadrupeds, and the segments or rings and their appendages of which the body of a worm, a centipede, etc., is composed. The latter is called *serial homology*. The parts which stand in such a relation to each other are said to be *homologous*, and one such part or organ is called the *homologue* of the other. In different plants the parts of the flower are homologous, and in general these parts are regarded as homologous with leaves.

HOMOPTERA. An order or sub-order of insects having (like the Hemiptera) a jointed beak, but in which the fore-wings are either wholly membranous or wholly leathery. The Cicadae, frog-hoppers, and aphides, are well-known examples.

HYBRID. The offspring of the union of two distinct species.

HYMENOPTERA. An order of insects possessing biting jaws and usually four membranous wings in which there are a few veins. Bees and wasps are familiar examples of this group.

HYPERTROPHIED. Excessively developed.

ICHNEUMONIDAE. A family of hymenopterous insects, the members of which lay their eggs in the bodies or eggs of other insects.

IMAGO. The perfect (generally winged) reproductive state of an insect.

INDIGEENS. The aboriginal animal or vegetable inhabitants of a country or region.

INFLORESCENCE. The mode of arrangement of the flowers of plants.

INFUSORIA. A class of microscopic animalcules, so called from their having originally been observed in infusions of vegetable matters. They consist of a gelatinous material enclosed in a delicate membrane the whole or part of which is furnished with short vibrating hairs (called cilia), by means of which the animalcules swim through the water or convey the minute particles of their food to the orifice of the mouth.

INSECTIVOROUS. Feeding on insects.

INVERTEBRATA, or INVERTEBRATE ANIMALS. Those animals which do not possess a backbone or spinal column.

LACUNAE. Spaces left among the tissues in some of the lower animals, and serving in place of vessels for the circulation of the fluids of the body.

LAMELLATED. Furnished with lamellae or little plates.

LARVA (pl. LARVAE). The first condition of an insect at its issuing from the egg, when it is usually in the form of a grub, caterpillar, or maggot.

LARYNX. The upper part of the windpipe opening into the gullet.

LAURENTIAN. A group of greatly altered and very ancient rocks, which is greatly developed along the course of the St Laurence, whence the name. It is in these that the earliest known traces of organic bodies have been found.

LEGUMINOSAE. An order of plants represented by the common peas and beans, having an irregular flower in which one petal stands up like a wing, and the stamens and pistil are enclosed in a sheath formed by two other petals. The fruit is a pod (or legume).

LEMURIDAE. A group of four-handed animals, distinct from the monkeys and approaching the insectivorous quadrupeds in some of their characters and habits. Its members have the nostrils curved or twisted, and a claw instead of a nail upon the first finger of the hind hands.

LEPIDOPTERA. An order of insects, characterized by the possession of a spiral proboscis, and of four large more or less scaly wings. It includes the well-known butterflies and months.

LITTORAL. Inhabiting the seashore.

LOESS. A marly deposit of recent (Post-Tertiary) date, which occupies a great part of the valley of the Rhine.

MALACOSTRACA. The higher division of the Crustacea, including the ordinary crabs, lobsters, shrimps, etc., together with the woodlice and sand-hoppers.

MAMMALIA. The highest class of animals, including the ordinary hairy quadrupeds, the whales, and man, and characterized by the production of living young which are nourished after birth by milk from the teats (*mammae, mammary glands*) of the mother. A striking difference in embryonic development has led to the division of this class into two great groups; in one of these, when the embryo has attained a certain stage, a vascular connection, called the *placenta*, is formed between the embryo and the mother; in the other this is wanting, and the young are produced in a very incomplete state. The former, including the greater part of the class, are called *placental mammals*; the latter, or *aplacental mammals*, include the marsupials and monotremes (*Ornithorhynchus*).

MAMMIFEROUS. Having mammae or teats (see MAMMALIA).

MANDIBLES, in insects. The first or uppermost pair of jaws, which are generally solid, horny, biting organs. In birds the term is applied to both jaws with their horny coverings. In quadrupeds the mandible is properly the lower jaw.

MARSUPIALS. An order of Mammalia in which the young are born in a very incomplete state of development, and carried by the mother, while sucking, in a ventral pouch (marsupium), such as the kangaroos, opossums, etc. (see MAMMALIA).

MAXILLAE, in insects. The second or lower pair of jaws, which are composed of several joints and furnished with peculiar jointed appendages called palpi, or feelers.

MELANISM. The opposite of albinism; an undue development of colouring material in the skin and its appendages.

METAMORPHIC ROCKS. Sedimentary rocks which have undergone alteration, generally by the action of heat, subsequently to their deposition and consolidation.

MOLLUSCA. One of the great divisions of the animal kingdom, including those animals which have a soft body, usually furnished with a shell, and in which the nervous ganglia, or centres, present no definite general arrangement. They are generally known under the denomination of 'shellfish'; the cuttlefish, and the common snails, whelks, oysters, mussels, and cockles, may serve as examples of them.

MONOCOTYLEDONS, OR MONOCOTYLEDONOUS PLANTS. Plants in which the seed sends up only a single seed-leaf (or cotyledon); characterized by the absence of consecutive layers of wood in the stem (endogenous growth), by the veins of the leaves being generally straight, and by the parts of the flowers being generally in multiples of three. (*Examples*, grasses, lilies, orchids, palms, etc.)

MORAINES. The accumulations of fragments of rock brought down by glaciers.

MORPHOLOGY. The law of form or structure independent of function.

MYSIS-STAGE. A stage in the development of certain crustaceans (prawns), in which they closely resemble the adults of a genus (Mysis) belonging to a slightly lower group.

NASCENT. Commencing development.

NATATORY. Adapted for the purpose of swimming.

NAUPLIUS-FORM. The earliest stage in the development of many Crustacea, especially belonging to the lower groups. In this stage the animal has a short body, with indistinct indications of a division into segments, and three pairs of fringed limbs. This form of the common freshwater cyclops was described as a distinct genus under the name of Nauplius.

NEURATION. The arrangement of the veins or nervures in the wings of insects.

NEUTERS. Imperfectly developed females of certain social insects (such as ants and bees), which perform all the labours of the community. Hence they are also called *workers*.

NICTITATING MEMBRANE. A semi-transparent membrane, which can be drawn across the eye in birds and reptiles, either to moderate the effects of a strong light or to sweep particles of dust, etc., from the surface of the eye.

OCELLI. The simple eyes or stemmata of insects, usually situated on the crown of the head between the great compound eyes.

OESOPHAGUS. The gullet.

OOLITIC. A great series of secondary rocks, so called from the texture of some of its members, which appear to be made up of a mass of small *egg-like* calcareous bodies.

OPERCULUM. A calcareous plate employed by many Mollusca to close the aperture of their shell. The *opercular valves* of cirripedes are those which close the aperture of the shell.

ORBIT. The bony cavity for the reception of the eye.

ORGANISM. An organized being, whether plant or animal.

ORTHOSPERMOUS. A term applied to those fruits of the Umbelliferae which have the seed straight.

OSCULANT. Forms or groups apparently intermediate between and connecting other groups are said to be osculant.

OVA. Eggs.

OVARIUM OR OVARY (in plants). The lower part of the pistil or female organ of the flower, containing the ovules or incipient seeds; by growth after the other organs of the flower have fallen, it usually becomes converted into the fruit.

OVIGEROUS. Egg-bearing.

OVULES (of plants). The seeds in the earliest condition.

PACHYDERMS. A group of Mammalia, so called from their thick skins, and including the elephant, rhinoceros, hippopotamus, etc.

PALAEOZOIC. The oldest system of fossiliferous rocks.

PALPI. Jointed appendages to some of the organs of the mouth in insects and Crustacea.

PAPILIONACEAE. An order of plants (see LEGUMINOSAE). The flowers of these plants are called *papilionaceous*, or butterfly-like, from the fancied resemblance of the expanded superior petals to the wings of a butterfly.

PARASITE. An animal or plant living upon or in, and at the expense of, another organism.

PARTHENOGENESIS. The production of living organisms from unimpregnated eggs or seeds.

PEDUNCULATED. Supported upon a stem or stalk. The pedunculated oak has its acorns borne upon a footstalk.

PELORIA OR PELORISM. The appearance of regularity of structure in the flowers of plants which normally bear irregular flowers.

PELVIS. The bony arch to which the hind limbs of vertebrate animals are articulated.

PETALS. The leaves of the corolla, or second circle of organs in a flower. They are usually of delicate texture and brightly coloured.

PHYLLODINEOUS. Having flattened, leaf-like twigs or leaf-stalks instead of true leaves.

PIGMENT. The colouring material produced generally in the superficial parts of animals. The cells secreting it are called *pigment-cells.*

PINNATE. Bearing leaflets on each side of a central stalk.

PISTILS. The female organs of a flower, which occupy a position in the centre of the other floral organs. The pistil is generally divisible into the ovary or germen, the style and the stigma.

PLACENTALIA, PLACENTATA, OR PLACENTAL MAMMALS. See MAMMALIA.

PLANTIGRADES. Quadrupeds which walk upon the whole sole of the foot, like the bears.

PLASTIC. Readily capable of change.

PLEISTOCENE PERIOD. The latest portion of the Tertiary epoch.

PLUMULE (in plants). The minute bud between the seed-leaves of newly germinated plants.

PLUTONIC ROCKS. Rocks supposed to have been produced by igneous action in the depths of the earth.

POLLEN. The male element in flowering plants; usually a fine dust produced by the anthers, which, by contact with the stigma effects the fecundation of the seeds. This impregnation is brought about by means of tubes (*pollen-tubes*) which issue from the pollen grains adhering to the stigma, and penetrate through the tissues until they reach the ovary.

POLYANDROUS (flowers). Flowers having many stamens.

POLYGAMOUS PLANTS. Plants in which some flowers are unisexual and others hermaphrodite. The unisexual (male and female) flowers, may be on the same or on different plants.

POLYMORPHIC. Presenting many forms.

POLYZOARY. The common structure formed by the cells of the Polyzoa, such as the well-known sea-mats.

PREHENSILE. Capabale of grasping.

PREPOTENT. Having a superiority of power.

PRIMARIES. The feathers forming the top of the wing of a bird, and inserted upon that part which represents the hand of man.

PROCESSES. Projecting portions of bones, usually for the attachment of muscles, ligaments, etc.

PROPOLIS. A resinous material collected by the hive-bees from the opening buds of various trees.

PROTEAN. Exceedingly variable.

PROTOZOA. The lowest great division of the animal kingdom. These animals are composed of a gelatinous material, and show scarcely any trace of distinct organs. The Infusoria, Foraminifera, and sponges, with some other forms, belong to this division.

PUPA (pl. PUPAE). The second stage in the development of an insect, from which it emerges in the perfect (winged) reproductive form. In most insects the *pupal stage* is passed in perfect repose. The *chrysalis* is the pupal state of butterflies.

RADICLE. The minute root of an embryo plant.

RAMUS. One half of the lower jaw in the Mammalia. The portion which rises to articulate with the skull is called the *ascending ramus*.

RANGE. The extent of country over which a plant or animal is naturally spread. *Range in time* expresses the distribution of a species or group through the fossiliferous beds of the earth's crust.

RETINA. The delicate inner coat of the eye, formed by nervous filaments spreading from the optic nerve, and serving for the perception of the impressions produced by light.

RETROGRESSION. Backward development. When an animal, as it approaches maturity, becomes less perfectly organized than might be expected from its early stages and known relationships, it is said to undergo a *retrograde development* or *metamorphosis*.

RHIZOPODS. A class of lowly organized animals (Protozoa), having a gelatinous body, the surface of which can be protruded in the form of root-like process or filaments, which serve for locomotion and the prehension of food. The most important order is that of the Foraminifera.

RODENTS. The gnawing Mammalia, such as the rats, rabbits, and squirrels. They are especially characterized by the possession of a single pair of chisel-like cutting teeth in each jaw, between which and the grinding teeth there is a great gap.

RUBUS. The bramble genus.

RUDIMENTARY. Very imperfectly developed.

RUMINANTS. The group of quadrupeds which ruminate or chew the cud, such as oxen, sheep, and deer. They have divided hoofs, and are destitute of front teeth in the upper jaw.

SACRAL. Belonging to the sacrum, or the bone composed usually of two or more united vertebrae to which the sides of the pelvis in vertebrate animals are attached.

SARCODE. The gelatinous material of which the bodies of the lowest animals (Protozoa) are composed.

SCUTELLAE. The horny plates with which the feet of birds are generally more or less covered, especially in front.

SEDIMENTARY FORMATIONS. Rocks deposited as sediments from water.

SEGMENTS. The transverse rings of which the body of an articulate animal or annelid is composed.

SEPALS. The leaves or segments of the calyx, or outermost envelope of an ordinary flower. They are usually green, but sometimes brightly coloured.

SERRATURES. Teeth like those of a saw.

SESSILE. Not supported on a stem or footstalk.

SILURIAN SYSTEM. A very ancient system of fossiliferous rocks belonging to the earlier part of the Palaeozoic series.

SPECIALIZATION. The setting apart of a particular organ for the performance of a particular function.

SPINAL CHORD. The central part of the nervous system in the Vertebrata, which descends from the brain through the arches of the vertebrae, and gives off nearly all the nerves to the various organs of the body.

STAMENS. The male organs of flowering plants, standing in a circle within the petals. They usually consist of a filament and an anther, the anther being the essential part in which the pollen, or fecundating dust, is formed.

STERNUM. The breastbone.

STIGMA. The apical portion of the pistil in flowering plants.

STIPULES. Small leafy organs placed at the base of the footstalks of the leaves in many plants.

STYLE. The middle portion of the perfect pistil, which rises like a column from the ovary and supports the stigma at its summit.

SUBCUTANEOUS. Situated beneath the skin.

SUCTORIAL. Adapted for sucking.

SUTURES (in the skull). The lines of junction of the bones of which the skull is composed.

TARSUS (pl. TARSI). The jointed feet of articulate animals, such as insects.

TELEOSTEAN FISHES. Fishes of the kind familiar to us in the present day, having the skeleton usually completely ossified and the scales horny.

TENTACULA or TENTACLES. Delicate fleshy organs of prehension or touch possessed by many of the lower animals.

TERTIARY. The latest geological epoch, immediately preceding the establishment of the present order of things.

TRACHEA. The windpipe or passage for the admission of air to the lungs.

TRIDACTYLE. Three-fingered, or composed of three movable parts attached to a common base.

TRILOBITES. A peculiar group of extinct crustaceans, somewhat resembling the woodlice in external form, and, like some of them, capable of rolling themselves up into a ball. Their remains are found only in the Palaeozoic rocks, and most abundantly in those of Silurian age.

TRIMORPHIC. Presenting three distinct forms.

UMBELLIFERAE. An order of plants in which the flowers, which contain five stamens and a pistil with two styles, are supported upon footstalks which spring from the top of the flower stem and spread out like the wires of an umbrella, so as to bring all the flowers in the same head (*umbel*) nearly to the same level. (*Examples*, parsley and carrot.)

UNGULATA. Hoofed quadrupeds.

UNICELLULAR. Consisting of a single cell.

VASCULAR. Containing blood-vessels.

VERMIFORM. Like a worm.

VERTEBRATA: or VERTEBRATE ANIMALS. The highest division of the animal kingdom, so called from the presence in most cases of a backbone composed of numerous joints or *vertebrae*, which constitutes the centre of the skeleton and at the same time supports and protects the central parts of the nervous system.

WHORLS. The circles or spiral lines in which the parts of plants are arranged upon the axis of growth.

WORKERS. See neuters.

ZOËA-STAGE. The earliest stage in the development of many of the higher Crustacea, so called from the name of *Zoëa* applied to these young animals when they were supposed to constitute a peculiar genus.

ZOOIDS. In many of the lower animals (such as the corals, medusae, etc.) reproduction takes place in two ways, namely, by means of eggs and by a process of budding with or without separation from the parent of the product of the latter, which is often very different from that of the egg. The individuality of the species is represented by the whole of the form produced between two sexual reproductions; and these forms, which are apparently individual animals, have been called *zooids*.

INDEX

THE WORLD'S CLASSICS

A Select List

HANS ANDERSEN: Fairy Tales
Translated by L. W. Kingsland
Introduction by Naomi Lewis
Illustrated by Vilhelm Pedersen and Lorenz Frølich

ARTHUR J. ARBERRY (Transl.): The Koran

LUDOVICO ARIOSTO: Orlando Furioso
Translated by Guido Waldman

ARISTOTLE: The Nicomachean Ethics
Translated by David Ross

JANE AUSTEN: Emma
Edited by James Kinsley and David Lodge

Northanger Abbey, Lady Susan, The Watsons, *and* Sanditon
Edited by John Davie

Persuasion
Edited by John Davie

WILLIAM BECKFORD: Vathek
Edited by Roger Lonsdale

KEITH BOSLEY (Transl.): The Kalevala

CHARLOTTE BRONTË: Jane Eyre
Edited by Margaret Smith

JOHN BUNYAN: The Pilgrim's Progress
Edited by N. H. Keeble

FRANCES HODGSON BURNETT: The Secret Garden
Edited by Dennis Butts

FANNY BURNEY: Cecilia or Memoirs of an Heiress
Edited by Peter Sabor and Margaret Anne Doody

THOMAS CARLYLE: The French Revolution
Edited by K. J. Fielding and David Sorensen

LEWIS CARROLL: Alice's Adventures in Wonderland *and* Through the Looking Glass
Edited by Roger Lancelyn Green
Illustrated by John Tenniel

MIGUEL DE CERVANTES: Don Quixote
Translated by Charles Jarvis
Edited by E. C. Riley

GEOFFREY CHAUCER: The Canterbury Tales
Translated by David Wright

ANTON CHEKHOV: The Russian Master and Other Stories
Translated by Ronald Hingley

JOHN CLELAND: Memoirs of a Woman of Pleasure (Fanny Hill)
Edited by Peter Sabor

WILKIE COLLINS: Armadale
Edited by Catherine Peters

JOSEPH CONRAD: Chance
Edited by Martin Ray

Victory
Edited by John Batchelor
Introduction by Tony Tanner

NORMAN DAVIS (Ed.): The Paston Letters

CHARLES DICKENS: Christmas Books
Edited by Ruth Glancy

Sikes and Nancy and Other Public Readings
Edited by Philip Collins

FEDOR DOSTOEVSKY: Crime and Punishment
Translated by Jessie Coulson
Introduction by John Jones

ARTHUR CONAN DOYLE: Sherlock Holmes: Selected Stories
Introduction by S. C. Roberts

ALEXANDRE DUMAS *père*:
The Three Musketeers
Edited by David Coward

ALEXANDRE DUMAS *fils*:
La Dame aux Camélias
Translated by David Coward

MARIA EDGEWORTH: Castle Rackrent
Edited by George Watson

GEORGE ELIOT:
Felix Holt, The Radical
Edited by Fred C. Thompson

Middlemarch
Edited by David Carroll

Scenes of Clerical Life
Edited by Thomas A. Noble

Selected Critical Writings
Edited by Rosemary Ashton

GUSTAVE FLAUBERT: Madame Bovary
Translated by Gerard Hopkins
Introduction by Terence Cave

BENJAMIN FRANKLIN: Autobiography and other Writings
Edited by Ormond Seavey

ELIZABETH GASKELL: Cousin Phillis and Other Tales
Edited by Angus Easson

My Lady Ludlow and Other Stories
Edited by Edgar Wright

J. W. VON GOETHE: Faust, Part One
Translated by David Luke

OLIVER GOLDSMITH: The Vicar of Wakefield
Edited by Arthur Friedman

THOMAS HARDY: A Pair of Blue Eyes
Edited by Alan Manford

HESIOD: Theogony *and* Works and Days
Translated by M. L. West

E. T. A. HOFFMANN: The Golden Pot and Other Tales
Translated and Edited by Ritchie Robertson

HOMER: The Iliad
Translated by Robert Fitzgerald
Introduction by G. S. Kirk

THOMAS HUGHES: Tom Brown's Schooldays
Edited by Andrew Sanders

HENRIK IBSEN: An Enemy of the People, The Wild Duck, Rosmersholm
Edited and Translated by James McFarlane

HENRY JAMES: The Ambassadors
Edited by Christopher Butler

M. R. JAMES: Casting the Runes and Other Ghost Stories
Edited by Michael Cox

JOCELIN OF BRAKELOND:
Chronicle of the Abbey of Bury St. Edmunds
Translated by Diana Greenway and Jane Sayers

GWYN JONES (Transl.):
Eirik the Red and Other Icelandic Sagas

BEN JONSON: Five Plays
Edited by G. A. Wilkes

MADAME DE LAFAYETTE: The Princesse de Clèves
Translated and Edited by Terence Cave

WILLIAM LANGLAND: Piers Plowman
Translated and Edited by A. V. C. Schmidt

J. SHERIDAN LE FANU: Uncle Silas
Edited by W. J. McCormack

MIKHAIL LERMONTOV: A Hero of our Time
Translated by Vladimar Nabokov with Dmitri Nabokov

MATTHEW LEWIS: The Monk
Edited by Howard Anderson

MARCUS AURELIUS: The Meditations of Marcus Aurelius
Translated by A. S. L. Farquharson
Edited by R. B. Rutherford

KATHERINE MANSFIELD: Selected Stories
Edited by D. M. Davin

CHARLES MATURIN: Melmoth the Wanderer
Edited by Douglas Grant
Introduction by Chris Baldick

HERMAN MELVILLE: The Confidence-Man
Edited by Tony Tanner

Moby Dick
Edited by Tony Tanner

PROSPER MÉRIMÉE: Carmen and Other Stories
Translated by Nicholas Jotcham

MICHELANGELO: Life, Letters, and Poetry
Translated by George Bull with Peter Porter

MOLIÈRE: Don Juan and Other Plays
Translated by George Graveley and Ian Maclean

E. NESBIT: The Railway Children
Edited by Dennis Butts

EDGAR ALLAN POE: Selected Tales
Edited by Julian Symons

JEAN RACINE: Britannicus, Phaedra, Athaliah
Translated by C. H. Sisson

ANN RADCLIFFE: The Italian
Edited by Frederick Garber

THE MARQUIS DE SADE:
The Misfortune of Virtue and Other Early Tales
Translated and Edited by David Coward

SIR WALTER SCOTT: The Heart of Midlothian
Edited by Claire Lamont

ANNA SEWELL: Black Beauty
Edited by Peter Hollindale

TOBIAS SMOLLETT: The Expedition of Humphry Clinker
Edited by Lewis M. Knapp
Revised by Paul-Gabriel Boucé

ROBERT LOUIS STEVENSON:
Treasure Island
Edited by Emma Letley

ANTHONY TROLLOPE: The American Senator
Edited by John Halperin

GIORGIO VASARI: The Lives of the Artists
Translated and Edited by Julia Conaway Bondanella and Peter Bondanella

VIRGINIA WOOLF: Orlando
Edited by Rachel Bowlby

ÉMILE ZOLA: Nana
Translated and Edited by Douglas Parmée

SURPRISE!
Yo
th
It's t
Japa
reads from right to left—so action, sound effects, and word balloons are completely reversed. This preserves the orientation of the original artwork—plus, it's fun! Check out the diagram shown here to get the hang of things, and then turn to the other side of the book to get started!

Vol. 40
Shojo Beat Edition

English Translation & Adaptation/Tomo Kimura
Touch-up Art & Lettering/Sabrina Heep
Design/Veronica Casson
Editor/Pancha Diaz

Printed in the U.S.A.

Published by VIZ Media, LLC
P.O. Box 77010
San Francisco, CA 94107

10 9 8 7 6 5 4 3 2 1
First printing, March 2018

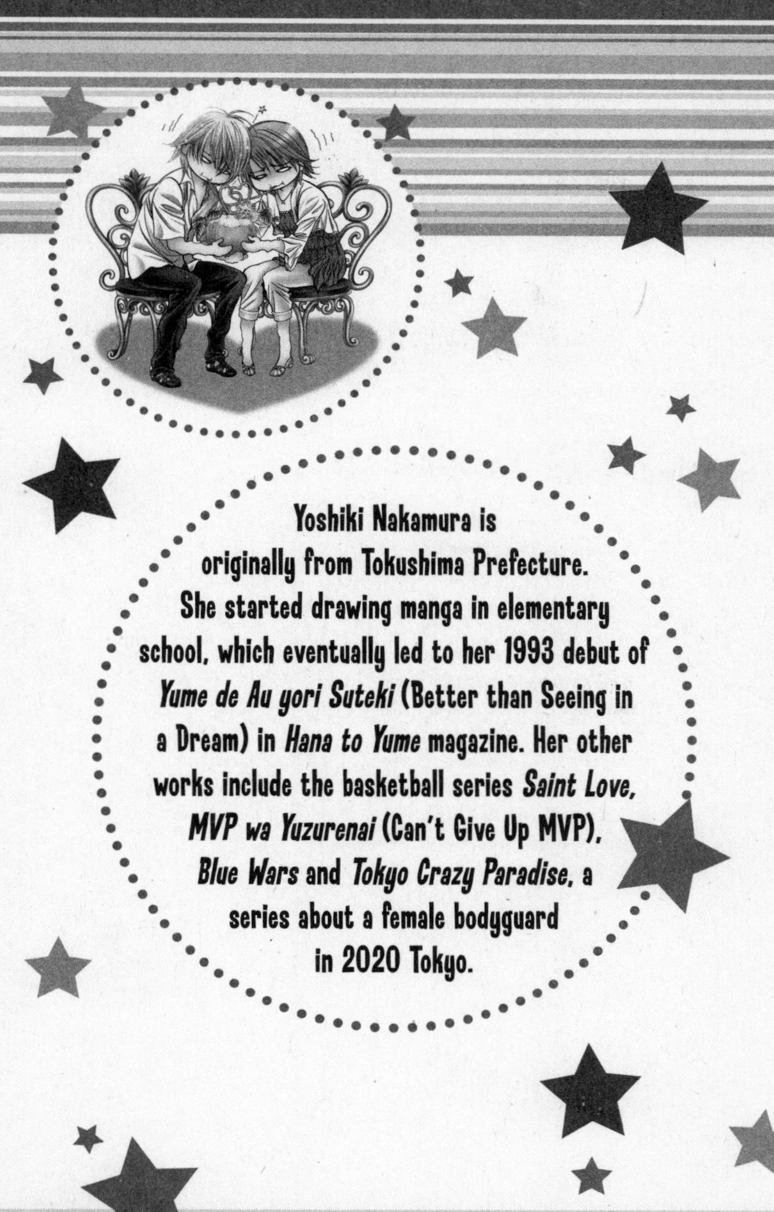

Yoshiki Nakamura is originally from Tokushima Prefecture. She started drawing manga in elementary school, which eventually led to her 1993 debut of *Yume de Au yori Suteki* (Better than Seeing in a Dream) in *Hana to Yume* magazine. Her other works include the basketball series *Saint Love, MVP wa Yuzurenai* (Can't Give Up MVP), *Blue Wars* and *Tokyo Crazy Paradise*, a series about a female bodyguard in 2020 Tokyo.

Skip-Beat! End Notes

Everyone knows how to be a fan, but sometimes cool things from other cultures need a little help crossing the language barrier.

Page 5: Painted sign
This says "Enjoying maple leaves" and "Hunting" in Japanese, but the first two characters can also be pronounced "Momiji." It is written in a traditional calligraphy style.

Page 39, panel 3: Edamame
Soybeans. The fresh beans have an outer skin through which the inner bean pops out.

Page 106, panel 7: White Day
White Day is a Japanese holiday that occurs exactly one month after Valentine's Day and is a chance for boys and men to give reciprocal presents to the ladies in their lives.

Page 131, panel 4: Kinboshi
In sumo, *kinboshi* refers to a lower-ranking wrestler winning against a grand champion.

?!!

MAYBE IF I DYE MY HAIR BLACK AND COME RIGHT BACK LIKE I DID WITH MIO...

NO.

THAT WON'T WORK.

HE WON'T WAIT FOR ME.

End of Act 242

SO JUST MAKE SURE THE PRODUCER LIKES YOU.

I...

I'VE...

NO MATTER WHAT ANYONE SAYS...

...I WON'T CAST PEOPLE I DON'T LIKE.

...ALREADY BEEN REJECTED BY MY LIFELINE!

WE ...

...MUST'VE BEEN REJECTED BECAUSE... WE DON'T... HAVE BLACK HAIR...

clench

I DON'T UNDERSTAND...
HOW COULD HE BE SO HIGH-HANDED?
I WANT TO SEE THE DIRECTOR.
NOTHING WILL CHANGE IF YOU SEE MORIZUMI.
THE SCRIPT-WRITER, THE DIRECTOR, THE REST OF THE CREW AND THE ENTIRE CAST...
...
SO...
...THIS MAN...
...ARE SELECTED...
...BY ME.
...IS THE PRODUCER!

...
D...
...
...YOU ASKED SUCH A STUPID QUESTION ...
THAT ...
...MEANS YOU DESERVE TO BE DISQUALIFIED.
DOES THAT MEAN... I'M DISQUALIFIED?
HUH?!
What?!
WHY ...
WHY?
YOU THREE KEEP YOUR MOUTHS SHUT...
...AND LEAVE THIS ROOM RIGHT NOW.
click
...

tap

shuu
snap
!
clak
clik
clik
?
clak
clik
stare

Conference

HUH?
WHAT'S GOING ON?
...
...MANAGING THIS GIRL, YASHIRO?
ARE YOU ...
YES.
I MANAGE REN AND KYOKO.
!
I SEE.
...
THEN MAY I ASK WHAT SORT OF ROLES SHE'S—
KA CHAK

EXCUSE ME FOR ASKING, BUT WHAT'S YOUR NAME?
UM.
I'M...
MY NAME...
...IS KYOKO.
I'M FROM LME.
WHA ?!

HUH?
SHE DIDN'T DYE HER HAIR FOR THIS AUDITION?
WHY "ALL OVER THE WORLD" ANYWAY?
I MEAN ...
WHAT DID SHE MEAN BY "WORLD"?
What part of the world?
AND ...
IF SHE DIDN'T DYE HER HAIR FOR MOMIJI, WHAT DID SHE MEAN BY "AIM HIGH"?
WHAT'S SHE AIMING FOR?
UH ...
glare
...
WHY'S SHE GLARING?
I FEEL ...
Now that I'm close to you.
...LIKE I'VE SEEN YOU SOMEWHERE.
BUT I CAN'T REMEMBER WHERE.

MS. MORIZUMI ?!
WHAT A SURPRISE.
Really?!
YOU LOOKED SO DIFFERENT, I DIDN'T RECOGNIZE YOU UNTIL YOU SAID YOUR NAME.
YOU DIDN'T ?
YOU DYED YOUR HAIR BLACK.
I LIKED THE WAY YOU LOOKED BEFORE, REALLY CUTE...
...BUT NOW YOU LOOK SHARP AND DIGNIFIED.
ASIAN BEAUTIES ARE POPULAR ALL OVER THE WORLD...
...SO I DECIDED TO AIM HIGH!
!
DON'T I LOOK LIKE AN ASIAN BEAUTY?

clatter
WELL.
WHY DON'T YOU SIT AND CALM DOWN?
Here.
SHOULD I BE SITTING DOWN?
I came in last...
SOME PEOPLE ARE STANDING...
NO WORRIES. IF THERE'S SOMEONE REALLY FAMOUS, WE NEED TO BE VERY CONSIDERATE...
...BUT WE'RE FINE HERE.
REAL...
...LY?
WELL...
IF YOU SAY SO—
MR. YASHIRO.
LONG TIME NO SEE.
I'M KIMIKO MORIZUMI.
WHAT?

YOU'RE RIGHT.
UH.
SO HE'S HER MANAGER?
WHAT ABOUT REN?
HE MANAGES BOTH OF THEM?
BUT SHE SEEMS LIKE A NOBODY.
Then why?
MAYBE I JUST DON'T KNOW HER...
BUT NO ONE ELSE SEEMS TO RECOGNIZE HER EITHER. SHE CAN'T BE A POPULAR ACTRESS.
RIGHT. I DOUBT...
...SHE COULD GET A STAR ROLE...
...OR A SPLASHY SUPPORT-ING ROLE ON TV.
BUT.
IF SHE'S A STAGE ACTRESS...
...I MIGHT NOT KNOW HER...

Now I see...
I'M THE ONLY ONE WITH LIGHT-BROWN HAIR...
Subdued dark brown
...
...
AAAAARGH
I screwed up.
I KNEW IT.
MY HAIR COLOR... IS A DISADVANTAGE...
...RIGHT, MR. YASHIRO?
smile
YES OR NO?!
HEY...
...KIMI.
LOOK.
WHAT?
I'M BUSY.
clik
clik
clik
clik
THAT'S MANAGER YASHIRO.

THEN ...
...SHE'S A NEWBIE.
UH.
BUT SHE'S GOT BLACK HAIR...
I get it...
BLACK HAIR IS MORE APPROPRIATE FOR A SAMURAI MOVIE...
...
...
IT'S HER ...
REALLY ?!
!
I WASN'T SURE. SHE LOOKS SO DIFFERENT FROM HER BLOG PHOTO.
Ack...
DARN...
I SHOULD'VE DYED MY HAIR BLACK TOO...

chak
EXCUSE M—
whisper
whisper
...
WHO'S THAT?
DON'T KNOW ...
whisper
NEVER SEEN HER.
...
...
...
whisper
HUH?
IS SHE AN AMA-TEUR?
whisper
...
THIS ISN'T A PUBLIC AUDITION.
COME ON...
AH.

THAT'S THE ROOM.
!
Oh!
Conference Room
Lotus in the Mud
Momiji audition waiting room.
He's a super-manager!
He's so dependable!
AH.
Conference Room
TIME TO GO.
YUP.
nok
nok
grab

YEAH.
THOUGH SOMETIMES THEY ARE...
...AND SOMETIMES THEY AREN'T.
tmp
tmp
SO HE NEEDS TO GO TAKE A LOOK?
tmp
BUT FIRST...
...I WANT TO TAKE A LOOK AT THE OTHER CANDIDATES.
tmp
THEY DIDN'T HAVE MUCH TIME TO HOLD THIS SECOND AUDITION, SO IT'S CLOSED. OTHERWISE THE FIRST SCREENING WOULD'VE BEEN TOO TIME-CONSUMING.
SO THE ONLY PEOPLE HERE FOR THIS AUDITION...
...ARE THOSE WHO MADE IT TO THE FINAL ROUND LAST TIME, AND THOSE BELONGING...
...TO AGENCIES THAT RECEIVED NOTICES ABOUT THIS AUDITION, LIKE YOU.
IF I RECOGNIZE ANY OF THE ACTRESSES...
...WE CAN FIGURE OUT STRATEGIES TO BEAT THEM.
...

clik
clak
clik
clak

2F ↗
Momiji audition waiting room.
Lotus in the Mud: Momiji
Wait in conference room B, second floor.

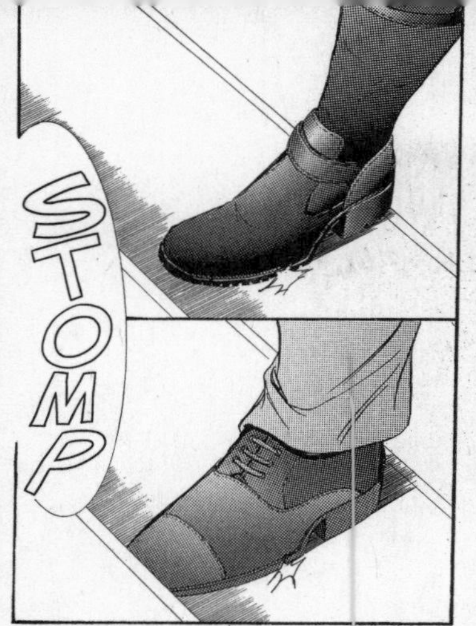
STOMP

Oh?
UH.
IS THE MANAGER'S WAITING ROOM UPSTAIRS TOO?
For these auditions.

ALL RIGHT.
I NEED TO GO TO THIS BUILDING.
UH. SURE.
Lotus in the Mud: Momiji
Wait in conference room B, second floor.
NEXT TIME WE MEET...
...I'LL BE CHI-DORI...
swf
...AND YOU'LL BE MOMIJI.
YEAH.
bump

Lotus in the Mud: Momiji
Wait in
conference room B,
second floor.
2F
Momiji audition
waiting room.
Lotus in the Mud: Momiji
Wait in
conference room B,
second floor.

Mo!
WHY'S SHE SO HYPER ?!
SHE SOUNDED SO DIS-COURAGED YESTER-DAY.
Ha ha ha
YEAH. I FELT THAT WAY TOO, AND WHEN I TOLD HER THAT...
I HAD A KINBOSHI RICE BALL THIS MORN-ING!
I'M READY NOW, SO I'LL GO ALL OUT!
I just need to get hyped up!
...SHE SAID.
She doesn't need to be that hyper!
I GET SO WORN OUT WHEN SHE GOES FROM NEGATIVE TO HYPER-POSITIVE.
SHE DOES THAT A LOT WHEN SHE'S WITH YOU, MS. KOTO-NAMI.
Do your best to accept her.
...
MAYBE ONCE EVERY FIVE TIMES.

THE TOWN ...
...IS ALSO USED FOR SHOOTS.
A town...
Lodging...
...
...
That ...
... means ...
It's a dream training camp where I can be with Moko from good morning to good night!
WE DON'T KNOW IF LOTUS WILL BE SHOOTING THERE.
I know you're enjoying imagining weird situations.
NO SCENES OVER THERE.
For sure.
WHAA—
THERE'LL BE LOTS OF LOCATION SHOOTS IF THE PLOT'S FAITHFUL TO THE NOVEL.
...
Loca ... tion ...
That means I can go on a dream (long) trip with Moko!
Woooookyaaaah!
I've never traveled with a friend! Is this what people call a buddy trip? We should get matching—
LET'S GO, MR. YASHIRO.
UH.
YEAH.

WOOOOOOOW.
THIS IS A HUUUGE STUDIO.
A NUMBER OF BIG MOVIE COMPANIES BUILT THIS TOGETHER.
bam
THE ONLY DISAD-VANTAGE IS THAT IT'S IN THE MIDDLE OF NOWHERE ...
...BECAUSE IT'S SO HUGE. BUT THEY'VE GOT GUEST LODGING.
THEY HAVE OTHER FACILITIES TO MAKE LIFE CONVENIENT...
...SO PART OF THE STUDIO IS LIKE A SMALL TOWN.
boop

EX-
CUSE
...
...ME.
clik
clak
clik
shut
clench

THAT ...
... LOVE ME GIRL ...
...HAS JOINED THE BATTLE FOR MOMIJI...
... I'LL ...
...SNEAK IN TO SEE WHAT HAPPENS.
THAT WOMAN ...
...WILL LOSE AGAINST A MEMBER OF THE LOVE ME SECTION THIS TIME...
...SO I WANT TO SEE HER BLOOD BOILING...
...IN ...
...FRUS-TRATION WITH MY OWN EYES.
twitch
...SO...
I SHALL GO NOW...
...BUT THE OTHER TWO WILL STAY HERE...
...SO CALL THEM IF YOU NEED ANYTHING.

...

MS. ERIKA.

YOU ...

...MAY HAVE ALREADY FORGOTTEN, SINCE YOU DON'T SEEM TO CARE ANYMORE...

...BUT TODAY ...

...THE SECOND AUDITION FOR MOMIJI...

...IS BEING BE HELD.

EX-
CUSE
US.
CALL
IF YOU
NEED
ANY-
THING.
trudge
trudge

clink

glug
glug
glug

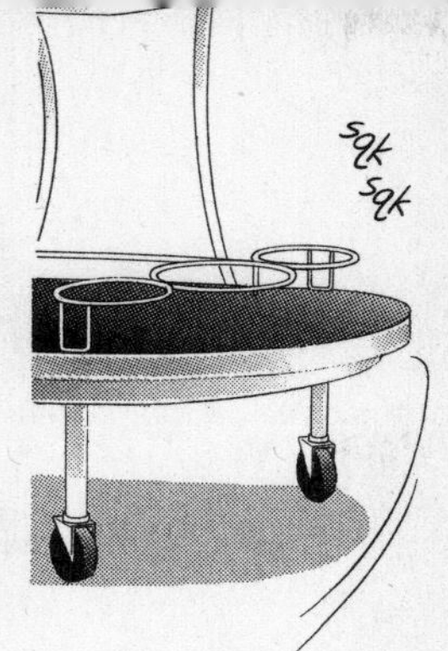
sqk
sqk

NOW ...
...MS. ERIKA.

WE'LL LEAVE YOU.
PLEASE EAT A LITTLE.

Skip·Beat!
Act 242: Lifeline

ONLY I CAN BE...
...MOMIJI.
BUT...
...IF THE EVIL WITCH...

End of Act 241

THANK YOU SO MUCH FOR YOUR...
Heh heh

heh heh
...REASSURING ENDORSEMENT.
TO BE HONEST...
...I ALSO FEEL THAT WAY.

NO MATTER...
...WHO SHOWS UP FOR THAT AUDITION...

Heh heh...

...NONE OF THEM...
...WILL PLAY MOMIJI AS REALISTICALLY...
...AS...
...I WILL.

DON'T WORRY ...
... ABOUT ...
...KIMIKO MORIZUMI.
I SAID SHE DIDN'T SUCK...
...BUT THERE WASN'T ANYTHING OUTSTANDING ABOUT HER EITHER.
I WATCHED HER ACT WHEN THE DRAMA WAS BEING SHOT, SO I KNOW WHAT I'M TALKING ABOUT.
SHE ...
...CAN NEVER BEAT YOU, KYOKO.
OH?
SHE CAN NEVER BEAT YOU.
Uh...
WAS ...
...MS. KOTONAMI TALKING ABOUT KIMIKO MORIZUMI?
Pffft
Heh heh heh
Yes.

VROOOOOOOOOM
Peek
KYOKO.
YES?

GOOD JOB.

VROOOOOOOM

I'M...
Good job, Bo!
Thank you!
...PROBABLY NOT THE ONLY ONE...
...WHO DIDN'T GET A GIFT FOR WHITE DAY.
SO...
...I'LL STOP...
...THINKING ABOUT IT.
ESPECIALLY...
...NOW...
...CUZ THERE'S SOMETHING ELSE...
...I NEED TO THINK ABOUT...
...FIRST.
!

Yes, it's possible.
twirrrl
tw irl
twir rrl
OH.
Itsumi's manager
SEE?
THERE IS SUCH A THING...
...AS SCREENING RULES FOR WHITE DAY GIFTS.
SO EVERYONE, ESPECIALLY ALL GUYS, WOULD FEEL THAT WAY.
SO.
EVERYONE AGREED THE RULES PROBABLY EXIST.
Two out of two people.
I WON'T THINK ABOUT IT.

THAT MEANS MEN MUST LIMIT THE RECIPIENTS OF THEIR WHITE DAY LARGESS ...
... OTHER-WISE...
...VALENTINE'S DAY BECOMES A BURDENSOME INCONVENIENCE.
...
YOU'RE SO RIGHT...
I've never thought about it...
DO YOU UNDER-STAND MY POINT OF VIEW?
YEAH, I GUESS IT'S POSSI-BLE.
Screen-ing rules for White Day Gifts.
!
...BUT THINGS ARE DEFINITELY SKEWED IN FAVOR OF GIRLS IN THIS CASE.
Really?!
RIGHT ?!
THOSE RULES SHOULD EXIST!
nod
YEAH, I THINK SO TOO.

DO YOU ...

...KNOW ALL THE RULES ...

...OF JAPANESE VALENTINE'S DAY FROM RULE 1 TO RULE 100?

GLAARE

WHAT ?

TH-THERE'RE 100 VALENTINE'S DAY RULES?

IF SOMEONE...

...RECEIVES SOMETHING OTHER THAN CHOCOLATES FOR NO OBVIOUS REASON...

...ARE THEY STILL OBLIGATED TO GIVE A GIFT ON WHITE DAY?

...CITIZENS OF THIS LAND HAVE BEEN TAUGHT TO GIVE CHOCOLATES ON VALENTINE'S DAY...

...SO IT'S ABSURD TO EXPECT A WHITE DAY GIFT IF YOU DIDN'T GIVE CHOCOLATES.

ESPECIALLY AS PEOPLE ARE EXPECTED TO GIVE SOMETHING DOUBLE THE PRICE OF WHAT THEY RECEIVED.

UH.
WHY...
...WOULDN'T YOU HAVE ANYTHING TO DO WITH HIS WHITE DAY GIFTS?
UH.
Well...
BE-CAUSE...
...JELLY IN-STEAD...
...I GAVE HIM...
...OF CHOCO-LATES—
BUT THE JELLY...
...WAS YOUR GIFT...
IT WAS...
...A HEARTFELT VALENTINE'S GIFT.
I...
...CAN'T BELIEVE MR. TSURUGA WOULD FORGET YOU...
THERE MUST'VE...
...BEEN SOME MISTAKE, AND YOU JUST NEVER RECEIVED HIS WHITE DAY GIFT?
...

KYOKO?
What's the matter? Your eyes seem empty.
White dée? I've got nothing to do with it, so I'm not feeling anything.
Emotionless robot eyes
HUH?!
BUT YOU BROUGHT HANDMADE CHOCOLATES.
Wha?!
The really cute ones.
YOU GAVE ME AND THE DIRECTOR SOME, SO I ASSUMED YOU GAVE THEM TO MR. TSURUGA TOO...
...DIDN'T GIVE CHOCOLATES TO MR. TSURUGA.
I...
WELL.
WELL...
I THOUGHT MR. TSURUGA...
...WOULD GET A TON OF CHOCOLATES...
...SO I GAVE HIM SOME JELLY INSTEAD.
UH... OKAY...
AND SO.
Yeah, I guess he'd get heartburn from all those chocolates.
I'M...
...OUTSIDE THE CIRCLE OF THOSE WHO RECEIVED MR. TSURUGA'S WHITE DAY GIFTS.
I MISSED BEING CHARM FRIENDS WITH YOU.
That's why...

YOU...
...MUST'VE GOTTEN ONE TOO.
THE WHITE DAY GIFT...
...FROM MR. TSURUGA.
MS. OHARA GOT A DIFFERENT CHARM THAN I DID...
...SO YOU MUST'VE GOTTEN A CHARM YOU'D LIKE...
UH.

...THINK it's weird ?
What?
HUH ?
YOU KNOW, THIS WAS JUST PART OF THE PACKAGE, RIGHT?!
"YOU KNOW... RIGHT ?!"
:
Huh?
To... use this like an accessory!
NO ...
I THINK IT'S FINE.
...
SHE'S ASSUM-ING I KNOW WHAT SHE'S TALKING ABOUT.
BUT...
U...
M...
UH...
Excuse me...
WHAT'S... THIS PACKAGE YOU'RE TALKING ...
...
THE...
... ABOUT ?
... WHITE DAY GIFT.
WHITE DAY GIFT ?

MS. MOMOSE?
OH?
KYOKO.
LONG TIME NO SEE.
IT'S BEEN A WHILE.
YOU'RE HERE FOR WORK?
YES.
IT'S A LOVE ME ASSIGNMENT.
Oh...
IS THAT SO.
WHAT ABOUT YOU, MS. MOMOSE?
I JUST FINISHED WORK. I'M WAITING FOR MY MANAGER TO PICK ME UP.
I HAD A HARD TIME GETTING IN TOUCH WITH HIM...
...BUT I FINALLY GOT A TEXT.
He'll be here in about 15 minutes.
jingle
THAT'S A CUTE CHARM.
The heart.
IT'S PLUMP AND SPARKLING, LIKE A PRINCESS HEART. ♡
JOLT
SO YOU LIKE THIS SORT–
Do you ...

beam
grin
grin
For
&
Kinboshi rice ball
For tomorrow's breakfast
+
And several other kinds
ALL RIGHT, KYOKO.
CALL ME WHEN YOU'RE DONE.
Y-yes!
Oh!
I WILL!
Promptly!
See you later!
HMM?
OH?
I THINK I RECOGNIZE THAT BACK...
COULD THAT BE...
sqk
sqk
sqk
sqk

OF COURSE...
...WE CAN.
...
THEN CAN WE GO THERE...
... FIRST—
Mo—ko
ooooooooo
I love youuuuu
FWIP
PLEASE.
UH.
SURE.
...
...
Kinboshi rice ball
Loved by people seeking victory, including students preparing for entrance exams.
Handmade Nihyo
Pickled Gold Star plum, a very sweet plum.

...FAIRLY?
fwip
...
MR. YASHIRO.
YES?
MAY I...
...ASK YOU...
...TO TAKE ME TO PURÉ BELLE?
THANK YOU.
YES.
OF COURSE.
THEN ...
MAY I...
...ALSO ASK YOU TO STOP BY NIHYO? IT'S A RESTAURANT.
!
fwip

crunch
crunch

...BUT YOU'LL BE FINE.

EVEN IF THE GIRL I MENTIONED SHOWS UP AT THIS AUDITION...

crunch
crunch

...SHE CAN NEVER BEAT YOU.

...

THERE'S SOMETHING...

...I SUDDENLY THOUGHT OF LAST NIGHT.

I WAS RELIEVED TO FIND OUT THE PRODUCER MIGHT BE A FAIR JUDGE...

BUT...

...WHAT IF THE PRODUCER AND THE DIRECTOR KNOW WHO MS. KOENJI IS?

MS. KOENJI SAID SHE HID HER IDENTITY...

KOENJI GROUP IS THE BIGGEST SPONSOR OF LOTUS IN THE MUD...

IF... ...THEY KNEW WHO SHE WAS...

...DID SHE WIN THE ROLE...

?
shp shp
I...
...saw a girl whose kneecaps were crushed into pieces by a truly evil witch of destruction...
...miserably collapse... That must be my fate tomorrow...
wah wah wah wah wah wah
shiver shake shiver shake
YOU WERE DAY-DREAMING AGAIN.

HEY.
THE BLISTERS ON YOUR HAND AREN'T SO VISIBLE ANYMORE.
HUH?
Uh...
YEAH.
I'VE BEEN USING...
...THE MOISTURE POWER PADS YOU TOLD ME ABOUT.
THEY'RE AMAZ-ING!
JUST AS I EXPECT OF TAIZEN PHARMA, "HEALS INSTANTLY!"
My new Blisters were flattened out over-night!
YOU'RE REALLY PASSION-ATE ABOUT BEAUTY!
UH... HIO GAVE ME THAT.
HUH?
HE TOLD ME "AN ACTRESS SHOULDN'T HAVE BLISTERS ON HER HAND."
Speak-ing in a mean way.
HE SAID IT WAS MY GIFT.
For White Day.
SO IF YOU'RE GONNA PRAISE ME, THANK HIO FOR THAT CURE-ALL.

SO.
WHEN MR. YASHIRO BECAME MY TEMPORARY MANAGER, I WAS SO OVER-WHELMED I TREMBLED...
...BUT THE SHOOT DATE FOR KIMAGURE MADE ME SHIVER IN FEAR FROM DESPAIR AND ANXIETY...
...
AND...
...SO?
With about ten years' worth of cold sweat.
I HAVEN'T TOLD HIM.
I TOLD HIM I CAN'T SPILL DETAILS SINCE...
...I'M DOING IT AS A LOVE ME MEMBER.
HE HASN'T ASKED SINCE...
...AND HE'S BEHAVING LIKE NORMAL.
HE UNDERSTANDS EXACTLY WHAT I DON'T WANT HIM TO ASK ME, THEN DOES WHAT'S NECESSARY.
I...
WELL... YOU WERE INITIALLY SENT IN ON A LOVE ME ASSIGN-MENT...
So you're not lying.
...THINK I UNDER-STAND WHY MR. YASHIRO WAS ASSIGNED TO BE...
...MR. TSURUGA'S MANAGER.
THE PRESI-DENT MUST'VE MADE THAT DECI-SION.
OH?

KYOKO IS HEADING OVER TO TBM...
...SO I CAN DROP YOU AT PURÉ BELLE IF YOU WANT.
...
SHARP INSIGHTS AND UNOB-TRUSIVE CARE.
UH ...
UM ...
So...
About Bo.
DID YOU TELL MR. YASHIRO?
I DON'T KNOW MUCH ABOUT IT, BUT BO AND MR. TSURUGA MET IN SUCH A TERRIBLE WAY...
...YOU CAN NEVER TELL HIM YOU'RE BO, RIGHT?
AND.
VVVVT
UH.
VVVVT
UH.
OF COURSE.
EXCUSE ME.
VVVT
Shp
Gloves
HEY ...
YOUR ...
...TBM JOB IS KIMAGURE, RIGHT?
YEAH.
RIGHT ...
IF MR. TSURUGA BELONGS TO THE UNDER-WORLD ...
...I MIGHT BE QUIETLY EXTER-MINATED. THAT'S HOW BAD THINGS ARE...
In many ways.
I told him I hate him... I know he had no idea what "tente-komai" means.
He must never find out.
THAT BAD?

...PURÉ BELLE AFTER PRAC- TICE.
!
Oh!
UM.
PURÉ BELLE IS THE BEAUTY CLINIC EVERYONE IN SHOW- BIZ IS TALKING ABOUT.
!
YES, IT IS...
BUT I DIDN'T SAY THE ONE IN MINATO- KU...
YOU'RE ALWAYS ON A DIET※...
...AND LOOKING FOR THE LATEST BEAUTY TREND.
※According to Kyoko
THEN YOU WOULDN'T GO ANYWHERE BUT THE MAIN SALON.
...
Uh...
...
She'd never go to the branches (though their prices are more reasonable) cuz she doesn't want people to think she's poor.
WOW... I'M IMPRESSED... MR. YASHIRO... FIGURED OUT WHICH PURÉ BELLE SHE GOES TO JUST FROM ITS NAME!
I WAS SURPRISED A MAN KNEW PURÉ BELLE IS A BEAUTY CLINIC!
I THOUGHT IT WAS A BAKERY WHEN MOKO TOLD ME ABOUT IT LAST YEAR...
Lala Belle is Kyoko's favorite Bakery.

UH.
?
ARE'T
YOU GOING
SOMEWHERE
IN MINATO-KU
AFTER THIS,
MS. KOTONAMI?
HUH?
I AM...
BUT
DID I
TELL
YOU
THAT?
Hmm?
KYOKO
...
...TOLD
ME
THIS
MORN-
ING.
MOKO!
HOW ABOUT
WE GO TO
NIHYO AND
HAVE THEIR
KINBOSHI
RICE BALL?!
For good luck! Good luck!
NIHYO?
Ah... the rice ball place.
NO.
I
CAN'T
MAKE
IT.
You hurled words of rejection without hesitating.
So blunt
I wanted a small feast to invite victory to our side...
I'M...
...
GO-
ING
TO...
...PURÉ
BELLE
AFTER
PRAC-
TICE.
!
Oh!

YOU TOO, KANA.
!
I'LL MAKE CHIDORI MINE...
...AND COME BACK WITH GOOD NEWS.
THANK YOU...
...FOR...
YES...
...TAKING YOUR PRE-CIOUS TIME TO TEACH US!

WELL ...
YOU STILL NEED TO BRUSH UP YOUR SKILLS ...
...BUT YOU LEARNED A LOT IN A SHORT WHILE.
NOW DO...
... WHAT-EVER IT TAKES ...
...TO CREATE YOUR OWN VERSION OF MOMIJI.
I WILL!

...
HEY ...
WHEN ...
...DID YOU LEARN TO DO THAT?!
...
Hee hee.
grin
IT'S ALL BECAUSE I USED...
...THE PRIVILEGE OF HAVING MY OWN MANAGER!

Pant
Pant
Pant
Pant

TROMP TROMP
KAK
TROMP TROMP
TROMP TROMP
KAK
sqk sqk
sqk
sqk
sqk
SHP
WHAP
LEAP
BAM

IT'S BEEN EIGHT DAYS...
...SINCE MR. YASHIRO BECAME ...
...MY MANA-GER.
Skip·Beat!
Act 241: The Source of Agony

End of Act 240

...A DRIVER'S LICENSE?
Wha?! Wha?!
I THINK YOU SAID...
"IF I CAN OVERCOME THIS SHORTCOMING, I'LL BE SOMEONE...
I think you did...
...FOR TWO WEEKS.
"...WORTHY OF REN TSURUGA."
"THE PERFECT-
TO...
...BECOME THE PERFECT MANAGER...
...I WENT TO A DRIVER'S LICENSE TRAINING COURSE...
THIS IS A LIMITED-TIME OPPORTUNITY TO HIRE...
...A SUPER-MANAGER.

...YOUNG LADY?
M...
MR. YASHIRO ?!
Wha...
Oh?
YOU HAVE ...

MAY I...
...DRIVE YOU TO YOUR DESTI- NATION...

... FOUR MIN-UTES.
DON'T WORRY ...
I CAN MAKE IT.
I CAN CATCH THE BUS...
STOMP
FWOOOO—SH
Shortcut
...IF I RISK MY LIFE!
Kyaaah!
HONK HONNK
su—u
!
!!
wave wave
?!
SECTION
LME production

MMM.
NOW EVERY-THING MAKES SENSE.
WELL.
snap
WHEN DOES THE BUS FOR KINGA MANOR LEAVE...
THIS IS BAD ...
I GOTTA CATCH THE NEXT BUS, OR I'LL BE LATE!
Pant Pant
6:17
Pant
COLLAPSE
Pant
Pant Pant
...
TH...
I'VE GOT...

Now I remember!

I DIDN'T GIVE HIM TRUFFLES!

I gave Mr. Tsuruga something different.

The Queen Rosa wine jelly!

HOW COULD I EXPECT A WHITE DAY GIFT WHEN I FORCED HIM TO ACCEPT SOMETHING I WASN'T EVEN ASKED TO GIVE?!

LOOK!

What I recieved in return for only two truffles!

From the Bridge brothers

From Ogata

From Yashiro

I'm like an evil loan shark!

Lend a small amount, then recover huge amounts.

YES...

I CAN'T IMAGINE A REALITY WITH A TSURUGA GIFT AMONG THESE OTHER GIFTS, WHEN THE FAIR PRICE OF MY VALENTINE'S DAY GIFTS IS ABOUT FIVE DOLLARS.

Her handmade chocolates

THEN WHY DID ALL THESE GUYS GIVE ME PRESENTS THAT'RE SO MUCH MORE EXPENSIVE THAN THE ONES I GAVE THEM?

IN RETURN FOR ONLY TWO TRUFFLES ...

TRUF...

WELL.
THAT'S ...
...BE CAUSE ...
Why's he...
...re-ferring to V.D?
... TODAY'S WHITE DAY.
... WHITE DAY.
UH...
WHY AM I FEELING DISAP-POINTED?!
I WASN'T EXPECTING ANYTHING IN RETURN!
SO I DIDN'T GET WHAT I HADN'T EXPECTED TO GET.
I SHOULDN'T BE SAYING "WHY NOT?!"
shake shake
HOW COULD I BE SO SHAME-LESS?!

...SO I'LL TAKE THIS HOME WHEN I HAVE TIME TO FULLY APPRECIATE IT.
MOST PEOPLE AREN'T ABLE TO GIVE THEIR GIFTS DIRECTLY ON WHITE DAY.
WHEN...
...THE PRESIDENT SUMMONED ME TO HIS OFFICE...
I thought I was safe!
...IT WAS...

SHE'S NOT QUITE SURE WHAT TO DO, SO I ASSIGNED HER AN EMPLOYEE ...
...WHO CAN GIVE HER ADVICE.
ka chak
I...
...HAVE MY SWORD TRAINING TODAY...

I WANT TO SURPRISE HIM.
WE'LL SEE HOW SUR-PRISED HE IS.
BY THE WAY.
DID KYOKO GET A NEW JOB OFFER?
YEAH.
THIS ONE'S A COMMER-CIAL.
SHE'S GOTTEN A COUPLE MORE OFFERS TOO.
I THINK THEY'RE ALL THANKS TO BOX "R"...
THE SECOND EPISODE RATINGS WEREN'T TOO GOOD...
...BUT KYOKO'S CHARACTER **REALLY** STANDS OUT.
AH, YES ...
MIO AND NATSU ARE SO DIFFERENT.
AND MS. MOGAMI HERSELF IS NOTHING LIKE THEM.
MS. MOGAMI WANTS TO CONCENTRATE ON HER ACTING...
...BUT WE CAN'T REJECT ALL OF THE NON-ACTING JOB OFFERS...

SUPER-
VISOR
SAWARA.
SO
YOU'RE
BACK IN
TOKYO
NOW,
YASHIRO
?

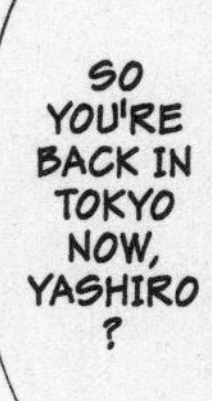
Yes.
FINALLY.

THANKS
FOR
TELLING
ME
WHERE
KYOKO
WAS
TODAY.
I was finally able to hand-deliver something I wanted to give her.
NO
PROB.

I WAS ABLE
TO FINISH
THINGS WAY
BEFORE REN
RETURNED
...
...SO
I'M
RE-
LIEVED.
nod
nod
YOU
HAVEN'T
TOLD REN
YET,
RIGHT?
NO.
Heh
heh

... GOTTEN THROUGH TO HER?
Or not...?
YOU'RE RIGHT.
WE'VE GOT CONNECTIONS NO ONE ELSE HAS!
I was a substitute manager for Mr. Tsuruga. I went on a secret mission to be his younger sister.
I GOT THOSE ASSIGNMENTS BECAUSE I'M A LOVE ME MEMBER!
MAKING PECULIAR CONNECTIONS IS ONE PERK OF BEING A MEMBER!
BAAAM
NOW SHE'S SAYING "PECULIAR" INSTEAD OF "SPECIAL"!
WELL... OF COURSE ...
She didn't get it.
SHE'S THE FIRST LOVE ME MEMBER AFTER ALL...
HEY, MS. MOGAMI. I NEED TO TALK TO YOU.
!
SURE.

THEY ALSO MAKE GOOD IMPRESSIONS...
...AND WORK AS A MARKETING TOOL.
THOUGH...
...PEOPLE DON'T SAY THAT ALOUD.
AND SO...
I'M GIVING YOU THIS VERY LATE WHITE DAY GIFT...
...BECAUSE YOU'RE SPECIAL, OKAY?
TH ...
Th-th-th-th-thank you so much!
Oh dear! I don't know what to do.
This is the first time a man has said I'm special!
So it took a while for my synapses to react!
...
HAVE I...

...MOST PEOPLE AREN'T ABLE TO GIVE THEIR GIFTS DIRECTLY ON WHITE DAY.
Especially Ren.
Uh...
YES...
Of course.
SO IT TAKES ABOUT A MONTH MORE FOR REN TO GIVE ALL HIS WHITE DAY GIFTS.
MANA-GERS AND THEIR CLIENTS...
...SINK OR SWIM TOGETHER...
REN HASN'T EXPLICITLY SAID IT...
...BUT I'M SURE THAT'S HOW HE DOES IT.
...SO I MAKE SURE I DO THINGS REN'S WAY.
NO ONE HATES GETTING GIFTS.

WHICH I FOUND OUT...
...THE OTHER DAY WHEN WE WERE RETURNING...

VROOOOOM

...FROM KYOKO'S PLACE.
SO I WAS FINALLY ABLE TO GIVE HER MY VERY LATE WHITE DAY GIFT.
I WAS OUT OF TOKYO FOR ABOUT FIVE DAYS...
...SO I WASN'T AROUND TO GIVE IT TO YOU ON THE 14TH...

...AND PEOPLE REACT JUST LIKE YOU DID...
...IF THE GIFT ISN'T GIVEN EXACTLY ONE MONTH AFTER VALENTINE'S DAY.

UH ...

THANK YOU...
I THOUGHT WHITE DAY GIFTS WERE ONLY GIVEN ON WHITE DAY.
To be honest...
I WASN'T EXPECTING A GIFT FROM YOU AT ALL...
WELL ...

IN THIS BUSI-NESS ...

Filigree
Pumpkin Carriage
Aroma Candle Holder
WOOOOOOOOOOOOOOOOOOOOOOOO
heeeeeee
JUST THE REACTION I WAS EXPECTING...
I'M SORRY IT TOOK SO LONG.
smile smile
beam beam
Floral aroma
Cuuutiful!
Hee hee hee hee
I DID HAVE IT READY ON WHITE DAY, BUT...
...I DIDN'T HAVE THE CHANCE TO GIVE IT TO YOU.
TO BE HON-EST.
NINETY-NINE PERCENT OF THE REASON WAS THAT I DIDN'T WANT TO GIVE HER MY GIFT BEFORE REN GAVE HIS.
And it's not something I can always carry around.
BUT MY THOUGHT-FULNESS WAS TOTALLY USELESS.

AH.
YOU'RE BACK.
NOW'S THE PERFECT TIME FOR THIS...
Well... um.
THAT'S WHY I CAME HERE TODAY.
KYOKO.
THIS IS...
...FOR YOU.
WHA ?
HUH?
...
U...
...m?
Uh... yeah...
YOU MUST BE WONDERING WHAT'S GOING ON.
Of course.
CUZ A MONTH HAS ALREADY PASSED.
?
A month?
THIS ...
...IS MY WHITE DAY GIFT TO YOU.
Whaaa?!

IF...
...I CAN SHOW THEM...
...HOW I WANT TO EMBODY MOMIJI...
...I STILL HAVE A CHANCE.
IT WORKED AT THE CURARA AUDITION.
THE COMPANY WANTED MS. KOENJI FROM THE VERY BEGINNING...
...BUT WE MANAGED A BRILLIANT UPSET VICTORY.
I WAS ABLE TO DO THAT BECAUSE I WAS WITH MOKO...
...BUT MY ACTING HAS GOTTEN BETTER SINCE THEN.
I WON'T BE AUDITIONING WITH MOKO, BUT SHE'S ON MY SIDE.
I'VE GOT NOTHING TO WORRY ABOUT.
I CAN DO IT!

THIS AUDITION IS RIIIIIGGED!
AAAAAAAH!
GLOOOIM
Uh...
WAIT, KYOKO.
Listen to me!
A PRODUCER HAS SAY OVER EVERYTHING.
Especially with movies.
HE CAN REJECT THE DIRECTOR'S PLAN...
...TO CAST KIMIKO MORIZUMI.
SO JUST MAKE SURE THE PRODUCER LIKES YOU.
...
WAIT...
MS. KOENJI WAS INITIALLY CHOSEN TO PLAY MOMIJI.
A NEW AUDITION IS BEING HELD...
...BECAUSE MS. KOENJI'S FATHER DEMANDED IT...
...BUT THE PRODUCER...
...IS FAIR ENOUGH TO CAST SOMEONE OUTSIDE THE MORIZUMI CLAN...
...AND FLEXIBLE ENOUGH TO NOT AUTOMATICALLY CAST THE RUNNER-UP.
Though the budget may be a factor.

Director/Screenwriter: Joji Morizumi

COMMENT

DID HE PRODUCE ...
... PURPLE DOWN?
NO.
THIS IS HIS OLDER BROTH-ER.
think
...
IS SHE...
...AUDI-TIONING FOR MOMIJI TOO?
I MEAN KIMIKO MORI-ZUMI.
I...
... THINK ...
... SO ...
YES ...
...

...SO I WOULD'VE RECOGNIZED HER NAME IF SHE'D REGULARLY APPEARED ON TV.

Lotus in the Mud Crew

...SO THE RUMORS ARE PROBABLY TRUE...
...
HAVE THERE BEEN ANY PROBLEMS WHEN SHE'S CAST...
...BE-CAUSE OF HER CONNEC-TIONS?
CUZ AN ACTRESS ...
...HAS TO BE ABLE TO ACT AS WELL AS THE OTHERS IN THE SHOW.
HMM ...
I DIDN'T THINK SHE SUCKED.
...
SOME-THING'S NOT RIGHT.
SHE COMES FROM THE RIGHT FAMILY, AND SHE HAS SOME TALENT.
BUT SHE DOESN'T SEEM TO GET MUCH MEDIA ATTEN-TION.
IS SHE MAINLY WORKING OUTSIDE TV?
I'M MORE INTERESTED IN THE BUSINESS NOW...

...REALLY WILLING TO KILL HER COMPETITION...

...TO GET A PART SHE WANTS?

NOW THAT I'VE READ HER BLOG, I DON'T THINK SHE IS.

WELL...

UH.

SHE COMES...

...FROM A SHOWBIZ FAMILY.

A SHOW-BIZ FAM-ILY?

SO SHE'S LIKE HIO?

HER FATHER WAS THE PRODUCER FOR PURPLE DOWN.

RUMOR IS PRODUCER MORIZUMI PERSON-ALLY CAST HER...

...BUT I DON'T KNOW IF THAT'S TRUE.

B U T...

DIRECTORS. CASTING DIRECTORS.

WHENEVER SOMEONE FROM THE MORIZUMI FAMILY IS INVOLVED, THERE SHE IS...

OH.
MR. YASHIRO.
...
WHA?
WHAT'RE YOU DOING?
THE BLOG YOU'RE READING.
DO YOU LIKE HER?
HUH?
NOT AT ALL.
I FOUND THIS WHEN I GOOGLED HER NAME.
SHE ...
...COSTARRED WITH MR. TSURUGA TWO YEARS AGO.
DO YOU REMEMBER HER?
IT WAS THE SECOND SEASON OF PURPLE DOWN.
UH...
I remember her.
YEAH.
WHAT WAS SHE LIKE?
IS SHE ...

...CO-STARRED WITH ...

Lovely Smiles
EVERY DAY
Kimiko Morizumi's Official Blog

Search

Home
Articles
Photos

Profile

Previous
Next

Dinner date with Ren

Kimiko Morizumi
Gender: Female
Subscribe

Recent Entries
I'm in it!♪
Surprise!
From my dearest.
New hair color
Sooo gorgeous.
In the rehearsal room now
I had so mubh fun!
Cherry Festival

He seems totally comfortable escorting women. I'm looking into his eyes, and we look pretty cool together.

...MR. TSURUGA ...

K...

KIMIKO...

... MORIZUMI?

That's what I'd like to say...

pretty cool together.

That's what I'd like to say...

But the entire cast and crew of the drama are here!

This is the wrap party of *Purple Down II*

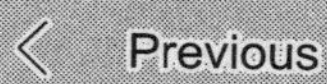

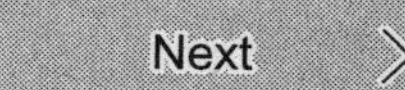

Dinner date with Ren

Category: Blog

Previous Next

Dinner date with Ren

He seems totally comfortable escorting women. I'm looking into his eyes, and we look pretty cool together.

Skip·Beat!
Act 240: Perfect Pickup Artist

End of Act 239

...NOT HER...

IS THAT ALL...

...YOU WANTED TO SAY...

...MS. KOENJI?

I DON'T...

...LIKE YOU.

grip

grip

clik

I DON'T LIKE YOU...

...BUT THIS TURN OF EVENTS...

shuu

BUT ...
...SHE KEPT HER PROMISE ...
...AND USED HER OWN SKILLS...
...TO WIN...
...HER ROLE...
...SO SHE COULD BATTLE MOKO...
...I ...
... WANT TO...
... TELL YOU ...
...ON A LEVEL PLAY-ING FIELD.
I DON'T WANT ANYONE TO KNOW...
...ABOUT MY CURRENT SITU-ATION.

OOPS.
fwmp
THIS IS HARD. EVEN THOUGH I DON'T HAVE TO TWIRL IT LIKE A BATON.
SUU
YEAH.
THE GRIP IS FLAT AND A BIT HEAVY.
It's different.
BUT I GOTTA BE ABLE TO HANDLE THIS BEFORE THE AUDITION BEGINS ...
WHAT ?
I APPLIED CUZ I WANT THAT ROLE!
...
YOU'RE AUDITIONING FOR MOMIJI ?!
GREAT! YOU ALREADY KNEW WHAT WE WANTED, LOVE ME GIRL!
Good!
YOU GO WIN THE ROLE! FOR MS. ERIKA!
Huh ?!
WHY'D I HAVE TO DO ANYTHING FOR MS. KOENJI ?!
I DON'T...
...LIKE MS. KOENJI...
...CUZ SHE KEPT DOING TERRIBLE THINGS TO MOKO.
I can even say I hate her.

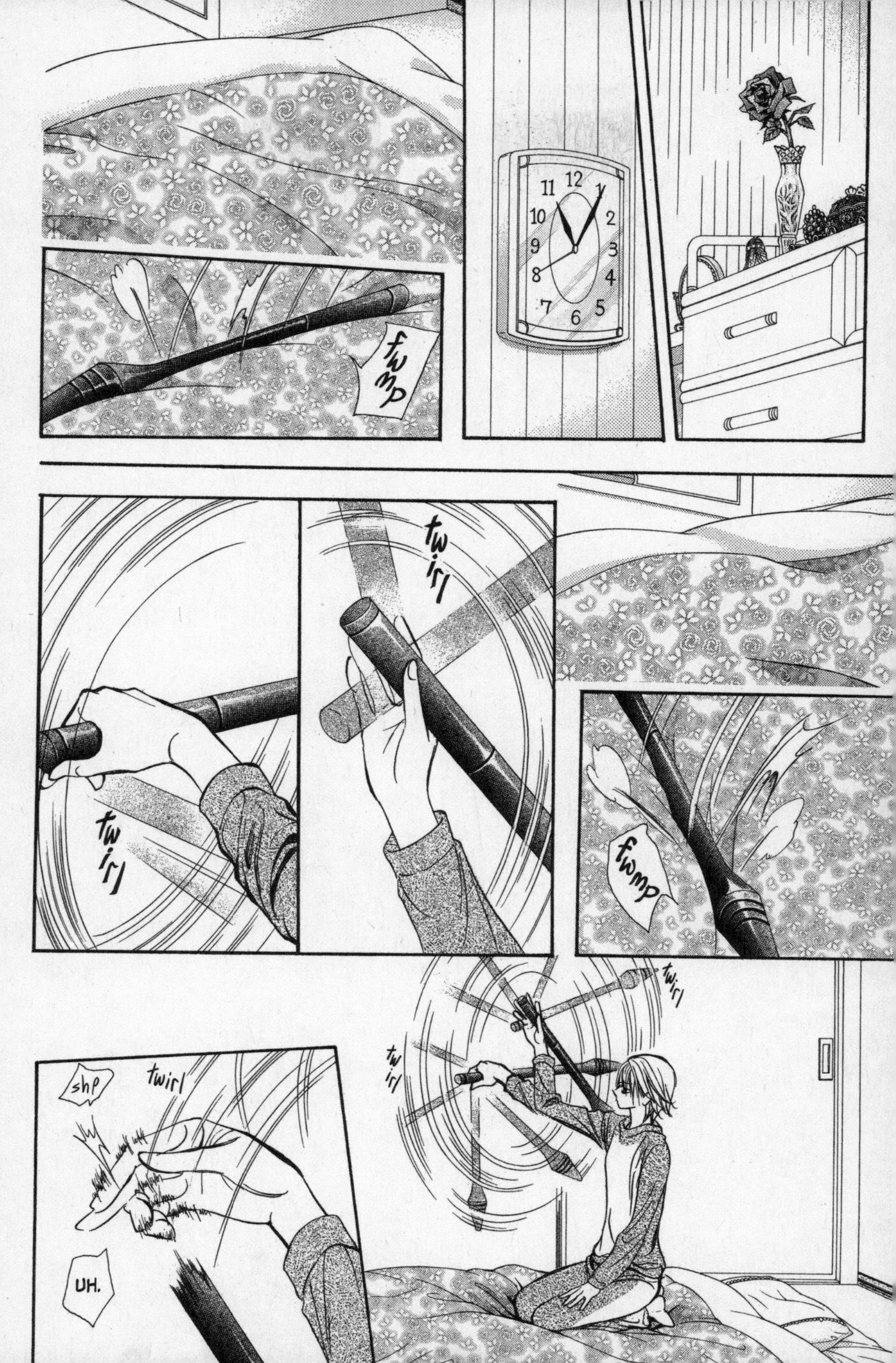
fwmp
twirl
twirl
fwmp
twirl
twirl
shp
twirl
UH.

WHAT'S THIS ABOUT SURPRISE ATTACKS?
When Ms. Kotonami was alone?
The three of you?
NNN?
STARE
WHEN...
...DID THIS HAPPEN?
WHAT?!
HUH?!
WHERE?
HOW?
FWOOOOOOOOOO
YOU REALLY DIDN'T KNOW ABOUT IT?!
WHY DON'T YOU TELL ME EVERY TINY DETAIL ABOUT IT.
We won't! They were tiny incidents! That's all in the past now!
だるまや
だるまや

NO?
"I WANT YOU TO CURSE THAT WOMAN WITHOUT MERCY AND KILL HER."
Why would she want you to play Momiji? How dare you.
I can't curse her!
Are you guys stupid?!
I DON'T HOLD GRUDGES AGAINST A TOTAL STRANGER!
WHY'RE YOU BEING SO MODEST?
YOU USED YOUR MYSTERIOUS POWERS TO PROTECT KANAE KOTONAMI ...
...EVERY TIME WE MADE SURPRISE ATTACKS.
Though you weren't with her.
?!
BUT ...
HMM.
YOU WINNING THE MOMIJI ROLE...
... SOUNDS LIKE A PLAN.
THAT WOMAN WILL END UP WINNING THE ROLE IF NO ONE COMPETES AGAINST HER...
...SINCE THE AUDITION IS ONLY BEING HELD TO QUELL MR. KATSUMASA'S ANGER.
TO TELL THE TRUTH, THAT WOM—
HOLD ON.

WE THOUGHT SHE'D GIVEN UP ON EVERY-THING...
...BUT SHE HADN'T.
...BUT WHEN WE SAW HER HEAD STRAIGHT TOWARD YOU, WE KNEW WHAT WAS HAPPENING.
OUR MS. ERIKA WOULDN'T GIVE UP MOMIJI SO EASILY.
SHE WANTED TO SPIT OUT THE FEEL-INGS SHE'D BOTTLED UP DEEP IN HER HEART.
THAT'S WHY SHE AP-PROACHED YOU.
MS. ERIKA ...
...IS VERY PROUD BY NATURE...
...SO SHE COULDN'T SAY IT...
...BUT ...
...WE'RE SURE SHE WANTED TO ASK YOU...
TO WIN THE ROLE ...
...OF MOMIJI IN HER PLACE?

BUT WE SAW A CHANGE IN HER TODAY.
OH NO.
I'M SORRY!
WHEN SHE SAW YOU...
tmp
tmp
tmp
roll roll roll
...SHE MOVED ...
... WILL-INGLY.
MS. ERIKA ...
... BECAME TOTALLY APATHETIC AFTER HER ACCIDENT ...

THE PROMISE SHE MADE WHEN SHE DECLARED WAR ON MOKO...
WE WOULD'VE PROTECTED MS. ERIKA, APPREHENDED AND PUNISHED THE CRIMINALS!
BUT ONE VAGUE EYE-WITNESS ISN'T ENOUGH!
...
THIS NEVER WOULD'VE HAPPENED IF WE HAD BEEN WITH HER!
DID MS. KOENJI SEE WHO PUSHED HER?
SHE WON'T SAY WHETHER SHE DID OR NOT.
SHE WAS TOTALLY INDIFFERENT EVEN WHEN WE TOLD HER THE AUDITION FOR MOMIJI WAS BEING HELD AGAIN.
?
SHE TOLD US SHE DOESN'T CARE ANYMORE AND REFUSES TO TALK ABOUT THE ACCIDENT.
SHE TRULY SEEMED NOT TO CARE.

AND THIS IS WHAT HAPPENS! IT'S NOT FAIR!
YOU...
...ARE STANDING IN A PEACEFUL AND BEAUTIFUL WORLD...
...WHERE YOU NEVER FALL DOWN AND GET COVERED IN MUD, EVEN IF YOU TRIP.
...
IF YOU WANT TO BEAT ME...
...STEP DOWN FROM YOUR WORLD.
AND GET COVERED IN MUD YOURSELF.
I'LL MAKE YOU REGRET ...
... THAT YOU EVER SAID...
... "STEP DOWN"!
MS. KOENJI...
SO...
...SHE KEPT HER PROMISE.

SHE HASN'T CHANGED.

HOW DARE YOU!

WE SWEAR...

MS. KOENJI IS STILL TAKING ADVAN-TAGE...
!
...OF HER FAMILY'S WEALTH.
...MS. ERIKA HASN'T PLAYED HER TRUMP CARD...
...SINCE YOU LOVE ME MEMBERS CRUSHED HER!

SHE'S DONE ALL SORTS OF THINGS TO HIDE HER IDENTITY ...
... SINCE THAT DAY.
MS. ERIKA EVEN REFUSES TO LET US ACCOM-PANY HER...
...WHEN SHE GOES TO AUDITIONS.

IT MUST'VE BEEN THAT WOMAN.
SHE MUST'VE DONE IT BECAUSE SHE LOST AGAINST MS. ERIKA...
WE CAN'T LET THAT WOMAN...
...MAKE MOMIJI HERS JUST BECAUSE THE PRODUCERS ...
...DECIDED MS. ERIKA WOULDN'T RECOVER IN TIME...
?
THERE'LL BE...
...ANOTHER AUDITION FOR MOMIJI, RIGHT?
YES.
MR. KATSUMASA...
MS. ERIKA'S FATHER MADE A FORMAL OBJECTION ...
...AND SAID THE KOENJI GROUP WOULD WITHDRAW...
... SPONSORSHIP OF LOTUS IN THE MUD...
...UNLESS ANOTHER AUDITION WAS HELD.

THE DOCTORS SAID...
...IT WAS PSYCHOGENIC PARALYSIS OF THE LOWER LIMBS.
SO ...
That means..
SHE CAN'T STAND BECAUSE OF HER EMOTIONAL TRAUMA ...
RIGHT.
I CAN WELL IMAGINE.
SHE MUST'VE BEEN TERRIFIED ...
THE CAR DRAGGED HER FOR YARDS.
!
WHAT?!
THE DRIVER SAID...
...MS. ERIKA SUDDENLY ROLLED OUT ONTO THE ROAD.
THE CAR MANAGED TO AVOID HITTING HER, BUT HER HANDBAG GOT CAUGHT IN THE BUMPER...
...AND THE CAR ENDED UP DRAGGING HER.
...
Eep...
THE PASSENGER CLAIMED ...
...THAT SOMEONE PUSHED MS. ERIKA ONTO THE ROAD.

TO MAKE A LONG STORY SHORT—

Hold it!

LET'S GO SOMEWHERE ELSE!

Right now!

mrmr mrmr

What's going on?

A GIRL'S FORCING TWO GROWN MEN TO GET DOWN ON THEIR KNEES...

mrmr mrmr

SOMETHING SERIOUS MUST BE GOING ON...

SHOULD WE CALL SECURITY?

fwsh fwsh fwsh fwsh

TO MAKE A LONG STORY SHORT...

EXTERNALLY, SHE HAS BRUISES AND ABRASIONS, BUT THEY AREN'T LIFE-THREATENING.

IT'S A MIRACLE, CONSIDERING HOW SERIOUS THE ACCIDENT WAS.

SHE **SHOULD** BE ABLE TO STAND AND WALK...

...BUT WHEN MS. ERIKA REGAINED CONSCIOUSNESS TWO DAYS AFTER THE ACCIDENT, SHE COULD COULDN'T DO EITHER.

?!
...
?!
?!
I...
I HEARD... THAT A YOUNG WOMAN CALLED...
I USED MY FALSETTO AND PRETENDED TO BE MS. ERIKA.
He's good at this.
?!
?!
?!
...
THE PERSON WHO TOOK MY CALL TOLD YOU ABOUT IT, RIGHT?
I ASSUMED THEY DIDN'T TAKE IT SERIOUSLY, SINCE EVENTUALLY THEY STOPPED RESPONDING TO ME.
Uh...
WELL... YES...
THEN...
...THE WOMAN...
...WHO WAS...
...INJURED IN A STAGED ACCIDENT AND CAN NEVER STAND AGAIN IS...

....
....
HUH ?!
...HUH ...?
RIGHT ?!
Oh!
SHE SAID THAT ...
...A GRUDGE NEEDS SETTLING ...
... NOW.
!
N... NO.
SO THE ONE WHO KEEPS CALLING LME WITH THAT REQUEST ...
WHA?
...IS MS. KOEN-
IT'S ME.
I called.

?!
BSHP
WE ASK YOU AGAIN.
ON BEHALF OF OUR MS. ERIKA...
...WE FORMALLY MAKE A REQUEST TO KYOKO OF THE LOVE ME SECTION...
...TO...
...PLEASE...
...SETTLE THIS GRUDGE!

SHE'S... WORLD-CLASS.
MS. ERIKA WAS BORN NOBLE AND DIGNIFIED...
EX-ACTLY.
THERE ...
...WAS NO WAY SHE COULD'VE ASKED ...
...A CRUDE, LOW-CLASS COMMONER WITH ABSOLUTELY NO SENSE OF BEAUTY FOR HELP.
EVERY SINGLE WORD THAT COMES OUT OF YOUR MOUTHS REALLY PISSES ME OFF...
"ASK FOR HELP" ?
WHAT HELP ?

tmp
PROMISE.
creak
UH.
LET ME PUSH YOUR WHEEL-CHAIR...
creak
MS. ERIKA ...
creak
...
DON'T TELL MOKO?
...
I WOULDN'T HAVE NOTICED HER...
...IF SHE HADN'T BROUGHT MIZUE BACK.
DID...
...HAVE SOMETHING ELSE...
...SHE WANTED TO TELL ME?
...MS. KOENJI ...
OF COURSE SHE WASN'T ABLE TO...

I...
...DON'T WANT ANYONE TO KNOW...
...ABOUT MY CURRENT SITU-ATION.
ESPE-CIALLY...
...NOT HER...
!
YOU MEAN MO...
...MS. KOTO...
...NAMI?
...
NOT HER?
TOSS
!
tmp
tmp

...
THERE'S ...
grab
clench
... SOMETHING- THING ...
...
MS. ERIKA ...
YES!
...
...I ...
... WANT TO...
... TELL YOU ...

SHE REACTED WHEN SHE SAW THIS LOVE ME PERSON, SO I THOUGHT WE COULD GET SOMETHING MORE OUT OF HER...
whisper
But...
NO... LET'S NOT GIVE UP YET...
whisper
SHE'D BEEN TOTALLY INDIF-FERENT UNTIL NOW.
THIS IS THE FIRST SIGN OF IMPROVE-MENT WE'VE WIT-NESSED.
WE SHOULD BE ABLE TO TAKE ADVANTAGE OF THIS SITUATION—
MS. ERIKA ...
GLOOM
LOVE ME GIRL.
!
YES?

THOSE ARE THE TRAGIC HABITS OF PEOPLE WHO HAVE NO MONEY...
...WHO USE ONE THING UNTIL IT CAN'T BE USED ANY-MORE.
MS. ERIKA ...

...MS. KOENJI...

...LOST HER ENERGY BECAUSE SHE WAS INJURED?

THIS GIRL DARES CALL HERSELF A CELEBRITY, BUT SHE'S STILL NOT WEARING ANY MAKEUP!

How could she be so care-less?!

Just finished her sword fight lesson, so she's only wearing sunscreen and lip balm.

A-AND HER CLOTHES AND BOOTS!

THIS IS EVEN WORSE!

SHE WAS WEARING THIS THE FIRST DAY OF THE CURARA AUDITION!

SHE'S OBVIOUSLY CARRIED THIS AROUND EVERY DAY SINCE THAT DAY!

And she seems to have worn it again just recently!

WE THOUGHT WEARING CLOTHES MORE THAN ONCE A YEAR WAS AN URBAN LEGEND, BUT WE NOW HAVE PROOF IT'S TRUE!

This isn't an urban legend, it's a galactic legend!

...IS TO MERCI-LESSLY HAMMER AND BREAK THE NAIL THAT STICKS UP.
SHE LOOKS ...
...COM-PLETELY DIFFERENT NOW!
IT'S NOT JUST HER LOOKS.
Her...
...LIFE FORCE ISN'T WHAT IT USED TO BE...
IF THERE WAS SUCH A THING AS FINGER-PRINTING SOMEONE'S LIFE FORCE, THEY'D NEVER MATCH!

M...
Ms. Koenji ?!
Huh?! No way!
YOU DIDN'T RECOGNIZE HER UNTIL YOU SAW US...
...WHILE WE RECOGNIZED YOU WITHOUT YOUR POISONOUS-LOOKING WORK UNIFORM.
YOU'VE BECOME EVEN MORE DIM-WITTED AND DISCOURTEOUS THAN BEFORE.
BUT...
B...
!
THOSE ARE...
ha ha ha ha ha
...THE TRAGIC HABITS OF PEOPLE WHO HAVE NO MONEY...
...WHO USE ONE THING UNTIL IT CAN'T BE USED ANYMORE.
SHE'S YOUR RIVAL...
...IN THAT WAY.
IF SHE'LL TELL YOU HONESTLY...
...THAT IS.
heh
hmph
THE FUN...
hah

...
YOU'RE
...

YOU'RE NOT...
... WEARING THAT CRAZY-COLORED UNIFORM ...
... TODAY?
Skip·Beat!
Act 239: Blue Leaves Falling

IT'S BEEN A WHILE ...
YOU'RE ...
YOU'RE NOT...
...LOVE ME GIRL.
...
UH...
... WEARING THAT CRAZY-COLORED UNIFORM ...
... TODAY?
End of Act 238

Foundation Of Momiji
☆ There's a Mozue too.
WHERE'S MIZUE?!
MIZUE! THE FOUNDATION OF MOMIJI!
DID SHE FLY OUT WHEN THAT CANNON-BALL SLAMMED INTO ME?!
OH NO! WHERE IS SHE ?!
WHERE'D SHE GO?!
creak
HELLO ...

Um...
OH NO!
SKKKSH
I-I'M SORRY. I'M SORRY. ARE YOU OKAY?!
Yes... UH ...
NOOOO!
flail flail
I'M FINE. I'M SORRY TOO.
I wasn't watching where I was going.
ARE YOU HURT—
NO, I'M FINE.
Really.
I HATE SHOTS!
Gyaaaaaah!
bow bow
HE'S ...
... TROUBLE ...
DON'T WORRY ABOUT ME.
Swf
?!

...EVERY-THING THAT HAPPENS TO ME...
...CAN BE CON-NECTED TO MY ACTING.
THAT WAS HER LESSON.
MS. AMAMIYA HERSELF DECIDED TO GO TO THE MAJISUKA SHOOTS...
...SO ...
...I DON'T THINK WHAT I SAID MADE ANY DIFFER-ENCE....
Uh...
H... HEY...
TATSU.
WAIT, DON'T RUN!
BAM

THAT'S ...

...HOW I FELT...

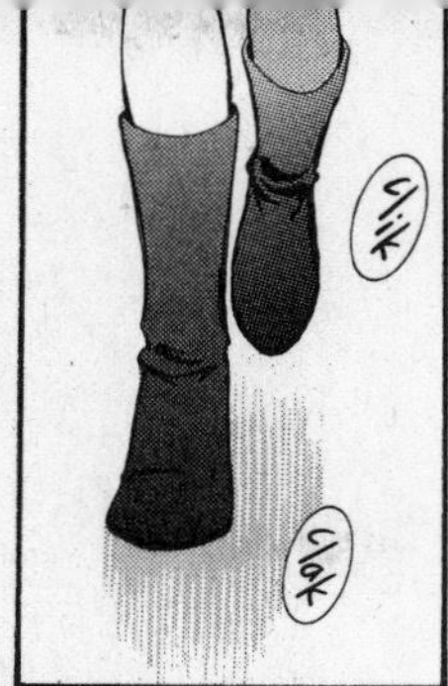

THERE WERE REASONS FOR HER APPEARING IN A VARIETY SHOW...
...BUT SHE HAD A HARD TIME ADJUSTING.
I HAD TO DRAG HER TO WORK EVERY TIME...
...SO I FIGURED CHIORI WOULDN'T SHOW UP IF I WEREN'T WITH HER.
ONCE YOU RUN AWAY FROM YOUR HARDSHIPS...
...YOU KEEP RUNNING AWAY, RIGHT?
YES.
...I CALLED YOU...
Heh heh
THAT'S WHY...
KYOKO ...
HAVE KYOKO COME TO THE HOSPI ...
...CUZ I WAS SURE ...
... SHE'D ...
... LISTEN TO YOU.

Th...
Full Marks
100%
Great Job
THANK YOU SO MUCH.
YOU'VE DONE A LOT FOR US.
NO... NO.
I WASN'T OFFICIALLY ASKED TO DO ANYTHING...
I feel like I'm forcing you to give me this reward...
Heh heh
DON'T SAY THAT.
YOU DID EVERYTHING PERFECTLY BEFORE I HAD A CHANCE TO ASK. OF COURSE YOU DESERVE THIS REWARD.
IF YOU WEREN'T AROUND...
...CHIORI WOULD NEVER HAVE RETURNED TO MAJISUKA.
NOW SHE HEADS FOR THE SHOOTS LOOKING PRETTY HAPPY.

DUH...
IT'S TOO HARD FOR ME...
To ask her out in a cool way...
...
I'LL WAIT ...
...UNTIL I'M AS TALL AS KANAE...
Number 202, Sawada. Enter exam room 4.
shf
shf
New Metropolitan General Hospital
sqk
sqk
sqk

HOW ABOUT THE TWO OF US JUST GO TO-GETHER ...
... NOW ...
CUZ ...
...I'M FEEL-ING LIKE...
...A TEA-ISH BREAK.
I SOUND SO STUPID!
Arrrgh! There's gotta be a better way to ask her out!
WHY DOES IT COME OUT SO AWFUL WHEN I TRY TO SOUND COOL!
DAM-MIT.
IT'S ALL BECAUSE I COULDN'T GET HER WEIRD WORDS OUT OF MY HEAD!
YOU'RE IN A TEA-ISH MOOD.
Taking it out on Kyoko.
HIO. I'M HEADING OUT.
THANKS FOR CHECKING UP ON MY PRO-GRESS.
... ...
I'll text you later.
SHWOO

!
JOLT
TOO BAD.
WHAT DO YOU MEAN ...?
THAT SHE ALREADY HAD AN APPOINT-MENT.
Huh ?!
WHY DO I—
YOU WANTED TO MAKE UP...
...FOR THAT HEAD BUTT ...
... RIGHT?
I TOLD HER ...
WE CAN HAVE TEA ANOTHER TIME.
Heh...
I GUESS.
I'LL ASK HER WHEN'S A GOOD TIME.
...

I NEED TO GO SEE SOME-ONE!
IF I SPEND TIME ELEGANTLY DRINKING TEA...
Argh!
snap
I'LL BE LATE!
I'M SORRY... I HAVE TO EXCUSE MYSELF...
NO PROB...
I mean...
WE CAN HAVE TEA NEXT TIME.
Huh?
NO, NO. GO HAVE YOUR TEA TODAY.
YOU'RE IN A TEA-ISH MOOD...
...SO YOU GO HAVE TEA WITH MOKO.
DON'T DO YOUR TRAINING IN FRONT OF OTHER PEOPLE!
Like with your Baton!
I won't~!
bounce
SORRY, MOKO. I'M LEAVING.
SEE YOU.
SEE YOU.
Uh, heeey.
Wait, Mizue!
...
SHE NAMED IT MIZUE?

Yes!
Thanks for treating us!
YOU WANT ME TO TREAT YOU?!
FINE.
Wha?
YOU ARE?!
clik clak clik clak
NO, THIS ISN'T RIGHT!
No no no!
THAT WAS SUPPOSED TO BE A JOKE!
Well...
HIO'S BEEN IN THIS BUSINESS MUCH LONGER THAN I HAVE...
...BUT HE'S YOUNGER THAN US!
A 7th-grader this year
UH.
OOPS!
DARN. SORRY, HIO.
I CAN'T GO WITH YOU!
WHAT?

OKAY.

I WILL.

ONE MORE THING.

WHAT YOU DID TO ME WHEN I WAS ON THAT SHOW...

...AND THAT TIME YOU PUNCHED ME...

DON'T YOU EVER MENTION IT AGAIN...

...CUZ I HAD MY PAYBACK WITH **THIS.**

Sheesh.

I DON'T WANT YOU TO THINK I'VE GOT A GRUDGE AGAINST YOU FOREVER.

crunch crunch crunch crunch

HEY.

I'M GONNA HAVE TEA SOME-WHERE AND THEN GO HOME, SO COME WITH ME.

Both of you.

crunch

!

GRANDPA'S TRAINING YOU...

...SO YOU CAN FIGHT LIKE THE NOVEL-VERSION OF MOMIJI.

SO STOP COM-PLAINING ...

...AND KEEP PRACTIC-ING.

O...
OWWWWWWWW.
...
HIO~
DO YOU THINK ...
... GRANDPA'S THAT STUPID ?!
GRANDPA...
...WOULD NEVER TEACH NONSENSE TO SOMEONE...
...WHO'S DOING THEIR ABSOLUTE BEST TO WIN A ROLE JUST BECAUSE HE HATES THEM.
GRANDPA'S BROAD-MINDED!
HE JUST BOILS OVER REAL FAST!
He's like that with everyone except me!
...
That means ...
HE SNAPS EASILY, BUT FORGETS QUICKLY?

I DOUBT MR. KOTETSU WOULD MAKE YOU DO SOMETHING TOTALLY USELESS WHEN YOU'RE SERIOUS ABOUT YOUR ACTING.
YOU REALLY THINK SO? ...
tmp tmp tmp tmp
...
BUT ...
... I'VE ...
... PUNCHED HIO WITH MY FIST.
AND I HARASSED HIM IN ALL SORTS OF WAYS WHEN HE WAS A GUEST ON OUR SHOW...
SO IF...
...HIO TOLD MR. KOTETSU ABOUT MY EVIL DEEDS ...
...UNDERSTAND WHY MR. KOTESTSU WOULD HATE ME AND NOT PROPERLY TEACH ME~
...I CAN ...
!
KA WHACK
WHOO--M

tmp
tmp
tmp
tmp

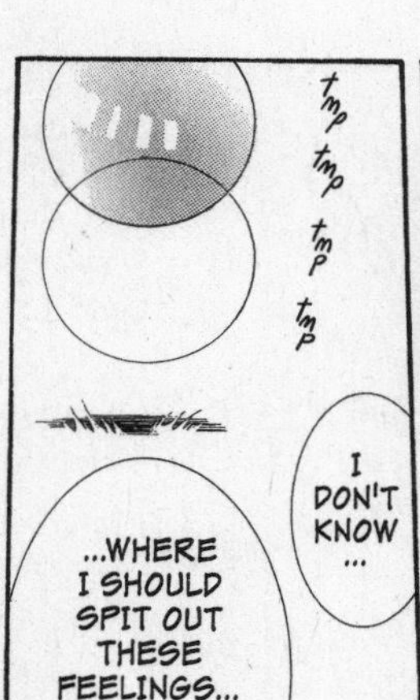

tmp tmp tmp tmp
I DON'T KNOW...
...WHERE I SHOULD SPIT OUT THESE FEELINGS...

I WANT TO LEARN SWORD FIGHTING...
Sword fight Sword fight Sword fight
tmp tmp tmp tmp tmp tmp tmp tmp
...

SHE HAS LESS THAN TEN DAYS UNTIL THE AUDITION, SO I CAN UNDER-STAND HER FRUSTRA-TION...

...FINALLY BEING ALLOWED TO TAKE MR. KOTETSU'S LESSONS ?!
WAS I TWIRLING TO PREPARE FOR IT?
SO WAS THIS...
...a "kinda seems cryptic but there's logic behind it" lesson ?!
Ball
Clubs ?
ARE THESE FOR RHYTHMIC GYMNASTICS?

SHUT UP!
FWSH
?!
Eep
Huh ?!
I DIDN'T SAY ANYTHING OUT LOUD.
I...
...COULD SENSE YOU BAD-MOUTHING HIO.
HE'S SCARY!
In so many ways!
Does Mr. Kotetsu have a black antenna too?
THIS IS HOW MUCH HE DOTES ON HIS GRANDSON...
WELL ...
BEING YOUNG IS WONDER-FUL.
Though you fumbled at first.
YOU'VE IM-PROVED FAST.
NOW IT'S TIME...
SWF
...FOR A NEW LESSON.
?!
AM I...

WHY...
...AM I BEING TREATED LIKE THIS?
What about my sword fight prac-tice?
FWOOSH
FWOOSH
Does he hate me? Is he shun-ning me?
FWOOSH
FWOOSH
FWOOSH
FWOOSH
Uh... no idea.
Stay back! You're wielding that baton like a weapon!
LOOK AT HOW HE'S TREAT-ING MOKO...
I'm sweat-ing so much I'm parched...
Ah.
OF COURSE.
I ASKED GRANDPA TO BE NICE TO KANAE.
HOW COULD YOU, HIO?! I WISH YOU'D ASKED HIM TO BE NICE TO ME TOO!
Oh-ho.
Hi.
You're here!
You look ten thou-sand times more lovely than this morning, my angel.
YOU'RE HALLU-CINATING, MR. KOTETSU...
Grandpa!!

SO.
tap
YOU ...
...JUST WATCH AND CHEER FOR KANA.
THEN ...
...I WAS GIVEN...
...THIS ITEM.
Whoa!
TWIRRL
RRRL
RRRRL
RRRRL
I HAVE BEEN TWIRLING EVERY SPARE MINUTE OF THE DAY...
I kept dropping it at first.
MAYBE YOU CAN ASSIST ME, SINCE YOU ARE MR. KOTETSU'S GRANDSON...
UH, WHY'RE YOU SPEAKING SO FORMALLY?

!
WHAM
FWIPP
KASHA—AK
UH ...
CLATTER

GAK
SHOVE
STOMP
GAK
GAK
KAKAKAK
ZSSSSSH
ZSSSH
STOMP
KASHAK

HMPH ...

YOU KNOW THE BASICS ...

...JUST LIKE KANA.

LME IS GOOD AT TRAINING THEIR PEOPLE.

I HAVEN'T COMPLETED THEIR SWORD FIGHT CURRICULUM LIKE MOKO HAS...

I'm Glad I remember some of it...

YOU DON'T NEED TO.
A STUNT-WOMAN WILL DO IT FOR YOU.
fwip
Counterpunch
WHA
CK
tmp tmp
I...
...THOUGHT HE LIKED ME NOW?! MOKO?!
peek
...
Doesn't know what to say.
sob
HERE.
DO THE MIDDLE STANCE.
swf

Four days ago
GOOD.
FIRST LET ME SEE YOU...
...DO BACK FLIPS!
Now!
JUMP LIKE A FRIED BEAN OVER TO THAT END!
EXCUSE ME, BUT I CANNOT!
I can move like an edamame that's been squeezed out!
Back-ward somer-sault
...
YOU WANT TO WIN THE ROLE OF A FEMALE NINJA, BUT YOU CAN'T EVEN DO BACK FLIPS...
YOU'RE HOPE-LESS.
Even young singer talentos can do back flips.
Sigh...
...
I-I'LL PRACTICE...

Peek
WHAT THE HELL ARE YOU DOING...
... KYOKO ?
Kyaaaaaaah!
MOOOKO! YOU lOOK SO COOl!
Woooooo
A-M-A-Z-I-N-G
Amazing! Kanae! Beaaautiful! Yah! Yeah!
tmp tmp tmpp
YOU SHOULD GO HOME IF YOU'RE NOT PRACTIC-ING.
What did you come here for?
EX-CUSE ME.
I'M NOT PLAY-ING.
Though it does look like I'm just fooling around.
twirrrrrrrl
MR. KOTETSU ...
...TOLD ME TO DO THIS.
GRAND-PA TOLD YOU?
YES.
IT WAS FOUR DAYS AGO, WHEN WE FIRST MET.

GAK
SWING
TMP
sak
sak
kak
kak
FWOOSH
kakak
HMPH.
SHE'S GOTTEN BETTER.
SHE CAN REALLY MOVE NOW.
GOOD JOB, GRAND-PA.
BY THE WAY...

STOMP
GAK
SHP

th-thump
th-thump

Skip·Beat!
Act 238: Beyond the Shock

LET'S HURRY TO THE TRAIN-ING HALL.
MR. KOTETSU GETS SCARY IF WE'RE EVEN A MINUTE LATE.
SURE!
End of Act 237

BUT HE ASKED YOUR NAME.
THAT'S HIS WAY OF TELLING YOU...
...THAT YOU'RE IN HIS GOOD GRACES.
I GUESS...
...HE LIKED YOUR LIVING SAMURAI MANNERS.

...AND MR. KOTETSU LIKES YOU NOW.

MR. KOTETSU...

...DOESN'T BOTHER TO REMEMBER THE NAMES OF HIS STUDENTS HERE.

HE ONLY CALLS THEM, "HEY," "YOU," "YOUNGSTER," "MISS," OR "GIRL".

WELL... OF COURSE HE HATES ME...
I ASKED FOR A LAST-MINUTE APPOINTMENT OUT OF THE BLUE, AND I EVEN HAD SOMEONE ELSE DO IT FOR ME.
HOW COULD I BE SO RUDE TO A SHOWBIZ HEAVY-WEIGHT?
...BUT I DIDN'T KNOW WHAT TO DO WHEN HE STARTED ACTING BRUSQUE.
MMM...
Yeah... right.
But...
HE'LL STILL TEACH ME...
I...
creak
creak
creak
creak
...ASSUMED HE WASN'T TOO OFFENDED, SINCE HIO WAS THE GO-BETWEEN...
creak
creak
I'M SO RELIE-
BUT YOU'VE MADE...
...A GOOD IMPRESSION...

I LOOK FORWARD TO YOUR LESSONS!
QUIET DOWN! YOU DON'T NEED TO SHOUT!
You think I'm a 500-year-old geezer who's hard of hearing?!
TURN DOWN THE VOLUME!
You fool!!
WHAA?!
creak
creak
creak
creak
...
I FEEL...
...LIKE I'M GETTING REVERSE COUNTERPUNCHED NO MATTER WHAT I SAY...
Reverse = Denies what he just said
Counterpunch = Then attacks me
WHEN I FIRST SAID MY GREETINGS...
LET ME FIRST APOLOGIZE FOR THIS SUDDEN REQUEST!
YOU SHOULD THANK ME INSTEAD.
I DIDN'T WILLINGLY AGREE. I REALLY HAD TO FORCE MYSELF AND GRUDGINGLY ACCEPT YOU AS A STUDENT.
IF I'D THANKED HIM FIRST...
YOU SHOULD APOLOGIZE INSTEAD.
After all, I'm spending time and effort...
...on a girl who appeared without warning.
...THAT WOULD'VE BEEN HIS RESPONSE...
Yeah...
I GUESS...
...THIS MEANS HE HATES ME...
...NO MATTER HOW OPTIMISTIC I TRY TO BE...

...MY PAST EXPERIENCES...
...LIKE THAT TIME WITH RURIKO...
SO.
WHAT'S...
...YOUR NAME?
!
Uh...
K-KYOKO.
MY REAL NAME...
...IS KYOKO MOGAMI.
THAT SO.
THEN...
...I'LL CALL YOU KYOKO.
(Wooden)
...
YES...
HE CALLS MOKO KANA...
I CAN'T HEAR YOU.
And you call yourself an actress?!
SPEAK FROM THE DIAPHRAGM.
Y...
Yes...

...THE NINJA ROLE YOU WANT...
... DOESN'T REQUIRE ANY OF THAT.
You're wasting your talents.
Your manners will be useful only if you play a town girl.
He did take a look at the novel.
Shock
...
He's right.
MOMIJI DOESN'T SPY ON A RESPECTED FAMILY!
IF SHE DISGUISES HERSELF AS A MAID OR A GEISHA AND SHOWS OFF HER ELEGANT AND SEXY SPY TECH-NIQUES...
...
Like Oden.
A female ninja in a long-running samurai drama.
THEY'RE PRESENT ...
...IN ALL OF YOUR MOVE-MENTS.
HOW-EVER...
...I MIGHT'VE ...
...BEEN ABLE...
...TO USE...

YES.
FOR A SAMURAI...
...THE TRUE BOW...
...IS A DEEP OBEISANCE RESERVED ONLY FOR THEIR LORD.
YOU'VE TAKEN...

swf
SIT DOWN.
THAT IS THE NAME OF THE BOW YOU MUST DO.
THE TRUE BOW.

PAR-DON ...
...ME.
SHU
SWF
PUSH
SHU
SUU
SWF
SWF
SWF

I TOLD YOU TO DO THE ORDINARY BOW!
DON'T STEP ...
...ON THE EDGES OF THE TATAMI MATS.

...IS THIS YOUNG MISS ...
...KANA'S SO-CALLED FRIEND WHO TOOK ADVANTAGE OF HIO TO BARGE IN HERE FOR SWORD FIGHTING LESSONS?
glare
NOW HE'S COLD AGAIN!
...
L...
HOW ABOUT WE GO INSIDE AND TALK.
...
...
tmp tmp
thud
PAR-DON ME.
LET ME FIRST APOLOGIZE FOR THIS SUDDEN REQUEST!
YOU SHOULD THANK ME INSTEAD.
I DIDN'T WILLINGLY AGREE. I REALLY HAD TO FORCE MYSELF AND GRUDGINGLY ACCEPT YOU AS A STUDENT.

You and you!
No
You just stepped on the tatami edging!
Now sit straight in that corner for an hour!
The rest of you go inside!
OH?
!
jolt
Oh-ho.
smile
Kanaaaaa.
Did I make you wait?
I didn't know you were already here.
HE'S ACTING SO NICE NOW.
NO.
WE JUST GOT HERE.
That so, that so.
Good, good.
AND...

...SO HE MUST'VE BEEN DOING JAPANESE ARCHERY OR SOME-THING.
J...
JAPAN-ESE...
...AR-CHERY ...
WOW! I WISH I COULD'VE SEEN HIM!
Awe-inspiring
HEEEEEEEEEY!
!

THE SAMURAI WORLD HAS SPECIFIC RULES TOO...
OH...
STUDENTS PRETEND THEIR SUPERIORS ARE PEOPLE OF HIGH STATUS.
WHEN SOMEONE ACTING AS A FEUDAL LORD COMES BY...
...PEOPLE REALLY PANIC.
SO THESE PEOPLE WILL BE ACTING IN A SAMURAI DRAMA?
YUP.
Don't stoop! Keep a straight line from your head down to your lower back!
Put strength in your abdominal and hip muscles!
WHACK
MR. KOTETSU USUALLY TEACHES HIS SWORD FIGHT CLASSES AT HIS OWN STUDIO...
...BUT HE COMES HERE WHEN SOMEONE ASKS HIM FOR HELP.
HE TEACHES ACTORS THE SAMURAI PROTOCOLS ...
...AND CHOREOGRAPHS SWORD FIGHTS DURING SHOOTS...
chirp chirp
...CUZ THIS PLACE IS OFTEN USED AS A LOCATION FOR SAMURAI DRAMAS.
I REMEMBER SEEING THIS PLACE IN SOME-THING.
MR. TSURUGA DID A COMMER-CIAL HERE.
I just heard about it.
...
OH.
IT WAS FILMED AT THE ARCHERY HALL...

HEEEEEEY!!
I've told you to do the ordinary bow when you pass your senior!
Move to the side! This is how your hands, waist and head should be!
WHAP
WHAP
KA WHA
CK
W-WE'RE SORRY. THANK YOU FOR COR-RECTING US!
...
UH ...
THE ORDI-NARY... BOW... WHA?
ARE WE... SUPPOSED TO KNOW THAT?
Like, does every Japanese citizen know this?
IT'S A SAMURAI BOW.
YOU DO THE ORDINARY BOW WHEN YOU ENCOUNTER YOUR SUPERIOR INSIDE A CASTLE.
THE WAY YOU BOW CHANGES DEPENDING ON THE STATUS OF THE OTHER PERSON.

THANKS FOR YESTER-DAY!
FOR GETTING PER-MISSION FOR MY TRAIN-ING.
HIO GOT YOU PERMISSION ...
...SO THANK HIM INSTEAD.
IS HE HERE TODAY ?
NO.
I ONLY HAVE A TWO-HOUR LESSON TODAY...
...SO HE COULDN'T MAKE IT.
HMM.
THAT MEANS ...
...HE'S ALWAYS HERE WHEN HE CAN MAKE IT.
Hee hee
HE'S SO NOBLE.
Though...
...I DON'T THINK YOU NEED ME TO TELL YOU THAT.
BY THE WAY.
ALWAYS MAKE SURE ...
...YOU OBSERVE GOOD MANNERS.
MR. KOTETSU IS VERY STRICT ABOUT IT.

Mornin.
The gem area again?
Yup.
Thanks.
...
...
MOKO!
!
SORRY.
AM I LATE?!
NO.
Pant pant
crunch crunch crunch
YOU'RE JUST ON TIME.

Oh!
SO MAYBE ...
...THIS MEANS ...
C'mon! Gimme good news! More good news!
I won't accept anything else! I'm starving for good news!
I'M FORCED TO MOVE FORWARD CUZ THIS CUTE FILTER IS MASKING A THR-
A STRONG WORD OF ENCOURAGE-MENT?
...
...
...
Comics Gallery
I...
I'LL DO MY BEST!
SHP
!
2:38 PM
Inbox
04/12 10:05 PM
UH ...
Kinga Manor

Don't worry.
You can do it, Ms. Mogami.
Manga adaptation of the book
Lotus in the Mud
...BUT IT'S REALLY EASY TO IMAGINE WHEN THERE ARE PIC-TURES!
Inbox
04/12 22:05
From Mr. Tsuruga
Thanks for texting me.
I'm about to leave.
Do your best at your audition.
Don't worry.
You can do it, Ms. Mogami.
ლ(° Д ° ლ)
...
...
...
WILL I BE ABLE TO...
...DO THIS ?
W...
...
WHY ...
...DID HE USE THIS EMO-TICON?
ლ(° Д ° ლ)
I COULDN'T HELP THINKING ABOUT IT LAST NIGHT...
More than just my surprise that Mr. Tsuruga used an emo-ticon ...
IT LOOKS SOME-THING ...
How dare you, rah!
Gimme some-thing better~!
... LIKE THIS.

WHIRL-WIND
FLASH
SH...
SHE'S SO COOOOOOOL!
TO BE HONEST, I COULDN'T ...
...REALLY PICTURE THE ACTION SCENES WHEN I READ THE BOOK...
th-thump
th-thump
th-thump
th-thump

...THOSE...
...LAMBS...
...SHOULD RUN WILD...
...AS MUCH AS THEY WANT!
CLANG
GAK

Heh...

I THINK...

...REN'S FINALLY COME TO TERMS WITH SOMETHING THAT WAS STILL BOTHERING HIM WHEN HE FIRST WENT TO GUAM...

IT'S NOT...

...JUST MY IMAGINATION, IS IT?

NO...

THEIR SPRITS...

...HAVE BECOME LIGHTER AND FREER.

BOTH REN'S...

...AND HERS.

...on flight 128 to Guam, proceed to security check-point...
A TEXT?
tap
IT'S FROM MS. MOGAMI.
!
Message
Ms. Mogami
tap
Message
Ms. Mogami
Hello.
You're leaving for the final shoot!
I hope you wrap up the film and your time as Actor X without a hitch.
I'm praying that everything goes well for TRAGIC MARKER and you.
MS. MOGAMI...
...
I'm praying
well for TRAGIC
you.
\(ˆoˆ)/\(ˆoˆ)/\(ˆoˆ)/
Banzai!
DOESN'T THIS EMOJI...
...mean...
...SOME-THING LIKE "MY LIFE IS OVER LOL"?
Not that I know much about emoticons...
She finally sent him a text with emoticons, but...
\(ˆoˆ)/\(ˆoˆ)/\(ˆoˆ)/
Over x 3

ka chak
clop
blah blah
mrmr
mrmr
clip
clop
clip
roll
roll
roll
roll
Lambda Airlines pas-sengers ...

...ISN'T AFRAID TO GRAB FOR WHAT SHE WANTS, EVEN IF SHE ENDS UP FALLING HARD.

EVERY TIME SHE MAKES A MOVE...

...SHE MANAGES TO WIN HERSELF MORE TERRITORY.

OKAY ...

...REN.

I'LL FOLLOW YOU SOON!

SO BE READY TO SEE ME AS SOON AS I ARRIVE!

WE'RE IN THE HOME STRETCH NOW.

YOU CAN DO THIS! I'LL DO EVERYTHING I CAN TO SUPPORT YOU!

THANKS.

I'M COUNTING ON YOU!

IN THE PLOT SUMMARY ON THE AUDITION ANNOUNCEMENT I FORCED MS. KOTONAMI TO TAKE.
MS. MOGMI WAS SO DISAPPOINTED WHEN SHE FOUND OUT HER PART BECOMES MS. KOTONAMI'S RIVAL...
VROOOOOOM
新空港道
SHIN-KUKO
成田 空港
Narita Airport
...BUT I THINK BEING ABLE TO WORK WITH MS. KOTONAMI WILL BE ENOUGH OF AN INCENTIVE...
...TO KEEP HER FROM BACKING OUT.
HMM ...
I WAS WORRIED ABOUT HER RELATIONSHIP WITH HER MOTHER...
...BUT SHE'S MANAGED TO LET A LITTLE LIGHT SHINE INTO THAT DARKNESS.
SHE ...

HUH?

AN AUDITION?

MS. MOGAMI IS AUDITIONING FOR THE UNGLAMOROUS TOO?

?

UNGLAMOROUS?

UH, FORGET IT.

It has nothing to do with you.

UH...

...SO, A FEMALE NINJA?

APPARENTLY.

AH... I THINK I DO REMEMBER A PART LIKE THAT BEING MENTIONED.

Skip·Beat!

Act 237: The Samurai and the Hunter

Skip·Beat!

Volume 40

CONTENTS

Skip·Beat!